Toxicity of Aquatic System and Remediation

The eco-friendly remediation technologies for the degraded environment are indeed the "need of the hour". Even though the regulatory mechanisms are in place to control the discharge of untreated contaminants into the natural environment, still, we could see a different picture; hence, remediation and restoration of the environment becomes an ardent requisite. The present-day fast pace of industrialization without proper disposal planning is impacting the water bodies adversely, generating the need for green management technologies. It is worth mentioning that these environment-friendly technologies are most cost-effective as well. The advancements in biotechnology have paved the way to mitigate the problem.

The primary audience of this book are the students and researchers who are working in the field of toxicology and bioremediation of aquatic environments. We have primarily focused in this book on bioremediation of aquatic system toxicity, considering this as an environment-friendly system and having the least adverse effects. Hence this book aims to bring forward together on a single platform the latest research in aquatic resource management, which includes the discussions and discourses on the degradation and the effect and the remediation.

This book includes a discussion on the different sources of contamination from industries or by the usage of commercial pesticides or even fertilizers. These contaminants, if discharged in their toxic form as effluent, cause harm to the aquatic systems and the subsoil and create the possibility of groundwater contamination. This book includes a discussion on the different routes of contamination and the food-chain transport possibilities of pesticide pollutants, which are very contemporary and required topics of research. It also includes relevant discussions on how to get rid of the toxicity.

Confluence of Research in Biotechnology for Human Welfare
Series Editor: Susmita Mukherjee
(*University of Engineering & Management, West Bengal, India*)

Natural Products: Alternate Therapeutic as Quorum Sensing (QS) Inhibitors
Edited by Moupriya Nag, Dibyajit Lahiri, Jaideep Banerjee and Taniya Roy Chowdhury

Toxicity of Aquatic System and Remediation: The Contemporary Issues
Edited by Susmita Mukherjee, Sanket J. Joshi, Sonali Paul, and Rita Kundu

For more information, please visit series page: www.routledge.com/Confluence-of-Research-in-Biotechnology-for-Human-Welfare/book-series/CRBHW

Toxicity of Aquatic System and Remediation

The Contemporary Issues

Edited by
Susmita Mukherjee,
Sanket J. Joshi, Sonali Paul,
and Rita Kundu

CRC Press
Taylor & Francis Group
Boca Raton London New York

CRC Press is an imprint of the
Taylor & Francis Group, an **informa** business

First edition published 2024
by CRC Press
2385 NW Executive Center Drive, Suite 320, Boca Raton FL 33431

and by CRC Press
4 Park Square, Milton Park, Abingdon, Oxon, OX14 4RN

CRC Press is an imprint of Taylor & Francis Group, LLC

Library of Congress Cataloging-in-Publication Data
Names: Mukherjee, Susmita (Professor of biotechnology), editor. | Joshi, Sanket J., editor. | Paul, Sonali, editor. | Kundu, Rita, editor.
Title: Toxicity of aquatic system and remediation : the contemporary issues / edited by Susmita Mukherjee, Sanket J. Joshi, Sonali Paul, and Rita Kundu
Other titles: Confluence of research in biotechnology for human welfare.
Description: Boca Raton, FL : CRC Press, 2024 | Series: Confluence of research in biotechnology for human welfare | Includes bibliographical references and index
Identifiers: LCCN 2023044548 (print) | LCCN 2023044549 (ebook) | ISBN 9781032286662 (hardback) | ISBN 9781032286709 (paperback) | ISBN 9781003297901 (ebook)
Subjects: LCSH: Water—Pollution. | Restoration ecology. | Phytoremediation. | Phycoremediation.
Classification: LCC TD420 .T693 2024 (print) | LCC TD420 (ebook) | DDC 363.739/4—dc23/eng/20231219
LC record available at https://lccn.loc.gov/2023044548
LC ebook record available at https://lccn.loc.gov/2023044549

ISBN: 978-1-032-28666-2 (hbk)
ISBN: 978-1-032-28670-9 (pbk)
ISBN: 978-1-003-29790-1 (ebk)

DOI: 10.1201/9781003297901

Typeset in Times
by Apex CoVantage, LLC

Contents

Preface xv
Editor Biographies xvii
List of Contributors xix

Chapter 1 Impact of Microplastics on the Environment and Its Mitigation: A Recent Review 1

Shankhadeep Chakraborty, Dip Mukherjee, Biswatosh Ghosh

1.1 Introduction 1
1.2 Global Distribution of Microplastics 2
1.3 Route and Fate of Microplastics 3
1.3.1 Major and Minor Routes 3
1.3.2 Fate of Microplastics 3
1.4 Extraction and Analysis of Microplastics 4
1.5 Effects of Microplastic Pollution on a Terrestrial and Aquatic Environment 5
1.5.1 Acting as Carrier of Contaminants 5
1.5.2 Effect on the Aquatic Ecosystem 6
1.5.3 Effects on the Terrestrial Ecosystem 6
1.5.4 Effects on Human Health 6
1.6 Microplastic Pollution Load Index (PLI) and Polymer Hazard Index (PHI) in the Context of India 7
1.7 Probable Control Measures and Mitigation Strategies 9
1.8 Conclusion 10
References 10

Chapter 2 Bioaccumulation of Cadmium in Fishes of Wetlands of West Bengal and the Effect on Human Health 13

Sonali Paul, Sharanya Paul, Shreya Bhattacharjee, Somava Nath, Susmita Mukherjee

2.1 Introduction 13
2.2 Heavy Metals in Polluting Wetland Ecosystems 14
2.3 Contamination of Cadmium in Wetlands 15
2.3.1 East Kolkata Wetlands (EKW) 15
2.3.2 Dankuni Wetlands (DW) 15
2.4 Accumulation of Cadmium and the Magnification in Fishes 16
2.5 Effects of Cadmium on Fishes 16
2.5.1 Behavioural Change 16

2.5.2 Endocrinological Effect 17
2.5.3 Genotoxic Effect 17
2.5.4 Oxidative Damage 18
2.6 Effects on Humans 18
2.7 Conclusion 19
References 20

Chapter 3 Aquatic Ecosystem Toxicity and Food Chain 25

Tilottama Roy

3.1 Introduction 25
3.2 Toxic Metal Contamination, Biomagnification, and Bioaccumulation 27
3.3 Mercury 28
3.4 Arsenic (As) 30
3.5 Selenium (Se) 30
3.6 Effects on Aquatic Vertebrates Like Fishes, Birds, and Marine Mammals 31
3.7 Microplastics in Aquatic Food Chains 31
3.8 Eutrophication 32
3.9 Contaminants of Emerging Concerns 32
3.10 Bioremediation in Aquatic Ecosystems: Zooremediation and Phytoremediation 33
References 34

Chapter 4 Mercury Toxicity in Humans: Sources, Pharmacokinetics, and Mitigation Measures 37

Sampriti Roy and Pritha Bhattacharjee

4.1 Introduction 37
4.2 Forms of Mercury 38
4.3 Sources of Mercury Exposure 40
4.4 General Routes of Exposure of Mercury in Humans 41
4.5 Toxicity 42
4.6 Acute and Chronic Exposure of Mercury with Effects 43
4.7 Pharmacokinetics of Mercury Exposure 44
4.7.1 Circulation 44
4.7.2 Brain 46
4.7.3 Thyroid 47
4.7.4 Gut 47
4.7.5 Kidney 48
4.7.6 Other Organs 48
4.7.7 Excretion 48

4.8 Treatment of Mercury Toxicity: Chelation Therapy 49
4.9 Mitigation Measures 49
4.10 Conclusion 51
References 51

Chapter 5 Toxic Effluent—Harmful Effects of Industries on Public Health 56

Sonali Paul, Anushka Ghosh, Arpan Banerjee, Sohini Roy, Subhabrata Paul, Susmita Mukherjee

5.1 Introduction 56
5.2 Types of Industries 56
5.2.1 Iron and Steel Industry 56
5.2.2 Textile Industry 57
5.2.3 Pulp and Paper Industry 57
5.3 Types of Industrial Effluents 57
5.3.1 Liquid Wastes 58
5.3.2 Solid Wastes 58
5.3.3 Chemical Wastes 58
5.3.4 Toxic Wastes 58
5.3.5 Heavy Metals 58
5.3.6 Effluent Discharge—Few Specific Industries 58
5.4 Impact on Ecosystem and Human Health 59
5.4.1 Implications of Pulp and Paper Industry Waste 59
5.4.2 Implications of Steel Industry Waste 60
5.4.3 Impacts of Textile Industry Waste 61
5.4.4 Impact of Heavy Metals on Human Health 61
5.5 A Comparative Study 64
5.6 Conclusion 67
References 68

Chapter 6 Toxicity of Surface Water Bodies and Public Health 70

Magapu Solomon Sudhakar and S. Senthil Kumar

6.1 Introduction 70
6.1.1 Clean Water Is Highly Indispensable for the Day-to-Day Life of Mankind 70
6.2 Water Bodies Getting Contaminated through Mining Waste 71
6.3 Contamination of Water Resources Like Groundwater and Lakes through Heavy Metals 73
6.4 Heavy Metal Contamination of Water Resources 77
6.5 Conclusion 79
References 81

Chapter 7 Anthracene in Aquatic Ecosystems: A Review 85

Susri Nayak, Sthitaprajna Nath Sharma, Smruti Prajna Pradhan, Subhashree Nayak, Lipika Patnaik

7.1 Introduction 85
7.2 Chemistry of Anthracene 86
7.3 Occurrence of Anthracene in Aquatic Environment 86
7.4 Bioaccumulation 92
7.5 Toxicity of Anthracene in Various Aquatic Animals 93
7.6 Conclusion 95
References 96

Chapter 8 Role of Allelochemicals in Alleviation of Stress in the Terrestrial and Aquatic Ecosystem 102

Pratik Talukder, Sounak Chanda, Meghna Mishra, Baishakhi Sinha, Anushka Dutta, Ayan Adhikary

8.1 Introduction 102
8.2 Plant Secondary Metabolites (PSMs) 103
8.3 Allelopathy 103
8.3.1 Root Exudation 104
8.3.2 Volatilization 105
8.3.3 Leaching 105
8.3.4 Decomposition 105
8.3.5 Insect Repellent from Naturally Derived Plant Products 106
8.3.6 Pest Control from Naturally Derived Plant Products 106
8.4 Allelochemicals as Bioherbicides 108
8.5 Straw Mulching 108
8.6 Role of Allelochemicals in Alleviation of Stress in the Aquatic Ecosystem 109
8.6.1 Planting Macrophytes 109
8.6.2 Direct Addition of the Extracted Allelochemicals 110
8.6.3 Algal Inhibition by Physiological Methods 110
8.7 Effect of Allelochemicals on the Environment 110
8.8 Current Studies and Future Prospect 110
8.9 Conclusion 111
References 111

Chapter 9 Comparative Assessment of Physico-Chemical and Biological Methods for the Treatment of Sanitary Landfill Leachate 115

Rachna Jain, Rita Mondal, Jhimli Bhattacharya, Akito I Sema, S. Saravanadevi, Dipanjali Majumdar

9.1 Introduction 115
9.1.1 Aeration 116
9.1.2 Advanced Oxidation Processes (AOPs): Fenton and Photo-Fenton 116
9.1.3 Bioaugmentation 116
9.1.4 Adsorption 116
9.1.5 Materials and Methods 116
9.1.6 Leachate Treatment 117
9.2 Results and Discussion 119
9.2.1 Physico-Chemical Characteristics of Raw Leachate ... 119
9.2.2 Application of Treatment Technologies 120
9.3.3 Degradation Kinetics 124
9.3.4 Comparative Efficiency of Different Technologies 125
9.3 Conclusion 125
Acknowledgements 126
References 127

Chapter 10 East Kolkata Wetlands: Unique Case of Treatment of Municipal Wastewater of Kolkata City 129

Susmita Mukherjee

10.1 Introduction 129
10.1.1 Evolution of EKW as a Wastewater-Fed Aquaculture System 130
10.1.2 Sewage Cleaning Process at EKW and STP 130
10.2 Materials and Method 131
10.3 Result 132
10.4 Discussion 134
10.5 Conclusion 134
References 135

Chapter 11 Role of Plant Polyphenolic Compounds in Alleviation of Metal-Induced Stress in Plants 136

Pratik Talukder, Sourish Bhattacharjee, Indrani, Elija Ghosh, Debankita Dutta

11.1 Introduction 136
11.2 Major Pathways Related to Polyphenol Production in Plants 137
11.3 Types of Major Polyphenol in Plants 138
11.3.1 Major Polyphenols in Stress Control 138
11.3.2 Biotic Stress 138
11.3.3 Abiotic Stress 139
11.3.4 Metal Stress and Its Negative Impact on Plants 139

11.4 Role of Macro and Microelements 139
11.4.1 Microelements 139
11.4.2 Macroelements 141
11.5 Role of Heavy Metals in Inducing Stress in Plants 141
11.5.1 Lead (Pb) 142
11.5.2 Mercury (Hg) 142
11.5.3 Cadmium (Cd) 142
11.5.4 Metalloids 142
11.6 Conclusion 143
References 143

Chapter 12 Aquatic Plant-Mediated Bioremediation Strategies 146

Falguni Barman, Shreya Nath, Swarnali Dey, Md. Tauhid Alam Ansari, Rita Kundu, Subhabrata Paul

12.1 Introduction 146
12.2 Inorganic Pollutants 147
12.2.1 Heavy Metals and Metalloids 147
12.2.2 Radioactive Wastes 154
12.3 Organic Pollutants 157
12.3.1 Petroleum Hydrocarbons and Polycyclic Aromatic Hydrocarbons (PAHs) 158
12.3.2 Pesticides and Fertilizers 161
12.3.3 Other Organic Compounds 162
12.4 Emerging Contaminants (ECs) 162
12.5 Conclusion 163
Acknowledgements 163
References 163

Chapter 13 Microbial Advancement towards Strategic Bioremediation Techniques: An Overview 169

Archisman Bhunia, Dibyajit Lahiri, Moupriya Nag

13.1 Introduction 169
13.2 Conventional Techniques of Remediation 170
13.3 Non-Conventional Techniques in Bioremediation 172
13.4 Players of Bioremediation 172
13.4.1 Phytoremediation 173
13.4.2 Microbial Remediation 174
13.5 Microbial Strategy Countering Toxic Pollutants 175
13.5.1 Factors Regulating the Biofilm Strategy 175
13.6 Microbial Counter Mechanism 175
13.6.1 Biodegradation 175
13.6.2 Biosorption 176
13.6.3 Bioaccumulation 176

13.6.4 Biomineralization 177
13.7 Techniques of Microbial Bioremediation 178
13.7.1 In Situ Bioremediation 178
13.7.2 Ex Situ Bioremediation 179
13.8 Conclusion and Future Perspectives 180
References 180

Chapter 14 Microbial Bioremediation: Importance, Features, Applications, and Future Perspectives 185

Soching Luikham and Jhimli Bhattacharya

14.1 Introduction 185
14.2 Methods 187
14.3 Principles of Bioremediation 187
14.3.1 Microbial Bioremediation of Inorganic Pollutants 188
14.3.2 Microbial Bioremediation of Organic Pollutants 188
14.4 Requirements to Support Microbial Bioremediation 189
14.5 Types of Microbial Populations for Remediation Processes 189
14.6 Various Factors Affecting Microbial Bioremediation 190
14.6.1 Physicochemical or Abiotic Factors Affecting Bioremediation Processes 191
14.6.2 Biological or Biotic Factors Affecting Bioremediation Processes 191
14.6.3 Extrinsic/Ex Situ Bioremediation 192
14.7 New Techniques on Bioremediation 193
14.8 Factors to Monitor Microbial Bioremediation 194
14.9 Conclusion 195
References 196

Chapter 15 Mitigation of Metal-Induced Stress in Plants with the Help of Microbial Association 198

Pratik Talukder, Anushka Ghosh, Arpan Banerjee, Subhabrata Paul

15.1 Introduction 198
15.2 Types of Stress in Plants 199
15.2.1 Types of Abiotic Stress in Plants 199
15.3 Plant Microbial Association 201
15.4 Different Mechanisms Involved in Biotic Stress Perception in Plants 202
15.4.1 Role of Microelements in Inducing Plant Stress 202
15.4.2 Role of Macroelements in Inducing Plant Stress 203
15.4.3 Role of Metalloids in Inducing Plant Stress 206
15.5 Mitigation of Metal-Induced Stress and Toxicity in Plants 207
15.6 Role of Microbial Association in Mitigation of Metal-Induced Abiotic Stress in Plants 208

15.6.1 Role of Bacteria in Mitigating Metal-Induced Stress in Plants 209
15.6.2 Role of Fungi in Mitigating Metal-Induced Stress in Plants 211
15.6.3 Role of Other Microbes in Mitigating Metal-Induced Stress in Plants 213
15.7 Conclusion 213
References 214

Chapter 16 Potential Use of Algae and Bacteria for Bioremediation of Wastewater: A Critical Review 219

Dipannita Parial and Satarupa Dey

16.1 Introduction 219
16.2 Use of Algae in Wastewater Treatment 220
16.2.1 Heavy Metal Bioremediation by Algae.... 221
16.2.2 Bioremoval of Emerging Contaminants Using Algae 224
16.2.3 Nutrients Bioremediation by Algae....226
16.3 Use of Bacteria in Wastewater Treatment 228
16.4 Microbes Associated with Wastewater....230
16.4.1 Role of Heterotrophic Bacteria....230
16.4.2 Role of Autotrophic Bacteria....230
16.5 The Utilization of Algae-Bacteria Consortia in Various Wastewater Streams....230
16.6 Application of the Algal Biomass.... 231
16.7 Challenges 232
16.8 Future Perspective 232
16.9 Conclusion 232
References 233

Chapter 17 Algae as a Source of Bioremediation and Resource Recycling.......237

S. Senthil Kumar and Magapu Solomon Sudhakar

17.1 Introduction 237
17.2 Macroalgae: A Powerhouse 238
17.2.1 Chemical Composition....238
17.3 Applications of Macroalgae....239
17.3.1 Advantages of Macroalgae as a Potential Source of Biofuel 239
17.4 Removal of Toxic Heavy Metals Using Macroalgae 241
17.5 Biosorption 243
17.6 Bioaccumulation....244
17.7 Bioremediation of Soil....245
17.8 Application of Seaweeds in Bioremediation of Nutrients 246

17.9 Bioremediation of Wastewater ... 247
17.10 Conclusion and Future Directions ... 248
References ... 249

Chapter 18 Blue-Green Algae in Wastewater Remediation ... 255

Susmita Mukherjee, Nivedita Chatterjee, Asmeeta Sircar, Shimantika Maikap, Abhilasha Singh, Sonali Paul

18.1 Introduction ... 255
18.2 Blue-Green Algae ... 255
18.3 Wastewater Definition: Industrial and Household ... 256
18.4 Remediation ... 257
18.5 How Blue-Green Algae Help in Remediation of Wastewater ... 257
18.5.1 Pathways/Mechanisms Involved ... 259
18.5.2 Applications and Benefits ... 260
18.6 Conclusion ... 261
References ... 261

Chapter 19 Advances in Nano-Biotechnology for Bioremediation Applications ... 264

Titir Guha, Sugata Mitra, Sumit Rez, Rita Kundu

19.1 Introduction ... 264
19.2 Conventional Technologies for Remediation ... 265
19.2.1 Physical Methods ... 265
19.2.2 Chemical Methods of Remediation ... 265
19.2.3 Biological Methods of Remediation ... 265
19.2.4 Limitations of Traditional Methods and the Need for Alternative Technologies ... 266
19.3 Applications of Nanotechnology for Remediation of Contaminants ... 266
19.3.1 Degradation of Polychlorinated Hydrocarbons ... 267
19.3.2 Degradation of Hydrophobic Compounds ... 267
19.3.3 Reductive Dechlorination by Oxidation of nZVI ... 267
19.3.4 Stabilization by Protective Coating on Nanoparticles ... 268
19.3.5 Nanoparticles in Water Purification ... 268
19.3.6 Enzyme-Linked Magnetic Nanoparticles ... 268
19.4 Application of Nanoparticles Along with Bioremediation for Removal of Hazardous Materials ... 269
19.4.1 Chlorinated Organic Contaminants ... 269
19.4.2 Petroleum Hydrocarbons ... 270
19.4.3 Heavy Metals Detoxification through Biogenic Synthesis of Nanoparticles ... 273
19.5 Remediation Using Nanoparticles in Aquatic Environments ... 273

19.5.1 Remediation of Industrial Wastewater with the Aid of Nano-Biotechnology 273
19.5.2 Enhancement of Bioremediation of Oil Spills via Nanotechnology 274
19.6 Future Perspectives and Conclusion 275
References 275

Preface

The field of science has developed in phases over the past centuries to meet the requirements of human beings. For survival, human beings depend on other living organisms, and here comes the importance and the role of ecosystem resources. In an ecosystem, organisms interact with each other, and by the mutual interactions, the ecosystem provides varied benefits for the sustenance of human beings. A healthy ecosystem is an absolute requirement for a healthy society.

Ecosystem resources are the natural requirement for human well-being. Whether we live in rural or urban areas, our dependence on ecosystem resources remains the same. For healthy living, we need fresh air to breathe, pure food to eat, safe drinking water, clothing, shelter, the basic needs, and so on; all are directly or indirectly through ecosystem resources.

Human beings have altered the ecosystem more rapidly and extensively to meet the growing demands in the course of economic development. These demands are considered important drivers of ecosystem degradation and biodiversity loss.

The obvious question comes, whether human beings are becoming less aware of their dependence on ecosystem and biodiversity. Any development at the cost of nature, can never be sustainable.

Economic development has resulted into rapid and most of the time unplanned industrialization and urbanization.which has impacted natural environment adversely. The rapid growth of the city and industrial units led to the unplanned dumping of wastes. Of course, a 100% pollution-free industry is a myth. We need to concentrate on industries with minimum discharge and a planned disposal system in place.

Ecosystem toxicity is a subset of environmental toxicity. It is the toxic effects on the ecosystem caused by natural or synthetic pollutants. The toxic effects become more deleterious when, through the food chain, bioaccumulation of the toxic chemicals leads to biomagnification in the higher trophic levels of the biosphere. However, in this book, we have restricted the discussion of ecosystem toxicity and remediation to the aquatic system. Surface water bodies are getting immensely polluted due to indiscriminate industrial discharge. Surface water toxicity is potentially affecting aquatic organisms and also human beings directly or through food chain contamination.

If in a city of West Bengal, India, smelting industries discharge their untreated wastes in the nearby ponds, polluting the pond water, it leads to ecosystem toxicity. Here, the bottom soil gets a deposition of lead, which is enough to pollute the groundwater, which affects the benthic organisms and also the people in the nearby area who use the water for their daily needs. Of course, the extent of toxicity varies depending on the position of the organism in the food web.

Hence there are enough reasons to restore the aquatic ecosystem from toxicity. Despite the regulatory mechanisms regarding non-contamination and opportunities for contaminant cleanup, restoration and remediation and its intersection with toxicology remains less explored. But sustainable and green remediation technologies are one of the major demands of the day. Bioremediation is an environment-friendly technique for decontamination of contaminated water bodies, including subsurface material. Biotechnological advancements have paved the way in developing tools to mitigate such problems using various biological systems that reduce the remnants of toxic materials present within the aquatic system. Research in the field of bioremediation obviously involves the usage of microbes (typically, heterotrophic bacteria and fungi) and green plants (termed as phytoremediation) to degrade or transform hazardous contaminants to materials such as carbon dioxide, inorganic salts, microbial biomass, and other by-products that may be less hazardous.

Heavy metals and metalloids like copper, cadmium, chromium, and arsenic are highly toxic for the aquatic ecosystem. The major potential environmental impacts of the heavy metals result from direct exposure of algae, benthic invertebrates, and embryos and fingerlings of freshwater fish and amphibians to these toxic pollutants. In microbial bioremediation, microbes utilize the contaminants and, by oxidation-reduction reaction, metabolize it, releasing energy. Both in situ (in the contaminated site) and ex situ (away from the site) bioremediation are part of this book. Bioremediation may sound simple but involves a series of complex processes facilitated by biological species. Toxic heavy metals which cannot be degraded by conventional methods are the targets of remediation by plants, phytoremediation. Bioremediation is also effective in handling xenobiotics. Different inorganic and organic pollutants coming out as effluent from different industries are highly toxic for water bodies. These can be treated by physical, chemical, or advanced oxidation processes, but these processes are not without limitations. Thus there is a need for remediation of these pollutants by some environment-friendly biological method.

The biological process is a slow but efficient process. It has emerged as a cost-effective, sustainable, and natural approach to treat a wide variety of chemical compounds and is an invaluable toolbox for wider application in the realm of environmental protection from toxic chemicals.

Susmita Mukherjee, Ph.D.
Editor

Editor Biographies

Susmita Mukherjee is Professor and Head of the Department of Biotechnology, University of Engineering & Management, Kolkata. She has done a BSc and MSc from University of Kalyani, India, and a PhD from Indian Institute of Engineering Science & Technology, Shibpur, India. Her research interest is on ecotoxicology and environmental remediation. She has worked on the East Kolkata Wetlands, the Ramsar site in West Bengal. Her major work is on arsenic remediation. She has more than 15 years of teaching experience. Supervised projects as PI. She is Associate Editor of the journal *Applied Biochemistry and Biotechnology*, Springer Nature. She has acted as Guest Editor in journals by Springer Nature, Elsevier, Taylor & Francis and edited books published by these publication houses. She has published many research papers in journals of international repute and has two patents published.

Dr. Sanket J. Joshi is a Professor and Deputy Director at Amity Institute of Microbial Technology, Amity University Rajasthan, Jaipur, India. He served as Deputy Director, Oil & Gas Research Center, and an Application Specialist, at Sultan Qaboos University, Oman, from 2013–2023. He holds BSc and MSc degrees from Sardar Patel University, India, and a PhD degree from M. S. University of Baroda, India—all in Microbiology. Prof. Joshi has 18 years of academic teaching and research experience, and 4 years of industrial R&D experience, in India and Oman. His current research interests encompass: Energy, Microbial products, and Environmental bioremediation. He has 193 scientific publications as papers in international journals (92), book chapters (23), conference proceedings (67), international books (10), and one Australian Innovation Patent to his credit. Prof. Joshi serves as an Academic/Associate/Guest Editor for some of the highly reputed journals. He is book series editor of 'Advances in Biotechnology and Bioengineering', Elsevier INC. He was awarded the "NRI Senior Scientist Award", from Microbiologists Society, India, during the academic year, 2019–2020. He has 4273 citations, h-index of 31, and i10 index of 60 (Google Scholar), and was listed among the 2% of the "Stanford University—Elsevier" list of highly cited scientists in the world-2022 and 2023, for excellence in scientific research in 'Energy' and 'Biotechnology' disciplines.

Dr. Sonali Paul is currently Professor, Department of Biotechnology, University of Engineering & Management, Kolkata. Having completed her graduate studies from Presidency College and her master's from University of Calcutta, she completed her doctoral degree from the Department of Biochemistry, University of Calcutta. Her present research interests include environmental toxicology and environmental biochemistry. She has published a number of papers in indexed journals of national and international repute. She has two published patents to her credit. She has supervised several projects as a joint supervisor who deals with understanding heavy metal toxicity effects on agricultural as well as medicinal plants and devising methods for remediation. She is serving as Assistant Editor of *International Journal* and is a member of varied scientific societies.

Prof. Rita Kundu is presently serving as Head and Professor, Department of Botany, University of Calcutta, working in the field of cell biology. Her research interest is to study and explore the potential of underutilized/unexplored natural/synthetic organic/inorganic compounds to regulate cell proliferation by inducing cell death either through apoptosis/autophagy or any other way, in cervical cancer cells *in vitro*. At the same time, she is also interested in identifying the bioactive compounds and their mode of action in regulating cell proliferation. She is also working on cadmium accumulation and its amelioration in rice grains. She also works on engineered nanocomposite mediated nutrient delivery and growth enhancement in rice plants to protect soil health due to overuse of fertilizers. She has received financial assistance from UGC, ICMR, WBDBT, WBDST, and UGC-UPEII for her research activities. She has published her findings in reputed journals. To date, eight students have submitted their work, and eight students are working on their doctoral degree.

Contributors

Ayan Adhikary
Department of Biotechnology
University of Engineering and Management, Kolkata
Kolkata, West Bengal, India

Arpan Banerjee
Department of Biotechnology
University of Engineering and Management, Kolkata
Kolkata, West Bengal, India

Falguni Barman
Department of Botany
University of Calcutta
Kolkata, West Bengal, India

Pritha Bhattacharjee
Department of Environmental Science
University of Calcutta
Kolkata, India

Shreya Bhattacharjee
Department of Biotechnology
University of Engineering and Management, Kolkata
Kolkata, West Bengal, India

Sourish Bhattacharjee
Department of Biotechnology
University of Engineering and Management, Kolkata
Kolkata, West Bengal, India

Jhimli Bhattacharya
Department of Chemistry
National Institute of Technology
Nagaland, India

Indrani Bose
Department of Biotechnology
University of Engineering and Management, Kolkata
Kolkata, West Bengal, India

Dr. Shankhadeep Chakraborty
Department- Oceanography
University- Techno India University
Kolkata, West Bengal, India

Sounak Chanda
Department of Biotechnology
University of Engineering and Management, Kolkata
Kolkata, West Bengal, India

Nivedita Chatterjee
Department of Biotechnology
University of Engineering & Management
Kolkata, West Bengal, India

Satarupa Dey
Department of Botany
Shyampur Siddheswari Mahavidyalaya
Howrah, West Bengal, India

Anushka Dutta
Department of Biotechnology
University of Engineering and Management, Kolkata
Kolkata, West Bengal, India

Debankita Dutta
Department of Biotechnology
University of Engineering and Management, Kolkata
Kolkata, West Bengal, India

Anushka Ghosh
Department of Biotechnology
University of Engineering and Management, Kolkata
Kolkata, West Bengal, India

Biswatosh Ghosh
Post Graduate Department of Zoology
Bidhannagar College
Kolkata, West Bengal, India

Elija Ghosh
Department of Biotechnology
University of Engineering and Management, Kolkata
Kolkata, West Bengal, India

Sanket Joshi
Amity Institute of Microbial Technology
Amity University Rajasthan
Jaipur, Rajasthan, India

S. Senthil Kumar
Department of Applied Biotechnology
University of Technology and Applied Sciences, Sur
Sultanate of Oman

Rita Kundu
Department of Botany
Calcutta University
Kolkata, India

Soching Luikham
Department of Chemistry
National Institute of Technology, Nagaland
Dimapur, Nagaland, India

Shimantika Maikap
Department of Biotechnology
University of Engineering & Management
Kolkata, West Bengal, India

Dipanjali Majumdar
CSIR – National Environmental Engineering Research Institute
Kolkata Zonal Centre
Kolkata, India

Meghna Mishra
Department of Biotechnology
University of Engineering and Management, Kolkata
Kolkata, West Bengal, India

Dr. Dip Mukherjee
Zoology, SBS Government College, Hili
University of Gour Banga
Higher Education Department, Govt. of West Bengal
Balurghat, West Bengal, India

Susmita Mukherjee
Department of Biotechnology
Institute of Engineering & Management
University of Engineering and Management
Kolkata, India

Moupriya Nag
Department of Biotechnology
University of Engineering and Management
Kolkata, India

Shreya Nath
Institute of Health Sciences
Presidency University (2nd Campus)
Kolkata, West Bengal, India

Somava Nath
Department of Biotechnology
University of Engineering and Management, Kolkata
Kolkata, West Bengal, India

Subhashree Nayak
Environmental science research laboratory
Department of Zoology
Ravenshaw University
Cuttack, Odisha, India

Susri Nayak
Environmental science research laboratory
Department of Zoology
Ravenshaw University
Cuttack, Odisha, India

Lipika Patnaik
Environmental Science Laboratory, Department of Zoology, COE in Environment and Public Health
Ravenshaw University
Cuttack, Odisha, India

Sharanya Paul
Department of Biotechnology
University of Engineering and Management, Kolkata
Kolkata, West Bengal, India

Sonali Paul
Department of Biotechnology
University of Engineering and Management
Kolkata, India

Subhabrata Paul
Institute of Health Sciences
Presidency University
New Town, Kolkata, India

Smruti Prajna Pradhan
Environmental science research laboratory
Department of Zoology
Ravenshaw University
Cuttack, Odisha, India

Sampriti Roy
Department of Environmental Science, University of Calcutta
Kolkata, West Bengal, India

Sohini Roy
Department of Biotechnology
University of Engineering and Management, Kolkata
Kolkata, West Bengal, India

Tilottama Roy
Missouri Western State University

Sthitaprajna Nath Sharma
Environmental science research laboratory
Department of Zoology
Ravenshaw University
Cuttack, Odisha, India

Abhilasha Singh
Department of Biotechnology
University of Engineering & Management
Kolkata, West Bengal, India

Baishakhi Sinha
Department of Biotechnology
University of Engineering and Management, Kolkata
Kolkata, West Bengal, India

Asmeeta Sircar
Department of Biotechnology
University of Engineering & Management
Kolkata, West Bengal, India

Magapu Solomon Sudhakar
Applied Biotechnology Department
University of Technology and Applied Sciences
Sultanate of Oman

Pratik Talukder
Department of Biotechnology
University of Engineering and Management
Kolkata, India

1 Impact of Microplastics on the Environment and Its Mitigation

A Recent Review

Shankhadeep Chakraborty, Dip Mukherjee, Biswatosh Ghosh

1.1 INTRODUCTION

With the rapid rate of growth in the sector of industrialization and urbanization, global use of plastics has also increased manifold. Due to low cost, versatility and ease of manufacturing, plastic is of tremendous use to date (Razeghi et al. 2021). Plastic is a generic term to mean an artificial polymer with a special property of binding together. About 20–90 metric tonnes (Mt) of plastic waste is observed to be produced each year. If this waste is disposed in the environment, over time it degrades and produces microplastics (Bajt 2021).

The term "microplastic" was first coined by Thompson Richard Charles in 2004. This term indicates a smaller-sized plastic component, especially in the marine environment (Wang et al. 2019). They range in size from 5 mm to 1 μm. A number of additives are added to microplastics to make a series of ornamental products nowadays. They are produced from two sources: primary and secondary sources, as shown in Figure 1.1. The primary source involves production with <5 mm dimension, whereas the secondary sources encompass a bunch of microplastics produced with >5 mm dimension. These are further broken down to even smaller pieces biologically or chemically and contaminate the soil and water of a biotic community. Although a number of literatures is available nowadays on the effect of microplastics on the marine environment, very limited studies are available on the impact of microplastic pollution in terrestrial environments and especially India. Thus, a review on this area will be helpful to understand the current trend of microplastic pollution in the environment, and future research in terms of remedial measures particularly in the Indian context may also be outlined from this study.

DOI: 10.1201/9781003297901-1

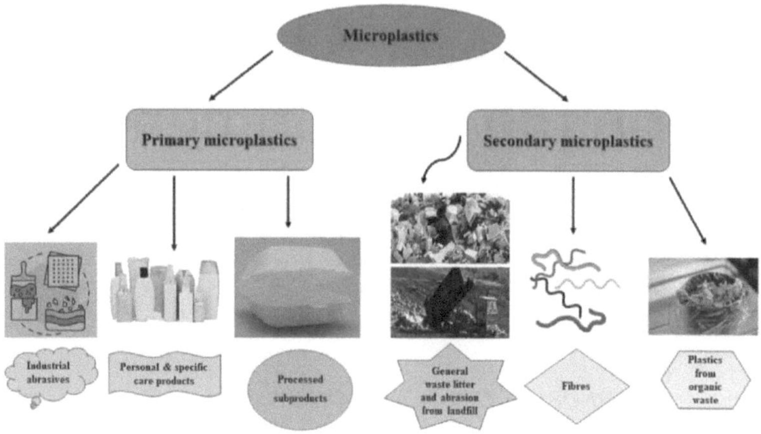

FIGURE 1.1 Major sources of microplastics.

Source: Adapted from Pinto et al. 2020.

1.2 GLOBAL DISTRIBUTION OF MICROPLASTICS

Although the size of microplastics is not very large, they are present in different spheres of the environment, namely estuaries, oceans, sediments, coastlines, soils and beaches. Table 1.1 summarizes the distribution of microplastics in the aquatic

TABLE 1.1
Global distribution of microplastics in aquatic environment

Place	Type of habitat	Type of sample	Findings	References
Gulf of Thailand, Japan	Marine	Surface sediment	Thailand: 100 pieces/kg dry weight	Matsuguma et al. 2017
Australia	Elwick Bay	Sediment and Dogshear Point in Derwent Estuary, Tasmania	2.43–4.2 fragments/gm of sediment	Willis et al. 2017
North Yellow Sea, China	Marine	Surface water and sediment	Surface water: 5.45 ± 282 items/m^3 Sediments: 37.1 ± 19.7 items/kg dry weight	
Western Lake Superior, Canada	Freshwater	Water	21,000–1,10,000 particles/km^2	Hendrickson et al. 2018
Goa beaches (Calanguti), India	Marine	Beach sediment	3,000 pellets from 6 kg dry weight	Veerasingam et al. 2016
Bay of Bengal, India	Estuary	Surface water	200–20,000 particles/m^3	Eriksen et al. 2014

TABLE 1.2
Global distribution of microplastics in terrestrial environment

Site	Country	Type of sample	Findings	References
Loess Plateau	China	Agricultural land surface soil	10.54 mg/kg	Zhang and Liu 2018
Plastic industry, Sydney	Australia	Soil	67,500 mg/kg	Fuller and Gautam 2016
Beach soil, Mumbai	India	Soil	220 items/kg	Tiwari et al. 2019
Synthetic fibre industry, Vienna	Austria	Sewage sludge and soil	Sewage sludge 11,497 items/gm Soil: 14,142 items/gm	Meixner et al. 2020

environment. Globally, microplastics are also distributed in the terrestrial environment to a large extent (Cole et al. 2011). Table 1.2 indicates the global distribution of microplastics in the terrestrial environment. Two factors are broadly associated with the generation of this pollutant, and these are (a) environmental factors like wind, tides, currents, etc. and (b) anthropogenic factors like mining, plastic production, etc.

1.3 ROUTE AND FATE OF MICROPLASTICS

As microplastics are lighter in weight, they have widespread dispersal as soon as they are released into an ambient environment. The transportation process is mostly carried by wind, rainfall, etc., and there is a route of transportation from the terrestrial to the aquatic environment by means of rivers, washing action of tunnels, agricultural and industrial discharges, etc.

1.3.1 Major and Minor Routes

A lot of studies reported that there are two ways of microplastic entry into the environment, namely major and minor routes. Major routes include river run-off, industrial effluents, etc., whereas the minor routes are cosmetics, like toothpastes containing a good number of microbeads. Whenever these minute particles enter the lotic environment, they get transported along with sediment into vast oceans and estuaries. After arrival in the marine environment, microplastics are propelled for a substantial distance by currents, waves, etc. It has also been reported that these pollutants move vertically because of biofouling, marine snow absorption and faecal pellet feeding (Cole et al. 2011).

1.3.2 Fate of Microplastics

Most of the microplastics are terrestrial in their origin. A study carried out by UNEP (2005) reported that nearly 70% of marine chunks are deposited into the sediment

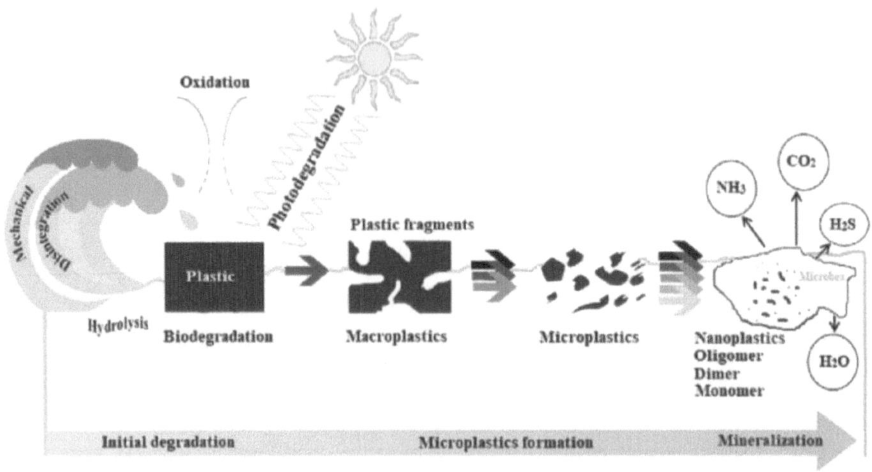

FIGURE 1.2 Transportation and fate of microplastics in the environment.

layer, and the remaining 30% of them are observed floating in the ambient aquatic layer. These micro molecules can be sedimented in nearby beaches, mudflats and coastal soils, and one very true fact is that these can be stored in the vegetative tissues of marine algae and gills, epithelium, etc. of tunicates and benthic invertebrates (Wong et al. 2020).

Another interesting fact is that the microplastics are stuck in the sediments, and after residing there for many years, they can get driven by wind either in their actual form or in a chemically converted state to elsewhere down the ridges, tunnels, etc. (GESAMP 2015).

1.4 EXTRACTION AND ANALYSIS OF MICROPLASTICS

Although there are several methods of microplastic extraction standardized from different sources, there needs to be one common method for extraction of microplastics. But it is not obvious that only one common methodology is good enough to get an optimum result. The general and basic scheme of extraction available in literature is illustrated in Figure 1.3. Physical sorting, drying, density separation, digestion, followed by filtration result in separation of diverse microplastics from soil, sediments, water, etc. The further analysis involves a microscopic study for physical interpretation followed by other biophysical and biochemical techniques like Fourier transform infrared spectroscopy (FTIR), gas chromatography–mass spectrometry (GC–MS), liquid chromatography–mass spectrometry (LC–MS), etc. to decipher their nature and types (He et al. 2018). The analytical methodology for diverse sources of sample collection with regards to behaviour of microplastics in different environments is still in its infancy, and more specific approaches are still wanting.

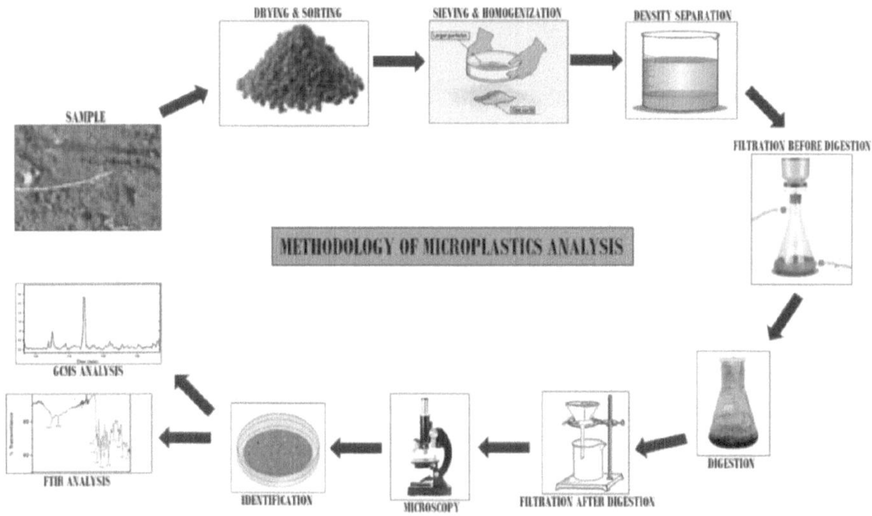

FIGURE 1.3 Experimental methodology of microplastic extraction and analysis from samples.

Source: Adapted from He et al. 2018.

1.5 EFFECTS OF MICROPLASTIC POLLUTION ON A TERRESTRIAL AND AQUATIC ENVIRONMENT

A chain of literatures reported the presence of microplastics in multidimensional niches including our diet, body tissue, vegetative tissue of plants, soil, water and even mid-oceanic ridges and underwater vents (WHO 2019). The food chain in land and water are therefore affected by these micropollutants.

1.5.1 Acting as Carrier of Contaminants

Analysis of the effects carried by microplastics shows that one of the primary risk factors associated with them is their stability and residual time. Furthermore, another risk is these can be impregnated into the soft tissues of terrestrial and aquatic invertebrates. As a prey-predator system always operates in nature, the accumulation factor gets increased along the food chain across the higher taxa of vertebrates (Corradini et al. 2019).

Microplastics have a special adsorbing property, based upon which they are reported to carry multiple organic and inorganic toxicants. Microplastics have an affinity with many polycyclic aromatic hydrocarbons (PAH), persistent organic pollutants (POPs), polychlorinated biphenyls (PCBs), organochlorine pesticides and dioxins as these have been identified on their surface from various locations around the beaches and islands.

1.5.2 Effect on the Aquatic Ecosystem

In the aquatic medium, a number of species are reported to ingest microplastics. Phytoplankton, zooplankton and benthic amphipods, polychaete and holothurians are found to ingest it due to its miniature dimension and ease of availability in the ambient media. The capacity of microplastic to absorb primary and secondary pollutants is another reason of grading it in the potential toxic pollutant category. It leads to biomagnification which results in perturbations of the ecosystem. In the aquatic environment, a number of food webs are present. So, as an interim predator as well as prey, fish is causing harm to other food chains by transferring and/or accumulating this toxic substance. Microplastics are found to reside in the stomach, respiratory and excretory tracts of organisms. Qiao et al. (2019) reported intestinal inflammation, alteration in gut microbiota and tissue metabolic profiles and oxidative stress in the zebra fish (*Danio rerio*) exposed to polystyrene. In a direct manner, microplastics are found to obstruct food entry, plug the digestive tract and elicit numerous types of physiological stress in zooplanktons, particularly the copepods. In addition, copepods also exhibited both species and stage specificity in response to microplastics (Bai et al. 2021).

1.5.3 Effects on the Terrestrial Ecosystem

Due to low levels of light and oxygen, microplastics can remain in the soil for more than 100 years. These microplastics have the capacity to interact with the soil microfauna and meiofauna, causing harmful implications (Corradini et al. 2019). Research conducted on springtails and earthworms suggested that they are capable of transiting microplastics in soil in both directions. Microplastics impact the burrows' structural changes, resulting in abnormal function of soil aggregation and operation (Huerta Lwanga et al. 2017).

Another very alarming fact is leaching to secondary toxic chemical bisphenol A from microplastics may result in endocrine system malfunction due to estrogenic effects. Hence, microplastics are creating a threat to the indispensable ecosystem services of the terrestrial organisms like plant pollinators and soil microorganisms.

1.5.4 Effects on Human Health

The portal of entry into the human body is multifarious for microplastics like cosmetics, meals, etc., and these particles are observed to induce several malfunctions in the internal milieu. A research study conducted on some people from Japan, Russia and the UK showed the presence of microplastics in stool (Harvey and Watts 2018). Seafood is said to be an alarming source of microplastic particles for humans nowadays. As human beings are at the apex of most food chains, the effect on human health is prominent. Few studies reported that the gastrointestinal tract is mostly affected by microplastics, causing obesity and cancers (GESAMP 2015). Figure 1.4 shows various consequences of microplastics on human health.

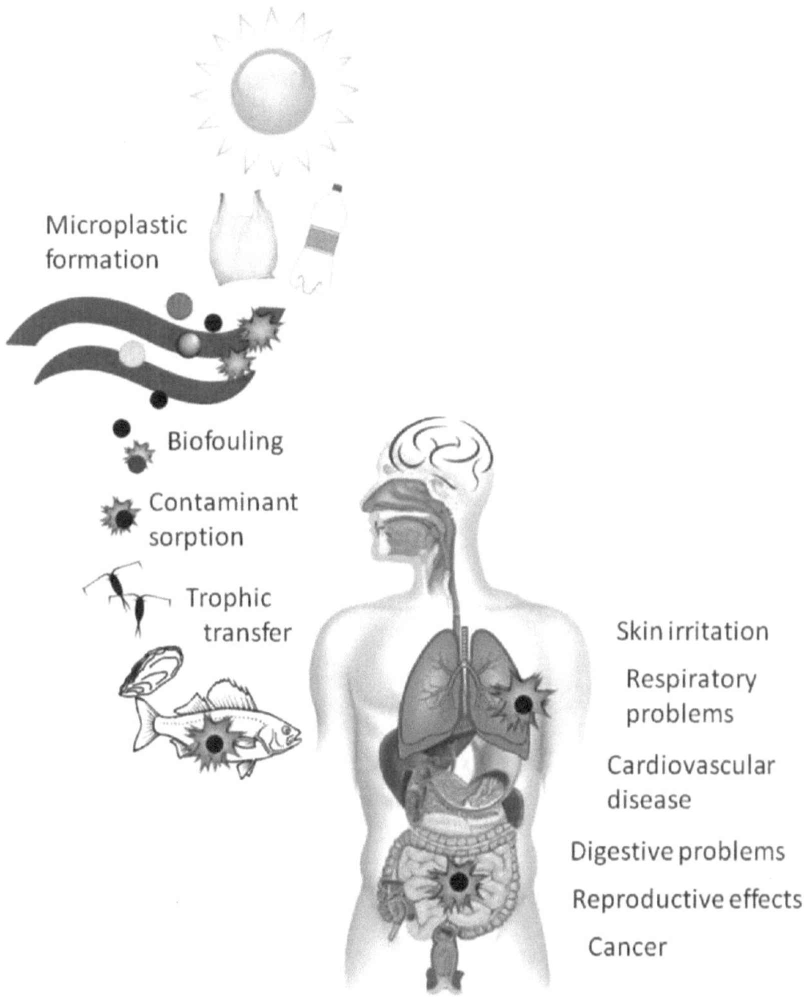

FIGURE 1.4 Detrimental effects of microplastics on human health.

Source: Courtesy: Carbery et al. 2018.

1.6 MICROPLASTIC POLLUTION LOAD INDEX (PLI) AND POLYMER HAZARD INDEX (PHI) IN THE CONTEXT OF INDIA

To determine the degree of microplastic pollution in surface sediments, an integrated Pollution Load Index (PLI) was calculated based on Tomlinson et al. (1980). PLI can be determined as follows:

$$CF_i = C_i \div C_{oi}$$

$$PLI = \sqrt{CF_i}$$

CF_i of the microplastic is the quotient of microplastic concentration from ambient sediment (C_i) and the background microplastic concentration (C_{oi}). The lowest microplastic concentration detected was considered as the background value.

The polymer hazard assessment is an accurate method to evaluate the potential hazard of microplastic. The formula of Polymer Hazard Index (PHI) is

$$PHI = \sum P_n \times S_n$$

where P_n is the percentage of polymer, and S_n is the hazard score of polymers, derived from Lithner et al. (2011) as given in Table 1.3.

The values of PLI and PHI across different coastlands of the Indian subcontinent are significant. The risk categories associated with the values are depicted in Table 1.4 as Hazard Level I-V (Ranjan et al. 2020).

TABLE 1.3
Primary hazard statements and scores of microplastic polymers from the indian coast

Polymer	Monomer	Density (g/cm³)	Primary hazard statements
Polyethylene (PE)	Ethylene	0.91–0.97	Extremely flammable gas, may cause drowsiness
Polypropylene (PP)	Propylene	0.89–0.92	Extremely flammable gas
Polystyrene (PS)	Styrene	0.28–1.04	Flammable liquid and vapor, harmful if inhaled
Polyethylene terephthalate (PET)	Ethanediol	1.37–1.38	Harmful if swallowed
Polyamide (PA)	Hexamethylene-diamine	1.14–1.15	Harmful if it comes in contact with skin or swallowed. May cause respiratory irritation. Can cause skin burns and eye damage.

TABLE 1.4
Terminology used to describe hazard level criteria for microplastic pollution

PHI	Hazard category	PLI	Hazard category	Risk category
0-1	I	<10	I	Minor
0-10	II	xx	x	Medium
10-100	III	10-20	II	High
100-1000	IV	20-30	III	Danger
>1000	V	>30	IV	Extreme danger

Source: Adapted from Ranjan et al. 2020

TABLE 1.5
Hazard categories in various coastal areas in india

Coastal site	PHI	PLI	Hazard category and statement
Maharashtra coast	100–1000	10–20	PHI-IV; PLI-II; High risk
Karnataka South coast	100–1000	<10	PHI-IV; PLI-I; Moderate to high risk
Central coast of Kerala	100–1000	<10	PHI-IV; PLI-I; Moderate to high risk
North coast of Kerala	>1000	10–20	PHI-V; PLI-II; Very high risk, action utmostly needed
Bengal coast (Sundarbans)	100–1000	<10	PHI-IV; PLI-I; High risk
Chennai coast	100–1000	<10	PHI-IV; PLI-I; High risk
Rameswaram coast	100–1000	10–20	PHI-IV; PLI-II; High risk
Northern Goa coast	>1000	10–20	PHI-V; PLI-II; Very high risk, action utmostly needed

Source: Adapted from a study done by Ranjan et al. 2020

Based on the report and ecological risk assessment (calculating PHI and PLI), India was categorized as Hazard Level IV to V. Table 1.5 indicates hazard categories in different coastal areas of India in relation to values of PHI and PLI.

1.7 PROBABLE CONTROL MEASURES AND MITIGATION STRATEGIES

There are a number of strategies to control microplastic pollution, as reported in various literatures (Mallik et al. 2021). The main focuses of the control methods are (a) control at source by legislation and mass awareness programs, (b) reduce, reuse and recycle (3Rs) microplastics as far as possible and, finally, (c) mitigation strategies by chemical and/or biological methods.

The restriction, as taken in the USA and EU countries, includes cut-off in production and use of microplastics in parallel to large-scale cleaning up of these in marine environments by floating systems that capture and intercept plastics (Auta et al. 2018).

Implementation of laws to control plastic pollution has already been taken up in many countries, although the 3R approach (i.e. reduce, reuse and recycle) is found to be much more potent in terms of measures. Improved solid waste infrastructure and proper management are found to decrease plastic debris entering the rivers and oceans. Reducing use of plastic products, reusing biodegradable forms of plastics (as depicted in Table 1.6) and recycling of some biodegradable plastic (except Styrofoam) can be effective measures.

TABLE 1.6
Biodegradability of major microplastics found in the marine environment

Polymer	Density (g/cm³)	Biodegrading agent
Low-density polyethylene	0.917–0.965	Bacteria, fungi, waxworms, mealworms
Polypropylene	0.90–0.91	Bacteria
Polystyrene	1.04–1.10	Bacteria, mealworms
Polyurethane	1.20	Fungi
Polyester	1.24–2.3	Bacteria

Source: Adapted from a study by Wu et al. 2017

The microbial degradation of petroleum-based microplastics like polyethylene (PE), polystyrene (PS) and polypropylene (PP) has been examined since the 1970s. Different strains of bacteria are found to be potent degraders of PE, PS and PP; *Bacillus* and *Enterobacter asburiae* degrade PE; *Exiguobacterium* sp. which degrade PS, were isolated from guts of pest insect larvae of *Plodia interpunctella* (Indian moth) and *Tenebrio molitor* (mealworms) respectively.

1.8 CONCLUSION

This study elucidates the proven origins of microplastics, their fate and transfer process into various types of environments. The primary and secondary microplastics' transportation process are anomalous, though secondary microplastics are more harmful than the primary ones. The degree and magnitude of damage these pollutants cause to the environment and human body is a big matter of concern. PLI and PHI analysis show considerable damage has already been done on the coastal food chain. Local people's awareness and motivational strategy implementation for NGO and private sectors related to mitigation are of immediate need. The government and policymakers should initiate implementation of strict laws regarding use and practices related to microplastics. Further research is necessary to improve the control measures and find out newer techniques.

REFERENCES

Auta, H.S., Emenike, C.U., Jayanthi, B., Fauziah, S.H. 2018. Growth kinetics and biodeterioration of polypropylene microplastics by *Bacillus* sp. And *Rhodococcus* sp. Isolated from mangrove sediment. *Marine Pollution Bulletin* 127:15–21.

Bai, Z., Wang, N., Wang, M. 2021. Effects of microplastics on marine copepods. *Ecotoxicology and Environmental Safety* 217:112243.

Bajt, O. 2021. From plastics to microplastics and organisms. *FEBS Open Biology* 11(4):954–966.

Carbery, M., O'Connor, W., Palanisami, T. 2018. Trophic transfer of microplastics and mixed contaminants in the marine food web and implications for human health. *Environment International* 115:400–409.

Cole, M., Lindeque, P., Halsband, C., Galloway, T.S. 2011. Microplastics as contaminants in the marine environment: A review. *Marine Pollution Bulletin* 62(12):2588–2597.

Corradini, F., Meza, P., Eguiluz, R., Casado, F., Huerta-Lwanga, E., Geissen, V. 2019. Evidence of microplastic accumulation in agricultural soils from sewage sludge disposal. *Science of the Total Environment* 671:411–420.

Eriksen, M., Lebreton, L.C., Carson, H.S., Thiel, M., Moore, C.J., Borerro, J.C., Galgani, F., Ryan, P.G., Reisser, J. 2014. Plastic pollution in the World's oceans: More than 5 trillion plastic pieces weighing over 250,000 tons afloat at Sea. *PLoS One* 9(111):9–13.

Fuller, S., Gautam, A. 2016. A procedure for measuring microplastics using pressurized fluid extraction. *Environmental Science and Technology* 50(11):5774–5780.

GESAMP. 2015. Sources, fate and effects of microplastics in the marine environment: A global assessment. In: P.J. Kershaw (Ed.) *IMO/FAO/UNESCO-IOC/UNIDO/WMO/IAEA/UN/UNEP/UNDP Joint group of experts on the scientific aspects of Marine Environmental Protection* (Rep. Stud. GESAMP No. 90), International Maritime Organization, p. 96.

Harvey, F., Watts, J. 2018. Microplastics found in human stools for the first time. *Plastics Food Pollution on the News, The Guardian* (last accessed on 24 October 2020).

He, D., Luo, Y., Lu, S., Liu, M., Song, Y., Lei, L. 2018. Microplastics in soils: Analytical methods, pollution characteristics and ecological risks. *Trends in Analytical Chemistry* 109:163–172.

Hendrickson, E., Minor, E.C., Schreiner, K. 2018. Microplastic abundance and composition in Western Lake Superior as determined via microscopy, Pyr-GC/MS, and FTIR. *Environmental Science and Technology* 52(4):1787–1796.

Huerta Lwanga, E., Mendoza Vega, J., Ku Quej, V., Chi, J.L.A., Sanchez Del Cid, L., Chi, C., Escalona Segura, G., Gertsen, H., Salánki, T., van der Ploeg, M., Koelmans, A.A., Geissen, V. 2017. Field evidence for transfer of plastic debris along a terrestrial food chain. *Scientific Reports* 7(1):14071.

Lithner, D., Larsson, A., Dave, G. 2011. Environmental and health hazard ranking and assessment of plastic polymers based on chemical composition. *Science of the Total Environment* 409(18):3309–3324.

Mallik, A., Xavier, K.A.M., Naidu, B.C., Nayak, B.B. 2021. Ecotoxicological and physiological risks of microplastics on fish and their possible mitigation measures. *Science of the Total Environment* 779:146433.

Matsuguma, Y., Takada, H., Kumata, H., Kanke, H., Sakurai, S., Suzuki, T., Itoh, M., Okazaki, Y., Boonyatumanond, R., Zakaria, M.P., Weerts, S., Newman, B. 2017. Microplastics in sediment cores from Asia and Africa as indicators of temporal trends in plastic pollution. *Archives of Environmental Contamination and Toxicology* 73(2):230–239.

Meixner, K., Kubiczek, M., Fritz, I. 2020. Microplastic in soil–current status in Europe with special focus on method tests with Austrian samples. *AIMS Environmental Science* 7(2):174–191.

Pinto, J., Da, C., Teresa, R., Armando, C. 2020. *The environmental impacts of plastics and microplastics use, waste and pollution: EU and national measures*. European Union PE.

Qiao, R., Sheng, C., Lu, Y., Zhang, Y., Ren, H., Lemos, B. 2019. Microplastics induce intestinal inflammation, oxidative stress, and disorders of metabolome and microbiome in zebrafish. *Science of the Total Environment* 662:246–253.

Ranjan, V.P., Joseph, A., Goel, S. 2020. Microplastics and other harmful substances released from disposable paper cups into hot water. *Journal of Hazardous Materials* 404(Pt B):124118.

Razeghi, N., Hamidian, A.H., Wu, C., Zhang, Y., Yang, M. 2021. Microplastic sampling techniques in freshwaters and sediments: A review. *Environmental Chemistry Letters* 18:1–28.

Thompson, R.C., Olsen, Y., Mitchell, R.P., Davis, A., Rowland, S.J., John, A.W., McGonigle, D., Russell, A.E. 2004. Lost at sea: Where is all the plastic? *Science* 304(5672):838.

Tiwari, M., Rathod, T.D., Ajmal, P.Y., Bhangare, R.C., Sahu, S.K. 2019. Distribution and characterization of microplastics in beach sand from three different Indian coastal environments. *Marine Pollution Bulletin* 140:262–273.

Tomlinson, D.L., Wilson, J.G., Harris, C.R., Jeffrey, D.W. 1980. Problems in the assessment of heavy-metal levels in estuaries and the formation of a pollution index. *Helgoländer Meeresuntersuchungen* 33(1–4):566–575.

UNEP. 2005. *Marine litter: An analytical overview*. United Nations Environment Programme.

Veerasingam, S., Saha, M., Suneel, V., Vethamony, P., Rodrigues, A.C., Bhattacharyya, S., Naik, B.G. 2016. Characteristics, seasonal distribution and surface degradation features of microplastic pellets along the Goa coast, India. *Chemosphere* 159:496–505.

Wang, J., Liu, X., Li, Y., Powell, T., Wang, X., Wang, G., Zhang, P. 2019. Microplastics as contaminants in the soil environment: A mini-review. *Science of the Total Environment* 691:848–857.

WHO. 2019. Microplastics in drinking water. *Water Sanitation Hygiene*, 20 August. Available at: https://www.WHO.int/water.sanitation.health/publication/microplastics-in-drinking-water/en (last accessed on 22 June 2020). https://www.who.int/publications/i/item/9789241516198 (last accessed on 23rd December 2023)

Willis, K.A., Eriksen, R., Wilcox, C., Hardesty, B.D. 2017. Microplastic distribution at different sediment depths in an urban estuary. *Frontiers in Marine Science* 4:1–8.

Wong, J.K.H., Lee, K.K., Tang, K.H.D., Yap, P.S. 2020. Microplastics in the freshwater and terrestrial environments: Prevalence, fates, impacts and sustainable solutions. *Science of the Total Environment* 719(137):5–12.

Wu, W.M., Yang, J., Criddle, C.S. 2017. Microplastics pollution and reduction strategies. *Frontiers of Environmental Science & Engineering* 11(1):6.

Zhang, G.S., Liu, Y.F. 2018. The distribution of microplastics in soil aggregate fractions in South Western China. *Science of the Total Environment* 642:12–20.

2 Bioaccumulation of Cadmium in Fishes of Wetlands of West Bengal and the Effect on Human Health

*Sonali Paul, Sharanya Paul, Shreya Bhattacharjee, Somava Nath, Susmita Mukherjee**

2.1 INTRODUCTION

Wetlands are recognized as one of the key life support systems on our planet, along with agricultural lands and forests (Storelli et al., 2005). They are not only habitats for many endangered plant and animal species but also play a crucial role in maintaining ecological integrity and providing goods and services that are essential for human well-being (Cavas et al., 2005). Wetlands are highly productive ecosystems, second only to tropical rainforests, and they perform various functions such as groundwater recharge, shoreline stabilization, and flood storage, which are particularly important (Mitsch et al., 2012). In West Bengal, India, there are diverse types of wetlands ranging from coastal and marine wetlands to inland freshwater lakes, rivers, dams, swamps, and constructed wetlands in irrigation schemes and sewerage treatment systems (Grosell, 2011). The wetlands are recognized as conservation areas, as national reserves parks, Ramsar sites (such as the East Kolkata Wetlands), and important bird sanctuaries (such as the Kulik bird Sanctuary).

Apart from their biodiversity importance, wetland resources are crucial for income generation, livelihoods, and the well-being of local communities (Barjhoux et al., 2012). However, due to a lack of effective management mechanisms and proper appreciation of their true value, wetlands have been degraded through unsustainable activities, conversion, and overexploitation of resources (Celik et al., 2004; Boyle et al., 2011). Pressures on wetlands have been exacerbated by catchment degradation and pollution, leading to the proliferation of invasive species (Sahu and Sikdar, 2011). It is important to recognize and protect the functions and benefits provided by wetlands, including their role in supporting agriculture, tourism, industry, biodiversity conservation, social, economic, and cultural activities, and to implement sustainable management practices to conserve these vital ecosystems for current and future generations (Costa et al., 2013; Rana et al., 2016).

The deposition of heavy metals in wetlands and estuaries is a worldwide issue. The industrial revolution significantly increased heavy metal pollution in the environment

DOI: 10.1201/9781003297901-2

(Rai et al., 2012). The release of metal-containing sludge into the ocean continues to disrupt aquatic ecosystems, while metals such as iron (Fe), copper (Cu), zinc (Zn), and manganese (Mn) are essential for living beings, but overexposure can cause adverse impacts (Rajeshkumar and Li, 2018). In contrast, non-essential elements like mercury (Hg), lead (Pb), arsenic (As), and cadmium (Cd) are harmful even in trace amounts (Atobatele and Olutona, 2015).

Environmental factors such as electrical conductivity, pH, temperature, sediment particle size, oxidation-reduction potential, as well as contaminants released from sediments, can cause secondary pollution in an aquatic environment (Cao et al., 2017). Rapid human activities near rivers and wetlands in developing countries have led to heavy metal release, impacting biota (Kumar et al., 2020). Studies have investigated heavy metal contamination in rivers, such as the Ganga River in West Bengal, attributed to urban waste, industrial effluents, and catchment area outflows (Kar et al., 2008).

High levels of Cd, Cu, Pb, and Fe in aquatic and terrestrial ecosystems can act as ecological poisons (Ali et al., 2019). Research has focused on metal pollution and bioaccumulation (Wu et al., 2017), with studies describing heavy metal pollution in sublittoral ecosystems (Zahran et al., 2015). Contamination assessments have been conducted to understand the ecological and health risks associated with heavy metals.

Fishes are commonly used as indicators of metal pollution due to their position at the top of the aquatic food chain and their sensitivity to heavy metal exposure (Mukherjee et al., 2022). Fish studies have explored trace metal bioaccumulation in various species (Raeisi et al., 2014). Freshwater mussels, being benthic in nature, are also significantly affected by heavy metal pollution, raising concerns about the health risks associated with consuming fish and mussels (Al-Hossainy et al., 2017).

The objective of this work is to understand the bioaccumulation of Cadmium in fish muscle, which leads to biomagnification and subsequent adverse impacts on human health.

2.2 HEAVY METALS IN POLLUTING WETLAND ECOSYSTEMS

Heavy metals are known to play a significant role in threatening ecosystems. They are often discharged from wastewater treatment plants, especially in cases where there is the input of industrial waste (Joystu et al., 2017). Unlike organic contaminants that can be decomposed during natural processes, heavy metals in the river environment can get accumulated, magnify, or be concentrated by organisms and build into organic complexes (Das and Kumar, 2022). Even when present in low concentrations in the overlying water column, heavy metals can accumulate in sediments, and this accumulation is dependent on various factors such as the nature of the sediment particles, the properties of the adsorbed compounds, and the prevailing physiochemical conditions (Avijit and Kamalakannan, 2023; Massolo et al., 2012). Several studies have evaluated the risk of heavy metal distribution in sediment, and it is important to understand the environmental risks associated with heavy metal contamination (Sarkar et al., 2002; Nimmi et al., 2022; Dash et al., 2021). Heavy metals enter the wetland environment through a variety of natural and anthropogenic mechanisms, including hydrological processes, natural erosion, environmental deposition, farming, non-point source pollutants, and industrial activities (Jayawardena et al., 2021; Das Sharma, 2019). Heavy metals can enter organisms by direct inhalation, ingestion, and skin contact.

2.3 CONTAMINATION OF CADMIUM IN WETLANDS

Cadmium (Cd) compounds primarily originate from industrial activities, including the consumption of fossil fuels, waste incineration, use of phosphate fertilizers, and disposal of sewage sludge. Additionally, recycling and cremation of municipal solid waste and toxic waste also contribute to the release of Cd into the environment, further increasing the risk of exposure to this toxic metal (Alloway and Steinnes, 1999). The wetlands serve as the reservoir for sewage sludge. They have a major role in purifying the water while providing a substratum for the breeding of fish which are used by locals as food. But as the sewage contains heavy metals such as cadmium, it leads to the entry of cadmium into the food chain. The two major wetlands in West Bengal that are studied in the paper are the East Kolkata Wetlands and the Dankuni Wetlands.

2.3.1 East Kolkata Wetlands (EKW)

The East Kolkata Wetlands (EKW), which has been designated as a Ramsar site, serves as a vast collection of ponds that receive untreated wastewater and raw sewage from various sources, including tanneries and leather industries, municipalities, and other industries in the city of Kolkata (Dasgupta, 2014). Cadmium (Cd) has been found to impact human health when it is inhaled or ingested through food consumption, such as mussels.

Field surveys were conducted in the vicinity of EKW to assess the consumption of mussels by local populations. The accumulation of cadmium in EKW can have severe ecological and health implications. It can affect soil quality, reduce the productivity of the wetlands, and contaminate the groundwater. Exposure to cadmium can also lead to several health issues, including kidney damage, osteoporosis, and cancer (Wu et al., 2016). The surveys aimed to identify patterns of mussel consumption among different economic status groups across various age groups. Additionally, the health of EKW ecosystem was assessed using L. marginalis, a well-established bioindicator species. Furthermore, the use of L. marginalis as a bioindicator species proved valuable in understanding the environmental health of EKW. As a bioindicator, L. marginalis can provide critical information about the overall health and condition of the ecosystem, helping to assess the impact of environmental factors, including Cd contamination on the wetland's ecological health (Alaee et al., 2014). The management strategies include measures to reduce cadmium inputs into the wetland system, improve wastewater treatment processes, and implement regulatory measures to minimize cadmium contamination in mussels.

2.3.2 Dankuni Wetlands (DW)

The Dankuni Wetlands (DW), spanning an area of approximately 30 km^2, is located in the lower Gangetic floodplains and represents one of the few remaining contiguous marshy stretches in the region. It is bounded by a national highway on the east and runs parallel to a busy rail track, making it easily accessible. The DW is characterized by a complex network of small ponds that are often utilized for small-scale aquaculture activities. These ponds are interspersed with patches of relatively higher dry land. The wetland landscape is dynamic, with water levels fluctuating seasonally and providing critical environments for a diversity of plants and wildlife. But the wetland areas are usually urbanized and industrialized and are therefore subject to the release of trace metals into the environment (Sen et al., 2023). Cadmium is one of the vital trace metals that are being generated in recent developing countries. Sediments are biologically essential components of aquatic habitats, as well as a reservoir of toxins that play a crucial role in the trophic status of any water body. Contamination of the aquatic environment by Cd has been accumulating in the fishes and plants living in the water bodies (Patra et al., 2007).

The wetlands located in West Bengal are significant as these provide a consistent source of water for multiple activities such as irrigation, fishing, groundwater recharge, supporting the livelihoods of many people. Unfortunately, due to the discharge of untreated or partially treated domestic and municipal sewage, effluents from a range of industries are carried by storm weather flow to contaminate the wetlands, and these get deposited and contaminate the bottom soil. Finally, they are transferred into the higher trophic levels through the wetland flora and fauna.

2.4 ACCUMULATION OF CADMIUM AND THE MAGNIFICATION IN FISHES

Bioaccumulation is a natural phenomenon occurring continuously in the ecosystem. It refers to the gradual increase in the concentration of a contaminant along a food chain to higher trophic levels. Hence the organisms at higher trophic levels such as human beings, are prone to accumulate more contaminants through their food, leading to magnification of the contaminants in their body, leading to adverse health impacts (Sarkar et al., 2016).

Bioaccumulation of heavy metals, especially Cd in fishes of wetlands, is a very common and concerning occurrence. The wetlands are contaminated by metal wastes in different ways like stormwater or surface run-off. Hence the fishes living in these wetlands are also affected, in turn leading to accumulation of the contaminants in their body (Huseen and Mohammed, 2019). Generally, there are three entry points for the metals in the fish's body – skin, gill, and alimentary tract (Jitar et al., 2014). In the case of cadmium, it has been seen that the Cd ions enter the fish's body through gill respiration majorly and gets accumulated in the liver from where it is being distributed to other tissues. Bioaccumulation of metals like Cd in fishes effect their health and the health of humans consuming them (Naeem et al., 2021). The examination of the fish can help to determine the level of water pollution as well as to assess the risk of how it may affect the humans consuming such contaminated fish (Ali and Khan, 2018a). It has been observed that carnivorous fishes can accumulate more contaminants, being at a higher position of the food web (Weber et al., 2013). Also, there are reports of biomagnification of Cd in omnivorous fishes (Bawuro et al., 2018).

2.5 EFFECTS OF CADMIUM ON FISHES

2.5.1 Behavioural Change

The presence of environmental contaminants can significantly impact the behaviour of fish in aquatic ecosystems. Studies have shown that cadmium, which is known to be a neurotoxicant, can alter various aspects of fish behaviour, including locomotion, attraction and avoidance, swimming performance, respiratory behaviour, social interactions, reproductive behaviour, feeding, and predator avoidance (Atchison et al., 1987).

For example, fish exposed to toxic levels of contaminants may exhibit irregular, erratic, and darting swimming movements, hyperexcitability, capsizing, and attaching to the surface (Shavaji and Asiya Nuzhat, 2015; Mishra et al., 2012). The behavioural abnormalities are often associated with disruptions in nervous and muscular coordination, inhibition of brain enzyme activity, and cytotoxic hypoxia, which can cause brain damage and loss of equilibrium (Rao et al., 2005).

Furthermore, research has also shown that fish reproductive health can be negatively impacted by contaminants in the aquatic environment (Crump and Trudeau, 2009). Overall, the presence of heavy metals and other contaminants in aquatic environments can have significant effects on the behaviour of fish, potentially leading to disruptions in their normal activities and ecological interactions.

2.5.2 Endocrinological Effect

Cadmium exposure can seriously damage the reproductive system of fish. Prolonged exposure to cadmium can result in structural and functional damage to ovarian and sperm tissues, leading to reduced reproductive ability, incomplete development of hatched embryos, and retarded growth and development of juvenile fish. Gracia, in 2013, conducted a study on the effects of Cadmium (Cd) on the function of the thyroid gland in tilapia (*Oreochromis niloticus*). The study found that acute exposure to sub-lethal concentrations of cadmium resulted in changes in endocrine status and carbohydrate metabolism in O. niloticus. Chronic exposure to cadmium was shown to disrupt reproduction in fish, inhibiting the induction of vitellogenin, delaying oogenesis in brown trout, enhancing luteinizing hormone (LH) secretion, and decreasing parameters such as gonadosomatic index (GSI) and ovulation in Prussian carp (Szczerbik et al., 2006). Cadmium can also disrupt the reproductive regulation system of fish by affecting the expression and secretion of hormones. It can affect the synthesis and secretion of hormones related to the hypothalamic-pituitary-gonadal (HPG) axis, which is a critical pathway involved in fish reproduction. Cadmium can directly inhibit the expression of oestrogen receptors, leading to disruptions in yolk production, egg development, and reproductive outcomes. It can also bind to oestrogen receptors in the form of oestrogen analogy, reducing the concentration of oestrogen receptors in the gonads, further impacting reproductive processes (Liu et al., 2022).

2.5.3 Genotoxic Effect

Fishes have been widely used as indicators for assessing the genotoxic and mutagenic effects of xenobiotics (chemical substances foreign to an organism) and physical agents. Various tests such as sister chromatid exchange (SCE) tests and the comet assay have been applied to different fish species to evaluate genetic alterations and genotoxicity. For example, studies have shown that exposure to heavy metals such as cadmium can induce molecular changes and genetic damage in fish. In a genotoxicity study on

(Cavas et al., 2005) common carp (*Cyprinus carpio*), Prussian carp (*Carassius gibelio*), and peppered cory (*Corydoras paleatus*), formation of micronuclei and binuclei in fish cells was found, caused by exposure to cadmium. A SCE test on various fish species indicated the effects of carcinogenic-mutagenic chemicals on the kidney cells of *Cyprinus carpio* (Rana et al., 2011). Thus, fishes have been shown to be effective indicators for assessing environmental toxicity through xenobiotics or heavy metals.

2.5.4 Oxidative Damage

The stress caused by cadmium in fishes in the aquatic environment results in a plethora of reactive nitrogen species (RNS) and oxygen species (ROS) radicals are highly reactive molecules which contain oxygen that are produced as natural by-products of cellular metabolism. An increase in the ROS production under stress conditions activates the antioxidant defense mechanisms in fish to maintain cellular homeostasis. The toxic effect of Cd on fish is mainly reflected in the oxidative stress caused by reactive oxygen species, and its excessive accumulation leads to structural changes in biological macromolecules, such as DNA and proteins (Dutta et al., 2022). At the same time, ROS tends to change peroxidation reaction of the fatty acids (unsaturated) which are present on the cell membrane lipid and produces a toxic compound known as MDA (malondialdehyde), which damages the cell membrane structure of the fishes (Dai et al., 2020).

The appraisal of in intracellular ROS caused by Cd mainly occurs through the following pathways:

> Cadmium, as it tends to inhibit the gaseous exchange transmission chain in the mitochondria, which results in higher production of the by-product ROS content. The content of reactive oxygen species is increased by Cd as it increases nitric oxide synthase's activity, and this hinders the level of expression of the related genes in the fishes (Wu et al., 2021). Cadmium decreases the activity of antioxidant enzymes (Chen et al., 2021; Ghosh et al., 2019), which reduces the antioxidant capacity of cells.

2.6 EFFECTS ON HUMANS

Cadmium intake generally occurs through food and water. It gets absorbed from the gastrointestinal tract in combination with thiol groups of cysteine and glutathione. Upon entering cells, cadmium binds to metallothioneins (MT), the proteins that contain a high number of cysteine residues and bind to zinc. It leads to the formation of a Cd complex (Cd-MT complexes) with metallothioneins (low molecular weight protein), which are released into the blood from the liver and then travel and get accumulated in different parts of the body (Roy et al., 2011). Although the exact function of MT is not known, it is believed that they play a role in distributing zinc to other proteins and protecting against cadmium exposure and oxidative stress. MT genes are activated by both zinc and cadmium, and the proteins are degraded by various pathways. While most of the bound zinc is released from MT, cadmium remains bound even at lysosomal pH, which may explain why Cd-MT has a longer half-life in vivo than Zn-MT (Yang and Shu, 2015). In addition to MT, glutathione also plays

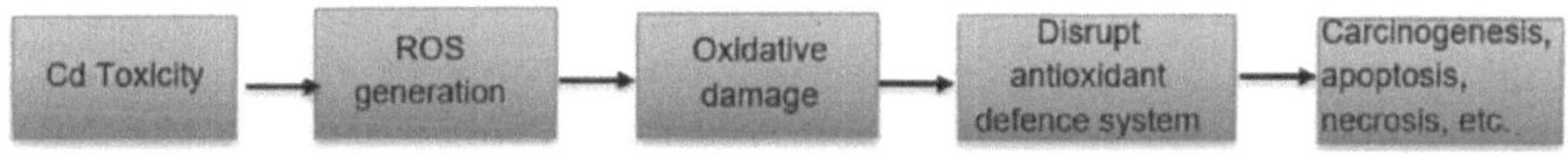

FIGURE 2.1 The effect of Cd toxicity on human health.

a role in binding to cadmium and may contribute to its excretion from the body in mammals. It can also get simply diffused through a system of membrane transporter proteins such as hZTL1 and ZNT1.

Cd causes kidney damage and long-term toxicity symptoms such as reduced renal function (Rahman and Singh, 2019). Cadmium affects the transport pathway of the kidney in which MT plays a huge role (Satarug et al., 2005). Cadmium can lead to glucosuria by decreasing the synthesis of Na+ glucose co-transporter in the proximal tubular brush border membrane, decreasing the abundance of sites for glucose reabsorption from the glomerular filtrate (Sarkar et al., 2013).

High concentrations of Cd were also found to be the cause of heart- and lung-related diseases (Satarug et al., 2011). Chronic exposure to low concentrations of Cd leads to an increase in blood pressure and cardiac output. In a study, it was seen that the strength of cadmium in blood pressure is found to be greatest among non-smokers, intermediate among former smokers, and small or absent among current smokers (Rani et al., 2014).

It can impact the cell cycle and its functions, such as cell differentiation, proliferation, and apoptosis, and is highly carcinogenic (Ali and Khan, 2018b). Cadmium carcinogenesis is due to a lack of effective excretion of Cd which leads to an increase in the concentration of cadmium in the body. It can cause oxidative stress which may result in disruption of the antioxidant defence mechanism, especially the action of superoxide dismutase, and glutathione reduces and reactive oxygen species (ROS) is formed. This ROS attacks the carbohydrates, proteins, and enzymes present in a cell and can cause irreversible damage. They cause DNA damage and mutation resulting in carcinogenesis (Genchi et al., 2020) as shown in Figure 2.1. The other reasons for carcinogenesis are aberrant gene expression, inhibition of DNA damage repair, and apoptosis causing genomic instability (Hartwig, 2010). It can also work as a co-genotoxic chemical causing proto-oncogene activation, that is, it activates the stress response genes which can lead to malignant transformation of cell lines. The activation of genes such as c-jun and c-myc and down regulation of the p53 tumor suppressor gene leads to the inhibition of apoptosis (Joseph, 2009). Suppression of the activity of the transcription factor NFkB also plays a role in this. It can cause breast cancer in females and prostate cancer in males (Strumylaite et al., 2019). Various preventive measures and nature-based solutions are being used to limit the discharge of heavy metals into the water system (Khound and Bhattacharyya, 2017).

2.7 CONCLUSION

Cadmium is a highly toxic metal that, when present in wetlands, causes contamination to aquatic organisms. Fishes, being at the top of the trophic level, consume the maximum. Through the food chain, it reaches humans and causes various toxic effects. In this work, various toxic effects of Cd on fish as well as human beings have been discussed. To save humanity from the deleterious effects of Cd contamination, a holistic approach has to be taken which will initiate from the source industries.

REFERENCES

Alaee, S., Talaiekhozani, A., Rezaee, S., Alaee, K., & Yousefian, E. (2014). Cadmium and male infertility. *Journal of Infertility and Reproductive Biology*, *2*(2), 62–69.

Al-Hossainy, A. F., Mohamed, A. E., Hassan, F. S., & Abd Allah, M. M. (2017). Determination of cadmium and lead in perch fish samples by differential pulse anodic stripping voltammetry and furnace atomic absorption spectrometry. *Arabian Journal of Chemistry*, *10*, S347–S354.

Ali, H., & Khan, E. (2018a). Bioaccumulation of non-essential hazardous heavy metals and metalloids in freshwater fish. Risk to human health. *Environmental Chemistry Letters*. https://doi.org/10.1007/s10311-018-0734-7

Ali, H., & Khan, E. (2018b). Trophic transfer, bioaccumulation, and biomagnification of non-essential hazardous heavy metals and metalloids in food chains/webs—Concepts and implications for wildlife and human health. *Human and Ecological Risk Assessment: An International Journal*, 1–24. https://doi.org/10.1080/10807039.2018.1469398

Ali, H., Khan, E., & Ilahi, I. (2019). Environmental chemistry and ecotoxicology of hazardous heavy metals: Environmental persistence, toxicity, and bioaccumulation. *Journal of Chemistry*, 2019.

Alloway, B. J., & Steinnes, E. (1999). Anthropogenic additions of cadmium to soils. *Cadmium in Soils and Plants*, *85*, 97–123.

Atchison, G. J., Henry, M. G., & Sandheinrich, M. B. (1987). Effects of metals on fish behavior: A review. *Environmental Biology of Fishes*, *18*, 11–25.

Atobatele, O. E., & Olutona, G. O. (2015). Distribution of three non-essential trace metals (Cadmium, Mercury and Lead) in the organs of fish from Aiba Reservoir, Iwo, Nigeria. *Toxicology Reports*, *2*, 896–903.

Avijit, D., & Kamalakannan, D. (2023). Geochemistry and mass balance of selected heavy metals in East Kolkata Wetlands, a Ramsar site of West Bengal, India. *Journal of Hazardous Materials*, *445*, 130574.

Barjhoux, I., Baudrimont, M., Morin, B., Landi, L., Gonzalez, P., & Cachot, J. (2012). Effects of copper and cadmium spiked-sediments on embryonic development of Japanese medaka (Oryzias latipes). *Ecotoxicology and Environmental Safety*, *79*, 272–282. https://doi.org/10.1016/j.ecoenv.2012.01.011

Bawuro, A. A., Voegborlo, R. B., & Adimado, A. A. (2018). Bioaccumulation of heavy metals in some tissues of fish in lake Geriyo, Adamawa State, Nigeria. *Journal of Environmental and Public Health*, *2018*, Article ID 1854892. https://doi.org/10.1155/2018/1854892

Boyle, D., Hogstrand, C., & Bury, N. R. (2011). Physiological response to a metal-contaminated invertebrate diet in zebrafish: Importance of metal speciation and regulation of metal transport pathways. *Aquatic toxicology (Amsterdam, Netherlands)*, *105*(1–2), 21–28. https://doi.org/10.1016/j.aquatox.2011.05.003

Cao, Q., Wang, H., Chen, X., Wang, R., & Liu, J. (2017). Composition and distribution of microbial communities in natural river wetlands and corresponding constructed wetlands. *Ecological Engineering*, *98*, 40–48.

Cavas, T., Garanko, N. N., & Arkhipchuk, V. V. (2005). Induction of micronuclei and binuclei in blood, gill and liver cells of fishes subchronically exposed to cadmium chloride and copper sulphate. *Food and Chemical Toxicology*, *43*(4), 569–574.

Celik, U., Cakli, S., & Oehlenschläger, J. (2004). Determination of the lead and cadmium burden in some northeastern Atlantic and Mediterranean fish species by DPSAV. *European Food Research and Technology*, *218*, 298–305.

Chen, J., Chen, D., Li, J., Liu, Y., Gu, X., & Teng, X. (2021). Cadmium-induced oxidative stress and immunosuppression mediated mitochondrial apoptosis via JNK-FoxO3a-PUMA pathway in common carp (Cyprinus carpio L.) gills. *Aquatic Toxicology*, *233*, 105775.

Costa, C., Antonucci, F., Boglione, C., Menesatti, P., Vandeputte, M., & Chatain, B. (2013). Automated sorting for size, sex and skeletal anomalies of cultured seabass using external shape analysis. *Aquacultural Engineering*, *52*, 58–64.

Crump, K. L., & Trudeau, V. L. (2009). Mercury-induced reproductive impairment in fish. *Environmental Toxicology and Chemistry*, *28*, 895–907.

Dai, Z., Cheng, J., Bao, L., Zhu, X., Li, H., Chen, X., . . . Huang, H. (2020). Exposure to waterborne cadmium induce oxidative stress, autophagy and mitochondrial dysfunction in the liver of Procypris merus. *Ecotoxicology and Environmental Safety*, *204*, 111051.

Das, G., & Kumar, A. (2022). Wetland flora of West Bengal for phytoremediation: Physiological and biotechnological studies—A review. *Biotechnological Innovations for Environmental Bioremediation*, *1*, 455–485.

Dasgupta, S. (2014). *Studies on the ecological status of East Kolkata Wetlands with special reference to effects of tannery effluents on fish and fish food organisms*. Kalyani: University of Kalyani, India.

Dash, S., Borah, S. S., & Kalamdhad, A. S. (2021). Heavy metal pollution and potential ecological risk assessment for surficial sediments of Deepor Beel, India. *Ecological Indicators*, *122*, 107265.

Das Sharma, S. (2019). Risk assessment and mitigation measures on the heavy metal polluted water and sediment of the Kolleru Lake in Andhra Pradesh, India. *Pollution*, *5*(1), 161–178.

Dutta, J., Zaman, S., Thakur, T. K., Kaushik, S., Mitra, A., Singh, P., . . . Datta, R. (2022). Assessment of the bioaccumulation pattern of Pb, Cd, Cr and Hg in edible fishes of East Kolkata Wetlands, India. *Saudi Journal of Biological Sciences*, *29*(2), 758–766.

Garcia-Santos, S., Fontaínhas-Fernandes, A., Monteiro, S. M., & Wilson, J. M. (2013). Effects of exposure to cadmium on some endocrine parameters in tilapia, Oreochromis niloticus. *Bulletin of Environmental Contamination Toxicology*, *90*, 55–59.

Genchi, G., Sinicropi, M. S., Lauria, G., Carocci, A., & Catalano, A. (2020). The effects of cadmium toxicity. *International Journal of Environmental Research and Public Health*, *17*(11), 3782.

Ghosh, S., Pal, P., Adhikary, J., Sarkar, A., & Das, B. K. (2019). Bioconcentration of cadmium (Cd) in rohu (Labeo rohita) collected from Sealdah, Garia and Sonarpur fish market in West Bengal. *Journal of Entomology and Zoology Studies*, *7*(2), 772–776.

Grosell, M. (2011). Intestinal anion exchange in marine teleosts is involved in osmoregulation and contributes to the oceanic inorganic carbon cycle. *Acta Physiologica*, *202*(3), 421–434.

Hartwig, A. (2010). Mechanisms in cadmium-induced carcinogenicity: Recent insights. *Biometals*, *23*, 951–960.

Huseen, H. M., & Mohammed, A. J. (2019). Heavy metals causing toxicity in fishes. *Journal of Physics: Conference Series*, *1294*(6), 062028.

Jayawardena, U. A., Wickramasinghe, D. D., & Udagama, P. V. (2021). Cytogenotoxicity evaluation of a heavy metal mixture, detected in a polluted urban wetland: Micronucleus and comet induction in the Indian green frog (Euphlyctis hexadactylus) erythrocytes and the Allium cepa bioassay. *Chemosphere*, *277*, 130278.

Jitar, O., Teodosiu, C., Oros, A., Plavan, G., & Nicoara, M. (2014). Bioaccumulation of heavy metals in marine organisms from the Romanian sector of the Black Sea. *New Biotechnology*. https://doi.org/10.1016/j.nbt.2014.11.004

Joseph, P. (2009). Mechanisms of cadmium carcinogenesis. *Toxicology and Applied Pharmacology*, *238*(3), 272–279.

Joystu, D., Choudhary, G., & Abhijit, M. (2017). Bioaccumulation of toxic heavy metals in the edible fishes of eastern Kolkata wetlands (EKW), the designated Ramsar Site of West Bengal, India. *International Journal of Aquaculture and Fishery Sciences*, *3*(1), 18–21.

Kar, D., Sur, P., Mandai, S. K., Saha, T., & Kole, R. K. (2008). Assessment of heavy metal pollution in surface water. *International Journal of Environmental Science & Technology*, *5*, 119–124.

Khound, N. J., & Bhattacharyya, K. G. (2017). Multivariate statistical evaluation of heavy metals in the surface water sources of Jia Bharali river basin, North Brahmaputra plain, India. *Applied Water Science*, *7*, 2577–2586.

Kumar, M., Patel, A. K., Shah, A. V., Raval, J., Rajpara, N., Joshi, M., & Joshi, C. G. (2020). First proof of the capability of wastewater surveillance for COVID-19 in India through detection of genetic material of SARS-CoV-2. *Science of the Total Environment*, *746*, 141326.

Kumar, V., Sharma, A., Pandita, S., Bhardwaj, R., Thukral, A. K., & Cerda, A. (2020). A review of ecological risk assessment and associated health risks with heavy metals in sediment from India. *International Journal of Sediment Research*, *35*(5), 516–526.

Liu, Y., Chen, Q., Li, Y., Bi, L., Jin, L., & Peng, R. (2022). Toxic effects of cadmium on fish. *Toxics*, *10*(10), 622.

Massolo, S., Bignasca, A., Sarkar, S. K., Chatterjee, M., Bhattacharya, B. D., & Alam, A. (2012). Geochemical fractionation of trace elements in sediments of Hugli River (Ganges) and Sundarban wetland (West Bengal, India). *Environmental Monitoring and Assessment*, *184*, 7561–7577.

McElroy, J. A., Shafer, M. M., Trentham-Dietz, A., Hampton, J. M., & Newcomb, P. A. (2006). Cadmium exposure and breast cancer risk. *Journal of the National Cancer Institute*, *98*(12), 869–873.

Mishra, A. S., Nautiyal, P., & Somvanshi, V. S. (2012). Length–weight relationship, condition factor and sex ratio of Uroteuthis (Photololigo) duvaucelii (d'Orbigny, 1848) from Goa, west coast of India. *Journal of Marine Biological Association of India*, *54*(2), 65–68.

Mitsch, W. J., Zhang, L., Stefanik, K. C., Nahlik, A. M., Anderson, C. J., Bernal, B., . . . Song, K. (2012). Creating wetlands: Primary succession, water quality changes, and self-design over 15 years. *Bioscience*, *62*(3), 237–250.

Mukherjee, A., Pal, S., Das, P., & Mukhopadhyay, S. K. (2022). Heavy metal exposure to a migratory waterfowl, Northern Pintail (Anas acuta), in two peri-urban wetlands. *Science of the Total Environment*, *851*, 158238.

Naeem, S., Ashraf, M., & Babar, M. E. (2021). The effects of some heavy metals on some fish species. *Environmental Science and Pollution Research*, *28*, 25566–25578. https://doi.org/10.1007/s11356–021–12385-z

Nimmi, N., Shiji, M., & Harikumar, P. S. (2022). Assessment of heavy metal contamination and ecological risk of core sediments in a coastal wetland of India. *Soil and Sediment Contamination: An International Journal*, *31*(7), 886–904.

Patra, S. K., Zaman, A., & Das, S. S. (2007). Heavy metal pollution of soils, waters and plants in the industrial area of Dankuni basin in West Bengal. In *National seminar on ecorestoration of soil and water resources towards efficient crop production*, Kalyani, India, 6–7 June 2007 (pp. 17–19). Crop and Weed Science Society.

Raeisi, S., Sharifi Rad, J., Sharifi Rad, M., & Zakariaei, H. (2014). Analysis of heavy metals content in water, sediments and fish from the Gorgan bay, southeastern Caspian sea, Iran. *International Journal of Advanced Biological and Biomedical Research*, *2*(6), 2162–2172.

Rahman, Z., & Singh, V. P. (2019). The relative impact of toxic heavy metals (THMs) (arsenic (As), cadmium (Cd), chromium (Cr) (VI), mercury (Hg), and lead (Pb)) on the total environment: An overview. *Environmental Monitoring and Assessment*, *191*, 1–21.

Rai, U. N., Prasad, D., Verma, S., Upadhyay, A. K., & Singh, N. K. (2012). Biomonitoring of metals in Ganga water at different ghats of Haridwar: Implications of constructed wetland for sewage detoxification. *Bulletin of Environmental Contamination and Toxicology*, *89*, 805–810.

Rajeshkumar, S., & Li, X. (2018). Bioaccumulation of heavy metals in fish species from the Meiliang Bay, Taihu Lake, China. *Toxicology Reports*, *5*, 288–295.

Rana, S., Jana, J., Bag, S. K., Roy, S. M., Biswas, J. K., Ganguly, S., . . . Jana, B. B. (2011). Performance of constructed wetlands in the reduction of cadmium in a sewage treatment cum fish farm at Kalyani, West Bengal, India. *Ecological Engineering*, *37*(12), 2096–2100.

Rana, V., Maiti, S. K., & Jagadevan, S. (2016). Ecological risk assessment of metals contamination in the sediments of natural urban wetlands in dry tropical climate. *Bulletin of Environmental Contamination and Toxicology*, *97*, 407–412.

Rani, A., Kumar, A., Lal, A., & Pant, M. (2014). Cellular mechanisms of cadmium-induced toxicity: A review. *International Journal of Environmental Health Research*, *24*(4), 378–399.

Rao, J. V., Begum, G., Pallela, G., Usman, P. K., & Rao, R. N. (2005). Changes in behavior and brain acetylcholinesterase activity in mosquito fish Gambusia Affinisin relation to sublethal exposure of chlorpyrifos. *International Journal of Environment Research*, *2*(3–4), 478–483.

Roy, U. S., Chattopadhyay, B., Datta, S., & Mukhopadhyay, S. K. (2011). Metallothionein as a biomarker to assess the effects of pollution on Indian Major carp species from wastewater-fed fishponds of East Calcutta wetlands (a Ramsar Site). *Environmental Research, Engineering and Management*, *58*(4), 10–17.

Sahu, P., & Sikdar, P. K. (2011). Threat of land subsidence in and around Kolkata city and east Kolkata wetlands, West Bengal, India. *Journal of Earth System Science*, *120*, 435–446.

Sarkar, A., Ravindran, G., & Krishnamurthy, V. (2013). A brief review on the effect of cadmium toxicity: From cellular to organ level. *International Journal of Biotechnology*, *3*, 17–36.

Sarkar, S. K., Bhattacharya, B., Debnath, S., Bandopadhaya, G., & Giri, S. (2002). Heavy metals in biota from Sundarban Wetland Ecosystem, India: Implications to monitoring and environmental assessment. *Aquatic Ecosystem Health & Management*, *5*(4), 467–472.

Sarkar, T., Alam, M. M., Parvin, N., Fardous, Z., Chowdhury, A. Z., Hossain, S., . . . Biswas, N. (2016). Assessment of heavy metals contamination and human health risk in shrimp collected from different farms and rivers at Khulna-Satkhira region, Bangladesh. *Toxicology Reports*, *3*, 346–350. https://doi.org/10.1016/j.toxrep.2016.03.003

Satarug, S., Garrett, S. H., Sens, M. A., & Sens, D. A. (2011). Cadmium, environmental exposure, and health outcomes. *Ciência & Saúde Coletiva*, *16*(5), 2587–2602. https://doi.org/10.1590/s1413-81232011000500029

Satarug, S., Nishijo, M., Ujjin, P., Vanavanitkun, Y., & Moore, M. R. (2005). Cadmium-induced nephropathy in the development of high blood pressure. *Toxicology Letters*, *157*(1), 57–68.

Sen, P., Das, N., Saha, N. C., Ghosh, S., Mondal, A., & Mandal, S. (2023). Modeling of cadmium bioaccumulation dynamics in Channa punctata (Bloch, 1793) of Dankuni wetland ecosystem and assessment of risk to human health. *Applied Biochemistry and Biotechnology*, 1–18.

Sfakianakis, D. G., Renieri, E., Kentouri, M., & Tsatsakis, A. M. (2015). Effect of heavy metals on fish larvae deformities: A review. *Environmental Research*, *137*, 246–255.

Shavaji, A. N. F. B. (2015). Behavioral and respiratory responses of the freshwater fish, umbaia carpio (Linnaeus) exposed to cadmium chloride. *International Journal of Pharmaceuticals and Biological Archives*, *6*(5), 12–16.

Storelli, M. M., Storelli, A., D'Addabbo, R., Marano, C., Bruno, R., & Marcotrigiano, G. O. (2005). Trace elements in loggerhead turtles (Caretta caretta) from the eastern Mediterranean Sea: Overview and evaluation. *Environmental Pollution*, *135*(1), 163–170.

Strumylaite, L., Kregzdyte, R., Bogusevicius, A., Poskiene, L., Baranauskiene, D., & Pranys, D. (2019). Cadmium exposure and risk of breast cancer by histological and tumor receptor subtype in White Caucasian women: A hospital-based case-control study. *International Journal of Molecular Sciences*, *20*, 3029.

Szczerbik, P., Mikolajczuk, T., Sokolowska-Mikolajczuk, M., Socha, J., Chyb, J., & Epler, P. (2006). Influence of long-term exposure to dietary cadmium on growth, maturation and reproduction of gold fish. *Aquatic Toxicology*, *77*(2), 126–135.

Weber, P., Behr, E. R., Knorr, C. D. L., Vendruscolo, D. S., Flores, E. M. M., Dressler, V. L., & Baldisserotto, B. (2013). Metals in the water, sediment, and tissues of two fish species from different trophic levels in a subtropical Brazilian river. *Microchemical Journal*, *106*(1), 61–66.

Wu, H., Liao, Q., Chillrud, S. N., Yang, Q., Huang, L., Bi, J., & Yan, B. (2016). Environmental exposure to cadmium: Health risk assessment and its associations with hypertension and impaired kidney function. *Scientific Reports*, *6*(1), 29989.

Wu, H., Wang, X., He, X., Zhang, S., Liang, R., & Shen, J. (2017). Effects of root exudates on denitrifier gene abundance, community structure and activity in a micro-polluted constructed wetland. *Science of the Total Environment*, *598*, 697–703.

Wu, Z., Zheng, R., Liu, G., Liu, R., Wu, S., & Sun, C. (2021). Calcium protects bacteria against cadmium stress via reducing nitric oxide production and increasing iron acquisition. *Environmental Microbiology*, *23*(7), 3541–3553.

Xie, J., & Shaikh, Z. A. (2006). Cadmium-induced apoptosis in rat kidney-epithelial cells involves modulation of nuclear factor-kappa-B activity. *Toxicological Sciences*, *91*, 299–308.

Yang, H., & Shu, Y. (2015). Cadmium transporters in the kidney and cadmium-induced nephrotoxicity. *International Journal of Molecular Sciences*, *16*(1), 1484–1494.

Zahran, M. A. E. K., El-Amier, Y. A., Elnaggar, A. A., Mohamed, H. A. E. A., & El-Alfy, M. A. E. H. (2015). Assessment and distribution of heavy metals pollutants in Manzala Lake, Egypt. *Journal of Geoscience and Environment Protection*, *3*(6), 107.

3 Aquatic Ecosystem Toxicity and Food Chain

Tilottama Roy

3.1 INTRODUCTION

In order for us to first understand aquatic ecosystem toxicity and its effects on food chains, we need to take a closer look into toxic chemicals and heavy metals flowing into the ocean as a result of human waste, as well as industrial and agricultural runoffs. These pollutants are usually intentionally released into rivers that then lead to their escape into the sea. These toxic materials have numerous repercussions on aquatic organisms including diseases, genetic mutations, reproductive and birth defects, behavioral changes, and even death in many dire circumstances. However, the effects and severity of the damage may vary substantially among different species. In many cases, those animals that occupy a position at the top of the food chain are most affected due to processes like bioaccumulation and biomagnification (Reid et al. 2020).

A large majority of the most dangerous toxins settle to the seafloor and are then ingested by organisms that inhabit the bottom sediments and feed off of them. These compounds, being indigestible, eventually accumulate within the animals that consume them, and increase in concentration as they pass along the food chain as animals eat them, and then are eaten in turn. This mechanism is known as biomagnification, and it describes the processes through which higher-level predators, such as fishes, birds, and marine mammals, build up greater and more dangerous amounts of toxic materials than animals lower on the food chain (Drouillard 2008).

An ecosystem is comprised of abiotic and biotic components that interact with each other to produce a stable system. Understanding the various relationships between the living organisms and their nonliving environment within a particular ecosystem is essential because these two components depend upon and impact each other in numerous ways (Crespo-Lopez et al. 2021; Vareda et al. 2019). Ecosystems operate on a daily exchange of energy which is transferred through food chains and more complex food webs. The energy exchanged within an ecosystem moves through a cyclical process combining biotic and abiotic components. A food chain can be described as a simple relationship between different organisms within a particular ecosystem. Photosynthetic organisms such as plants, green algae, and cyanobacteria usually lie at the bottom of the food chain and act as primary producers. These organisms act as nutrient sources for the primary consumers that are usually herbivores. From here onwards, a chain of consumers interact with each other strategically at different levels for their nutrient resources. After the death of the final consumer in the food chain, the bodies are degraded by different types of microbial decomposers,

DOI: 10.1201/9781003297901-3

which usually include fungi and bacteria. Food chains can often take the shape of complex food webs since most organisms usually consume more than one type of plant or animal. Apart from the transfer of nutrients, a variety of different contaminants can also transfer through the various levels of a food chain/food web. Food webs, which are usually interconnected networks of food chains, help us understand how changes to ecosystems, including removing a top predator or adding nutrients or contaminants, can affect the different members inhabiting them, either directly or indirectly. Usually, a variety of phytoplankton including cyanobacteria and algal protists combine together to constitute the basic structure of aquatic food webs. They are, in turn, consumed by primary consumers like zooplankton, small fish, and crustaceans. Humans consume aquatic life from every section of this food web.

Aquatic ecosystems include a variety of water bodies like oceans, lakes, rivers, streams, estuaries, and wetlands. Within these ecosystems, the various biotic components such as fishes, plants, herps and mammals, as well as a wide variety of microorganisms, depend on their surrounding water for survival (Taylor et al. 2007). The various aquatic ecosystems across the world can be categorized into two broad groups – marine ecosystems and freshwater ecosystems. Marine ecosystems form some of the largest aquatic ecosystems in the world, which can be further divided into oceans, estuaries, coral reefs, coastal ecosystems, etc. Freshwater ecosystems are further classified as 1) lentic ecosystems that encompass slow-moving water bodies like pools, ponds, and lakes; 2) lotic ecosystems that contain faster moving water, for example streams and rivers; and 3) wetlands containing areas where the soil remains partially saturated or inundated. Wetlands provide a means of connecting land and water and help the surrounding environment in multiple ways including providing a natural filtration system, reducing pollution, controlling flooding, and acting as breeding grounds for numerous aquatic organisms. Rivers, lakes, and estuaries serve as important transportation, recreation, and wildlife hubs. Aquatic ecosystems are usually very fragile and can be easily disturbed by pollution.

A watershed is an area of land covered by flowing water that eventually travels and joins a larger common source of water such as a lake or a pond (Kolok et al. 2009). We are all inhabitants that are linked to a watershed or drainage basin. Watersheds can be as large as the Ganga basin in India that includes the 11 basin states, namely Uttar Pradesh, Madhya Pradesh, Rajasthan, Bihar, West Bengal, Uttarakhand, Jharkhand, Haryana, Chhattisgarh, Himachal Pradesh, and Delhi. However, watersheds can also represent an area as minute and as small as a farm with a pond. A watershed is a good example of how the abiotic and biotic components within an ecosystem depend upon each other. Alterations to a watershed tend to have a negative impact on the living components within that watershed. All living organisms inhabiting a particular aquatic ecosystem share the same watershed.

In recent years, human actions have become a major cause of alterations to watersheds, brought about by population expansion and the resultant urbanization, leading to paved roadways and building constructions. The altered watersheds, in turn, affect the flow of water over the land and may eventually lead to numerous harmful pollutants to move directly into the water and contaminate it. This, in turn, has a massive negative impact on aquatic organismal life that depend on the water for their survival. For example, some fish feed on organisms in the water. Polluted water may cause these organisms to die, leaving the fish without any food. In this chapter, some major threats to aquatic ecosystems causing toxicity at different levels of the food chain will be discussed.

3.2 TOXIC METAL CONTAMINATION, BIOMAGNIFICATION, AND BIOACCUMULATION

It is essential for us to understand the distribution of toxic metals in aquatic ecosystems, so that we can assess their impacts on the environment as well as the associated health risks that they pose for humans (Hallare et al. 2011; Newman 2015). A major toxic effect on natural waters comes from microbial processes that can act as dominating factors controlling the distribution of specific metals. Although our current understanding of microbe-metal interactions is limited due to the complexity of both the microbiology and chemistry of natural systems, experimental results indicate the potential for significant interaction, at least within soil and sediment ecosystems. There is, however, a substantial amount of information available on specific interactions between microorganisms and iron and manganese and on the importance of these interactions in the biogeochemical cycling of these elements. The following discussion will focus on the cycling of toxic metals and the potential role of microbe-metal interactions in these processes.

Freshwater aquatic ecosystems form the habitats for least 10% of the world's known fauna. This includes 30% of all vertebrates and plays a crucial role in the global water cycle (Reid et al. 2020). Unfortunately, between 1970 and 2014, human activities have led to the decline of populations inhabiting the aquatic ecosystem by a whopping 83% (Reid et al. 2018). Utilization of natural resources such as minerals, fossil fuels, biomass, soil, wood, and water has increased from 27 to 92 billion tons during the past five decades, due to an increase in the demand and consumption of the world population. This has been one of the major factors for the loss of biodiversity and global environmental pollution (IRP 2019). Anthropogenic activities, in particular, majorly contribute to the natural dynamics of metals, resulting to the increase in their availability in aquatic systems in particular (Crespo-Lopez et al. 2021). Metals and metalloids like mercury, selenium, and arsenic are heavily utilized in industries like mining, agriculture, and forestry, contributing to an increase in their buildup in the environment naturally (Johnson et al. 2020). Intense anthropogenic activities have also resulted in alterations in their biogeochemical cycles, which, in turn, have led to their accumulations in the surrounding aquatic environments. Hg and Arsenic possess the potentials to accumulate at different trophic levels within the food chain, thereby causing harm to a broad range of organisms.

Biomagnification can be described as the phenomenon where the chemical concentration in an organism exceeds the concentration of its food, and eventually its persistence leads to harmful effects within the organism (Drouillard 2008; Newman 2015). Food web biomagnification more specifically describes the trophic enrichment of contaminants within food webs and refers to the progressive increase in chemical concentrations with the increasing trophic status within animals (Mackay et al. 2016). For hydrophobic organic contaminants, the terms biomagnification and food web biomagnification are usually synonymous and denote a phenomenon where the chemical potential in an animal exceeds its diet and environment, increasing through successive trophic levels. Human intervention and activities have led to the introduction of a variety of chemicals to the food web over time. Some naturally occurring metals such as cadmium, mercury, lead, and zinc have been inadvertently released by humans into our environment, and we are yet to conclude their harmful effects on our ecosystems. A variety of plastics, particularly microplastics, are gradually emerging as major threats to our ecosystem. Bioaccumulation is the buildup of chemicals in an organism that is above the level in its surrounding environment. This accumulation has, in recent days, become a major concern for our environment. Hydrophobic inorganic chemicals (HOCs) and metals are usually absorbed from aquatic or marine environments.

3.3 MERCURY

Mercury and Arsenic contamination pose a global threat to a diverse group of organisms, starting from humans to wildlife and plants. Their entry into the aquatic environments lead to majority of them getting accumulated in various fishes (Crespo-Lopez et al. 2021; Kato et al. 2020). Fish are often observed to be the organism that is most affected by such toxic metals and tend to accumulate majority of them into their systems. Since fishes are one of the major sources of protein-rich food consumed by humans globally, their toxicity eventually poses a direct harm to human health. Since biota absorb as well as accumulate chemicals, monitoring these chemicals in aquatic organisms has now become an essential part of a variety of programs that assess the chemical status of water bodies.

Effects of metal poisoning and biomagnification are numerous, and in wild organisms they can range from decreased reproductive capacity, reduced embryo viability, teratogenesis (malformations of the embryo that are congenital), and reduced enzymatic processes (Azevedo et al. 2021; Coelho et al. 2013). Although majority of the studies in terms of metal toxicity in aquatic ecosystems are focused on the effects of mercury at higher trophic levels of food webs, there is still some amount of data that has been reported to show the effects of As and Se. In aquatic organisms, metal toxicity can disrupt planktonic growth, as well as lead to growth and developmental problems in crustaceans and mollusks, and decrease photosynthetic activities, thereby further affecting a variety of ecological processes and the overall productivity in aquatic ecosystems. Excessive levels of Se have been shown to cause deformities in fishes, birds, and mammals. The accumulation of mercury above the tolerance level in humans affects the central nervous system, causes cytogenetic damage, cardiovascular and renal malfunctioning, as well as mental health issues. Mercury (Hg) is a persistent metal that enters aquatic food chains through natural and anthropogenic sources. Hg that enters our oceans, lakes, and rivers is converted to Methylmercury (MeHg) by aquatic organisms and bioaccumulates in aquatic food webs, majority of which is retained within fishes and shellfishes. Humans and wildlife are exposed to MeHg primarily through the consumption of contaminated fishes, especially large predatory fish species such as tuna, swordfish, shark, and whale. In humans, MeHg has been shown to cause neurotoxicity. The fetus is more sensitive to those effects than the adult (U.S. EPA 1997). Studies done by the World

Health Organization has shown that in countries such as Columbia, China, Brazil, Greenland, etc., where fishing is predominant, 17 out of every thousand children are affected with mental disorders caused by the consumption of contaminated fishes containing a high level of Hg.

Recent research shows that biomagnification of Hg is substantially influenced by a number of factors including mechanisms of speciation, food preferences, seasonality of the ecological dynamics, as well as physicochemical properties such as the total concentration of chemicals like Phosphorus and Nitrogen. This is more prominent in temperate biomes. Biomagnification, or the increase in the concentration of contaminants at every trophic level of the food chain, is usually caused by predatory organisms that absorb a contaminant more than they can excrete or eliminate from their bodies. As a result of this, the concentration of contaminants in their bodies may increase to a level that is more than that in their prey species. Research done in networks comprising fish, plankton, and benthic invertebrates in Lake Ontario, Canada, has shown variations in Hg concentrations ranging between 1.7, 2.4, and 0.2 throughout the year. This trend was especially predominant in fishes manifested with seasonal variation. A recent study done by Chiang and colleagues in 2021 has shown that food webs that are predominantly composed of marine macroinvertebrates, as well as fishes and seabirds (but excluding marine mammals) from Western Patagonia and the Western Arctic Peninsula, had an increase in the slope of trophic increase. This further indicated that biomagnification rates of Hg may be similar also to the great lakes, and the slope increases appeared to remain consistent across different Antarctic, tropical, and temperate ecosystems.

Another environmental factor that can explain Mercury concentrations in food chains is temperature (Lavoie et al. 2013). Several mechanisms have been suggested as related factors, including warmer temperatures that cause greater growth and stimulation of aquatic organisms, leading to a lower amount of Hg per unit of body mass as compared to areas with colder temperatures, where suppression of growth is more significant. Morphological structures as well as environmental characteristics, including concentration of gases like oxygen, as well as the pH of the water, can also affect the biomagnification of mercury. These are key factors that lead to the assimilation and mobility of Hg in a variety of organisms, especially in a variety of macrofauna. Other factors that can cause considerable effects include age, sex, rapid growth rate particularly in animals, as well as dietary changes during their growth and development. Some of these factors are linked to consumption of larger prey species, particularly more in organisms that are mature, leading to a larger amount of dietary intake of prey species, followed by an increase in MeHg concentrations that arise from the consumption of larger contaminated prey species. This pattern provides considerable support for the hypothesis that MeHg concentrations in aquatic food webs result from the combined effects of dietary exposure that is influenced by the dynamics of the particular ecosystem which is responsible for the dietary exposure and metabolic mechanisms. This further leads to their assimilation, transformation, and elimination, as well as environmental factors specific to particular geographic areas impacting the mobility of contaminants. In addition, it has been noted that in temperate freshwater lakes, Hg methylation and biomagnification can be directly affected by the levels of sulfate in the water. It may be safe to suggest that in both temperate and tropical ecosystems, bioaccumulation and biomagnification levels are dependent on the combined influences of warmer temperatures, productivity of chemicals, such as nitrogen, phosphorous, etc., of the ecosystem, availability of food, as well as the high influx of Hg-enriched organic matter.

3.4 ARSENIC (AS)

Aquatic environments may also contain Arsenic (As), present as a mixture of Arsenic and Arsenate, which is another major metal that affects the food chain and poses considerable harm to wildlife and human health. Arsenate is more prevalent in aquatic ecosystems (Kato et al. 2020). Although not much is known about the trophodynamics of Arsenic in freshwater ecosystems, natural concentration in aquatic environments range from 0.5 $\mu g\ L^{-1}$ to more than 5000 $\mu g\ L^{-1}$. In freshwater systems, typical concentrations are less than 10 $\mu g\ L^{-1}$. Moreover, the concentrations are also often determined by the availability, source, and geochemistry of the water body. Phytoplankton are the primary producers in all aquatic ecosystems and also play a crucial role in biotransformation, thereby leading to bioaccumulation and trophic transfer. Studies have also shown excessive concentrations in a variety of aquatic invertebrates including polychaetes and crustaceans, as well as in algae. A wide range of drivers for Arsenic bioaccumulation have been reported in aquatic ecosystems. In marine ecosystems, the concentration is correlated to the enrichment of Arsenobetaine and the degradation of food containing a high level of arseno-sugar, whereas in freshwater ecosystems, it's dependent on the rate of its biodilution. At higher trophic levels, the concentration was lower in pelagic dietary fishes in summer and higher in benthic fishes over winter. It is also a possibility that organisms at lower trophic levels are impacted to a lesser degree as a result of biodilution along food webs.

3.5 SELENIUM (SE)

Excessive concentration of Se, an essential element, has an adverse effect on a variety of different organisms (Jasonsmith et al. 2008; Savery et al. 2013). Selenium is introduced in a particulate form to aquatic ecosystems through a variety of planktonic organisms. From there onwards it gets incorporated into organic matter and makes its way through the food chain across a variety of inverts, ultimately to fishes. Studies have shown that Se has the potential to biomagnify only in fishes, and it may be correlated to the complexity of the food webs they are present in. Such food webs usually form networks that comprise plankton, algae, inverts, macrophytes, and fishes. Although debatable, and requiring further research, Selenium (Se) has been shown to have a reducing effect against MeHg toxicity. Selenium is believed to possess the ability to modify the biokinetic mechanisms of Hg through the action of Selenomethionine, Selenocysteine, and Selenite. A study done by Arcagni and colleagues in 2013 also suggested a direct correlation between the reductive effect of Se on MeHg, and the trophic structure, ecological processes, chemical characteristics, as well as specific detoxification mechanisms developed by some organisms in specific fishes. This study explored the relevant factors that act as drivers of biomagnification of Hg, As, and Se in aquatic food webs.

3.6 EFFECTS ON AQUATIC VERTEBRATES LIKE FISHES, BIRDS, AND MARINE MAMMALS

Metals and metalloid contamination pose a serious threat to wildlife and humans (Ruffle et al. 2019; Rumbold et al. 2018). The most affected among all aquatic organisms are fishes. Arsenic in fishes leads to enzymatic reduction and hyperglycemia, also creating a significantly damaging impact on the immune system. Hg can lead to multiple negative impacts, including damages to tissues, genes and protein synthesis, reduced reproductive capacity, as well as induction of congenital defects in the embryo. Se, although not as toxic by itself, can negatively impact and cause reproductive issues, including decreased survival of the embryo and fertility, overall reproductive success, reduced growth, and increased mortality rates in the juveniles.

In various seabirds and waterbirds, the bioaccumulation of Hg can damage their flight capacities, as well as lead to a deterioration in their overall health, especially depending on their feeding habits (Ullah et al. 2014). Hg has a high affinity to keratin and ramphoteca, which are rich in sulfur-containing amino acids. In seabirds, Hg has also been shown to affect the gonadotropin-releasing hormone, leading to a loss of reproductive success in males and females. Se also affects reproductive success in birds through loss in overall body weight, hatching problems in eggs, as well as a variety of morphological deformities including birth defects leading to defects in eyes and beaks in adults. As has often shown to cause reduction in red blood cell formation and endocrine disruption.

Although available literature is scarcer for mammals as compared to other aquatic vertebrates, studies have shown Hg toxicity can damage reproductive organs, as well as have an adverse effect on nervous and excretory systems (Wolfe et al. 1998). Arsenic has been known to cause a homeostatic imbalance, as well as disrupt endocrine functions. An excessive concentration of Se, on the other hand, has shown to reduce uptake of vitamins and minerals, causing their deficiency, further leading to cardiomyopathy and problems in the circulatory system.

3.7 MICROPLASTICS IN AQUATIC FOOD CHAINS

Anthropogenic activities are one of the leading causes for releasing plastic products into our ecosystems, due to their enormous consumptions in every aspect of our daily lives (van Raamsdonk et al. 2020). As a result of this high demand, the annual production of plastics has increased drastically and is expected to reach 33 billion tons by the year 2050. A variety of plastic products are disposed of daily into the environment, and the pollution of plastics as well as microplastics (MPs) caused by their degradation in the environment has now become a crucial environmental pollution issue that has garnered the attention of scientists, governmental and non-governmental organizations, and the public media all over the world. Large quantities of plastic are dumped in beaches, lakes, navigation channels, and other water bodies, with an even bigger volume in low-income countries that lack proper waste disposal facilities. MPs are generated through the degradation of larger plastic items. The food chain exposure process is based on human consumption of aquatic organisms, animals or plants, that may have consumed the plastic through MP-loaded water or from feeding off of other organisms. There are many studies that have reported the presence of MPs in a wide range of organisms starting from the lowest levels of the food chain, such as zooplanktons, to the highest levels in both invertebrates (crustacea, mollusks) and vertebrates (fishes, turtles, etc.). Güven et al. (2017) studied the Mediterranean coast (Turkey) for MPs in fishes, and their results revealed a total of 1,822 and 1,337 MP particles were recovered from stomachs and intestines respectively. Furthermore, it was also reported that MPs reached 92.3% of the total number of plastic items (meso/microplastics) in fishes collected from coastal and freshwater in China. There are currently over 1000 marine species that have been reported to have interacted with plastic debris, while about 800 of these have shown interaction with MPs. Aquatic organisms such as fishes, sandhoppers, sea turtles, crustaceans and mussels are some of the animals that have been primarily affected by MPs. MPs are usually transferred to organisms through ingestion, leading to an increased chemical exposure and toxicity.

3.8 EUTROPHICATION

Eutrophication is now considered as one of the leading global causes of toxicity to numerous freshwater and marine ecosystems. It is described as excessive plant and algal growth that stems from the increased availability of one or more limiting growth factors required for photosynthesis (Schindler 2006), such as sunlight, carbon dioxide, and nutrient fertilizers. Eutrophication occurs naturally over centuries as lakes age and are filled in with sediments (Carpenter 1981). The adverse effects of anthropogenic activities have led to an acceleration in the rate and extent of eutrophication through sewage discharge from factories and sewage treatment plants, as well as from contaminated water containing nitrogen and phosphorus flowing directly into various aquatic ecosystems. These factors together contribute to what is also known as cultural eutrophication. This, in turn, leads to toxicity of water sources, fisheries, and recreational water bodies (Carpenter et al. 1998). Aquaculture scientists and pond managers often start the process of eutrophication of water bodies through the addition of fertilizers in order to enhance primary productivity and increase the density and biomass of recreationally and economically important fishes through bottom-up effects on higher trophic levels (Boyd and Tucker 1998). Examples of some of the prominent consequences of cultural eutrophication include cyanobacterial blooms, contamination of potable water sources, degradation of recreational opportunities, contamination of fisheries, and death of livestock. Algal blooms often lead to the limitation of light penetration in aquatic ecosystems, reducing growth of various organisms, causing faunal death in littoral zones. These, in turn, lead to lower success of predators that need light to pursue and catch prey.

3.9 CONTAMINANTS OF EMERGING CONCERNS

Contaminants of emerging concerns or CECs are found everywhere in surface waters and have become one of the major threats to various aquatic ecosystems globally (Salimi et al. 2017). These include pharmaceuticals including different kinds of medications like antibiotics, antihistamines, analgesics etc., as well as plasticizers and flame retardants. Major sources of CECs causing toxicity to aquatic environments include small- and large-scale water treatment plants and hospitals. Some CECs also get released into aquatic ecosystems through leaching of landfills, surface runoff, manures and fertilizers used in agriculture, as well as atmospheric deposition. The distribution of CECs after their release depends on their physicochemical properties and shows the capacity to bioaccumulate in aquatic food webs, with the potential to be transformed by biotic or abiotic processes. CECs like flame retardants, personal care products, and synthetic hormones lead to endocrine dysfunctions causing a variety of health effects and reduction of reproductive potential. Reduction in reproductive potential further leads to a decrease in population and eventually has an adverse effect on the trophic cascade. A recent study by Lagadic et al. (2017) showed endocrine dysfunction in snails through tributyltin exposure, and a similar effect was also observed in vertebrates, including fishes. Behavioral effects have also been observed in fishes as a consequence of CEC exposure.

3.10 BIOREMEDIATION IN AQUATIC ECOSYSTEMS: ZOOREMEDIATION AND PHYTOREMEDIATION

Scientists and environment protection agencies across the world have come up with various methodologies to counteract some of the earlier-mentioned causes that lead to toxicity at various trophic levels within aquatic ecosystems. Here we will briefly discuss the principles of bioremediation and phytoremediation, including zooremediation. Bioremediation mostly uses microorganisms, leading to the degradation of the environmental contaminants into less toxic forms (Vidali 2001). This process makes use of bacteria, fungi, or plants that occur naturally to detoxify substances hazardous to human health and/or the environment. Although bacteria comprise the most common group of organisms used for bioremediation, phytoremediation, or the use of plants and algae (phycoremediation), as well as zooremediation, or the use of animals, are gradually become popular methods to mitigate various pollutants contaminating aquatic ecosystems. Due to ethical concerns and also because a large number of aquatic organisms are traditionally cultured or harvested commercially, eventually leading to their consumption by humans, zooremediation (Gifford et al. 2007) has long been a method not explored well. However, many animals in aquatic ecosystems naturally hyperaccumulate, stabilize, or degrade pollutants and are now being studied by scientists to remedy environmental pollution in aquatic habitats. Zooremediation can be achieved through multiple ways, namely 1) through the extraction of pollutants from an area by means of capturing the wild populations; 2) or via the animals which can be introduced to a particular water body, cultured, and harvested, or, lastly, 3) through aquaculture and supplementation or maintenance of wild animal populations that possess the capability to stabilize or degrade the pollutants. Due to the simple life histories of many animal species, they are able to resist toxicity and possess the ability to generate an economic return following remediation mechanisms. Studies have revealed that oysters, mussels, clams, fish, polychaetes, and sponges can often act as suitable bioremediators. Phytoremediation (Salt et al. 1998) is an eco-friendly and advantageous technique where plants are utilized to remove, detoxify, or immobilize environmental contaminants. The mechanism involved absorbs pollutants through roots and then accumulates them within body tissues, further decomposing and transforming them to milder and less toxic forms. Phytoremediation can be used effectively in cleaning up water contamination. Various species of aquatic plants from the families Ranunculaceae, Lemnaceae, Cyperaceae, Haloragaceae, Hydrocharitaceae, Potamogetonaceae, Typhaceae, Najadaceae, Pontederiaceae, Juncaceae, and Zosterophyllaceaehave been verified and recognized for their efficiency to accumulate inorganic and organic contaminants from water (Prasad and Freitas 2003).

REFERENCES

Arcagni, M., Campbell, L., Arribére, M., Marvin-DiPasquale, M., Rizzo, A., Ribeiro, S. 2013. Differential mercury transfer in the aquatic food web of a double basined lake associated with selenium and habitat. Sci. Totl Environ. 170–180. https://doi.org/10.1016/j.scitotenv.2013.03.008

Azevedo, L.S., Pestana, I.A., Almeida, M.G., Ferreira da Costa Nery, A., Bastos, W.R., Magalhães Souza, C.M. 2021. Mercury biomagnification in an ichthyic food chain of an amazon floodplain lake (Puruzinho Lake): influence of seasonality and food chain modeling. Ecotoxicol. Environ. Saf. 207, 2–10. https://doi.org/10.1016/j.ecoenv.2020.111249

Boyd, C.E., Tucker, C.S. 1998. Pond Aquaculture Water Quality Management. Kluwer.

Carpenter, S.R. 1981. Submersed vegetation: an internal factor in lake ecosystem succession. Am. Nat. 118, 372–383.

Carpenter, S.R., Caraco, N.F., Correll, D.L., Howarth, R.W., Sharpley, A.N., Smith, V.H. 1998. Nonpoint pollution of surface waters with phosphorus and nitrogen. Ecol. Appl. 8(3), 559–568.

Chiang, G., Kidd, K.A., Díaz-Jaramillo, M., Espejo, W., Bahamonde, P., O'Driscoll, N.J., Munkittrick, K.R. 2021. Methylmercury biomagnification in coastal aquatic food webs from western Patagonia and western Antarctic Peninsula. Chemosphere 262, 128–360. https://doi.org/10.1016/j.chemosphere.2020.128360

Coelho, J.P., Mieiro, C.L., Pereira, E., Duarte, A.C., Pardal, M.A. 2013. Mercury biomagnification in a contaminated estuary food web: effects of age and trophic position using stable isotope analyses. Mar. Pollut. Bull. 69, 110–115. https://doi.org/10.1016/j.marpolbul.2013.01.021

Crespo-Lopez, M.E., Augusto-Oliveira, M., Lopes-Araújo, A., Santos-Sacramento, L., Yuki Takeda, P., Macchi, B. de M., do Nascimento, J.L.M., Maia, C.S.F., Lima, R.R., Arrifano, G.P. 2021. Mercury: what can we learn from the Amazon? Environ. Int. 146, 106–223. https://doi.org/10.1016/j.envint.2020.106223

Drouillard, K.G. 2008. Biomagnification. In: Jørgensen, S.E., Fath, B.D. (Eds.), Encyclopedia of Ecology. Academic Press, pp. 441–448. https://doi.org/10.1016/B978-008045405-4.00377-3

Gifford, S., Dunstan, R.H., O'Connor, W., Koller, C.E., MacFarlane, G.R. 2007. Aquatic zooremediation: Deploying animals to remediate contaminated aquatic environments. Trends Biotechnol. 25(2), 60–65.

Güven, O., Gökdağ, K., Jovanović, B., Kıdeyş, A.E. 2017. Microplastic litter composition of the Turkish territorial waters of the Mediterranean Sea, and its occurrence in the gastrointestinal tract of fish. Environ. Pollut. 223, 286–294. https://doi.org/ 10.1016/j.envpol.2017.01.025

Hallare, A.V., Seiler, T.B., Hollert, H. 2011. The versatile, changing, and advancing roles of fish in sediment toxicity assessment-a review. J. Soils Sediments 11, 141–173. https://doi.org/10.1007/s11368-010-0302-7

IRP. 2019. Global resources outlook 2019: natural resources for the future we want. In: Oberle, B., Bringezu, S., Hatfeld-Dodds, S., Hellweg, S., Schandl, H., Clement, J., Cabernard, L., Che, N., Chen, D., Droz-Georget, H., Ekins, P., Fischer-Kowalski, M., Flörke, M., Frank, S., Froemelt, A., Geschke, A., Haupt, M., Havlik, P., Hüfner, R., Lenzen, M., Lieber, M., Liu, B., Lu, Y., Lutter, S., Mehr, J., Miatto, A., Newth, D., Oberschelp, C., Obersteiner, M., Pfster, S., Piccoli, E., Schaldach, R., Schüngel, J., Sonderegger, T., Sudheshwar, A., Tanikawa, H., van der Voet, E., Walker, C., West, J., Wang, Z., Zhu, B. (Eds.), A Report of the International Resource Panel. United Nations Environment Programme, Nairobi, Kenya.

Jasonsmith, J.F., Maher, W., Roach, A.C., Krikowa, F. 2008. Selenium bioaccumulation and biomagnification in lake Wallace, New South Wales, Australia. Mar. Freshw. Res. 59, 1048–1060. https://doi.org/10.1071/MF08197

Johnson, R.C., Stewart, A.R., Limburg, K.E., Huang, R., Cocherell, D., Feyrer, F. 2020. Lifetime chronicles of selenium exposure linked to deformities in an imperiled migratory fish. Environ. Sci. Technol. 54, 2892–2901. https://doi.org/10.1021/acs.est.9b06419

Kato, L.S., Ferrari, R.G., Leite, J.V.M., Conte-Junior, C.A. 2020. Arsenic in shellfish: a systematic review of its dynamics and potential health risks. Mar. Pollut. Bull. 161, 111–693. https://doi.org/10.1016/j.marpolbul.2020.111693

Kolok, A.S., Beseler, C.L., Chen, X.H., Shea, P.J. 2009. The watershed as a conceptual framework for the study of environmental and human health. Environ. Health Insights 3, 1–10. https://doi.org/10.4137/ehi.s1925

Lagadic, L., Katsiadaki, I., Biever, R., Guiney, P.D., Karouna-Renier, N., Schwartz, T., Meador, J.P. 2017. Tributyltin: Advancing the science on assessing endocrine disruption with an unconventional endocrine-disrupting compound. In: De Voogt, P. (Ed.), Reviews of Environmental Contamination and Toxicology, vol. 245. Springer, pp. 65–127.

Lavoie, R.A., Jardine, T.D., Chumchal, M.M., Kidd, K.A., Campbell, L.M., 2013. Biomagnification of mercury in aquatic food webs: A worldwide meta-analysis. Environ. Sci. Technol. 47, 85–94. https://doi.org/10.1021/es403103t

Mackay, D., Celsie, A.K.D., Arnot, J.A., Powell, D.E. 2016. Processes influencing chemical biomagnification and trophic magnification factors in aquatic ecosystems: implications for chemical hazard and risk assessment. Chemosphere 154, 99–108. https://doi.org/10.1016/j.chemosphere.2016.03.048

Newman, M.C. 2015. Fundamentals of Ecotoxicology the Science of Pollution. CRC Press/Taylor & Francis. https://doi.org/10.1017/CBO9781107415324.004

Prasad, M.N.V., Freitas, H.M.D.O. 2003. Metal hyperaccumulation in plants – Biodiversity prospecting for phytoremediation technology. Electron. J. Biotechnol. 6, 3. http://dx.doi.org/10.2225/vol6-issue3-fulltext-6

Reid, A.J., Carlson, A.K., Creed, I.F., Eliason, E.J., Gell, P.A., Johnson, P.T.J., Kidd, K.A., MacCormack, T.J., Olden, J.D., Ormerod, S.J., Smol, J.P., Taylor, W.W., Tockner, K., Vermaire, J.C., Dudgeon, D., Cooke, S.J. 2018. Emerging threats and persistent conservation challenges for freshwater biodiversity. Biol. Rev. 94, 849–873. https://doi.org/10.1111/brv.12480

Reid, A.J., Carlson, A.K., Hanna, D.E.L., Olden, J.D., Ormerod, S.J., Cooke, S.J. 2020. Conservation challenges to freshwater ecosystems. In Encyclopedia World's Biomes. pp. 270–278. https://doi.org/10.1016/b978-0-12-409548-9.11937-2

Ruffle, B., Baird, S., Kirkwood, G., Breidt, F.J. 2019. Estimation of fish consumption rates based on a creel angler survey of an urban river in New Jersey, USA. Human Ecol. Risk Assess. 26, 944–967. https://doi.org/10.1080/10807039.2018.1546549

Rumbold, D.G., Lange, T.R., Richard, D., DelPizzo, G., Hass, N. 2018. Mercury biomagnification through food webs along a salinity gradient down-estuary from a biological hotspot. Estuar. Coast Shelf Sci. 200, 116–125. https://doi.org/10.1016/j.ecss.2017.10.018

Salimi, M., Esrafili, A., Gholami, M., Jonidi Jafari, A., Rezaei Kalantary, R., Farzadkia, M., Kermani, M., Sobhi, H.R. 2017. Contaminants of emerging concern: a review of new approach in AOP technologies. Environ. Monit. Assess. 189(8), 1–22.

Salt, D.E., Smith, R.D., Raskin, I. 1998. Phytoremediation. Annu. Rev. Plant Biol. 49(1), 643–668.

Savery, L.C., Evers, D.C., Wise, S.S., Falank, C., Wise, J., Gianios, C., Kerr, I., Payne, R., Thompson, W.D., Perkins, C., Zheng, T., Zhu, C., Benedict, L., Wise, J.P. 2013. Global mercury and selenium concentrations in skin from free-ranging sperm whales (Physeter macrocephalus). Sci. Total Environ. 450–451, 59–71. https://doi.org/10.1016/j.scitotenv.2013.01.070

Schindler, D.W. 2006. Recent advances in the understanding and management of eutrophication. Limnol. Oceanogr. 51(1 part 2), 356–363.

Taylor, C.A., Schuster, G.A., Cooper, J.E., DiStefano, R.J., Eversole, A.G., Hamr, P., Hobbs, H.H., Robison, H.W., Skelton, C.E., Thoma, R.F. 2007. A reassessment of the conservation status of crayfishes of the United States and Canada after 10+ years of increased awareness. Fisheries 32, 372–389.

Ullah, K., Zaffar, M., Naseem, R. 2014. Heavy-metal levels in feathers of cattle egret and their surrounding heavy-metal levels in feathers of cattle egret and their surrounding environment: a case of the Punjab province, Pakistan. Arch. Environ. Contam. Toxicol. 66, 139–153. https://doi.org/10.1007/s00244-013-9939-8

U.S. EPA. 1997. Exposure Factors Handbook. Final Report. U.S. Environmental Protection Agency, Washington, DC, EPA/600/P-95/002F a-c, 1997.

van Raamsdonk, L.W., van der Zande, M., Koelmans, A.A., Hoogenboom, R.L., Peters, R.J., Groot, M.J., Peijnenburg, A.A., Weesepoel, Y.J. 2020. Current insights into monitoring, bioaccumulation, and potential health effects of microplastics present in the food chain. Foods 9(1), 72.

Vareda, J.P., Valente, A.J.M., Durães, L. 2019. Assessment of heavy metal pollution from anthropogenic activities and remediation strategies: a review. J. Environ. Manag. 246, 101–118. https://doi.org/10.1016/j.jenvman.2019.05.126

Vidali, M. 2001. Bioremediation. an overview. Pure Appl. Chem. 73(7), 1163–1172.

Wolfe, M.F., Schwarzbach, S., Sulaiman, R.A. 1998. Effects of mercury on wildlife: a comprehensive review. Environ. Toxicol. Chem. 17, 146–160. https://doi.org/10.1897/1551-5028(1998)017<0146:EOMOWA>2.3.CO;2

4 Mercury Toxicity in Humans

Sources, Pharmacokinetics, and Mitigation Measures

Sampriti Roy and Pritha Bhattacharjee

4.1 INTRODUCTION

Mercury is a heavy metal that is toxic in nature and is widely distributed in the environment. Most of the mercury exposures to humans reported so far are the result of consumption of fish that has been exposed to mercury in the aquatic environment, along with inhalation of atmospheric mercury and use of dental amalgams. Methyl mercury contamination is known for having caused public health disasters in Minamata Bay, Japan (Ministry of the Environment, Japan, 2002) and in Iraq (Clarkson et al., 1981). However, the present clinical impact of smaller amounts of mercury exposures is still controversial.

Mercury has been found to occur naturally in the earth's crust. Human activity has often been stated to be one of the main causes of mercury releases into the environment. These human activities are varied, ranging from residential coal burning for the purpose of cooking, waste incinerators, coal-fired power stations, several industrial processes, and gold, along with others. Once the element is in the environment, particularly the aquatic environment, there are a few particular species of bacteria that can transform mercury into methylmercury. Methylmercury has been found to be bio-accumulated and biomagnified in fish, shellfish, and other associated organisms in the higher trophic order.

It must be noted that all humans are exposed to at least some amounts of mercury in their lifetime, and this exposure can be categorized into chronic and acute. Most people, often through chronic exposure (continuous or intermittent long-term contact), are exposed to low levels of mercury. However, when people are exposed to high levels of mercury in a short period of time, acute mercury toxicity occurs. This could happen due to industrial accident or cases of acute poisoning (WHO, 2017). In this chapter, we will look at the forms in which mercury occurs in nature, the different sources of mercury, and the fate of mercury once it enters the human system. We will also discuss treatment options currently available for mercury and mitigation measures that must be implemented for reduction of mercury in the environment.

DOI: 10.1201/9781003297901-4

4.2 FORMS OF MERCURY

The several forms in which mercury exists can be grouped into the following:

i. Inorganic mercury: Includes mercury vapour (Hg^0), metallic mercury, and mercuric (Hg^{2+}) or mercurous ($Hg_2{}^{2+}$) salts
ii. Organic mercury: Includes those compounds in which mercury is bonded to a structure which contains atoms of carbon

Table 4.1 demonstrates further the various physical properties of different forms of inorganic and organic mercury.

TABLE 4.1
Physical properties of different forms of mercury

	Inorganic mercury			Organic mercury
Parameters	**Elemental mercury (Hg^0)**	**Mercurous chloride [Mercurous ($Hg_2{}^{2+}$) mercury]**	**Mercuric chloride [Mercuric mercury (Hg^{2+})]**	**Methylmercury (MeHg)**
1. Structure		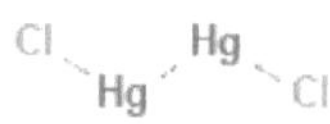	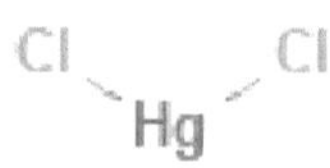	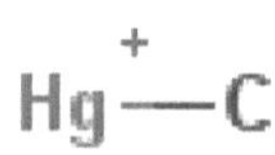
2. Synonyms	Hydrargyrum, quicksilver, metallic mercury, metallic silver	Calomel, mercury monochloride	Dichloromercury, mercury(II) chloride, sublimate mercury bichloride	Methylmercury II
3. Molecular Weight (g/mol)	200.59	472.09	271.50	216.6326
4. Physical Description	A silvery liquid that is heavy and mobile as well as odourless (Source: ICSC)	Mercurous chloride is an odourless white solid. It is found to sink in water.	Mercuric chloride is observed to be an odourless white crystalline solid.	Found in the solid state
5. Solubility	0.28 μmoles/l of water at a temperature of 25°C; partly soluble in lipids	Solubility in water is 4 mg/l at 25°C (practically insoluble)	5 to 10 mg/mL at 72°F	The nature of the anion is responsible in determining the solubility of methylmercury compounds in water – a property that has been found to vary greatly. Most compounds are soluble in polar solvents like water, but much less soluble in the case of non-polar solvents.

TABLE 4.1 *(Continued)*
Physical properties of different forms of mercury

	Inorganic mercury			Organic mercury
Parameters	**Elemental mercury (Hg^0)**	**Mercurous chloride [Mercurous ($Hg_2{}^{2+}$) mercury]**	**Mercuric chloride [Mercuric mercury (Hg^{2+})]**	**Methylmercury (MeHg)**
6. Corrosivity	The high mobility and tendency to disperse that is exhibited by mercury, and the simplicity with which it forms alloys (as in the case of amalgams) with electrical contact metals and many laboratory metals, can cause severe instances related to corrosion in laboratories.	Moderate	Corrosion rate has been observed to be • 0.0003 mm/yr (1% concentration) at 100°C; • mm/yr (5% concentration) at 100°C; • 0.001 mm/yr (10% concentration) at 100°C	Methylmercury is able to cause corrosive action. It has been found to damage any tissue it comes in contact with when it is presented to the same in adequately high concentrations.
7. Stability	It is slightly volatile at room temp. When in the pure state, it does not tarnish upon exposure to air at room temperature. However, when heated to temperature near the boiling point, it is observed to slowly be oxidized to the form of mercuric oxide (HgO).	No rapid reaction with air or with water.	Slightly soluble in water.	• During a three-day period, if the conditions are so that light is absent, MeHgCl does not decompose to the form Hg^{2+}, even when there is presence of 25% acids. • MeHgCl has proven to be particularly hard to be decomposed because it is not easily decomposed by heating in the presence of 50% acid concentrations; however, with UV light, the decomposition of MeHgCl to Hg^{2+} takes place immediately, depending on the intensity of UV light. • In the presence of sulfhydryl group or -SH groups (L-cysteine), there is observed a partial stability and complexation of mercury.

(Continued)

TABLE 4.1 (*Continued*)
Physical properties of different forms of mercury

	Inorganic mercury			Organic mercury
Parameters	Elemental mercury (Hg^0)	Mercurous chloride [Mercurous (Hg_2^{2+}) mercury]	Mercuric chloride [Mercuric mercury (Hg^{2+})]	Methylmercury (MeHg)
8. Toxicity	High toxicity	Moderate to low toxicity	High to moderate toxicity	High to moderate toxicity

(*Source:* NCBI, n.d.)

4.3 SOURCES OF MERCURY EXPOSURE

In nature, mercury exists primarily as elemental mercury or as a sulphide. It has been found in the earth's crust at 0.5 parts per million approximately (Dewi, 2022).

The sources of mercury can be classified into the following:

i. **Atmospheric mercury**:
Atmospheric exposures of mercury can occur through volcanic activity or through outgassing from rocks (Bernhoft, 2012).

ii. **Mercury in water:**
Divalent mercury, Hg(II), is responsible for causing the most concern, while mercury toxicity in the aquatic environment is Methylmercury (MeHg). The primary source of methylmercury is methylation of inorganic Hg(II), carried out by anaerobic bacteria when the latter enters a natural aquatic environment (Zhang et al., 2014). The concern arises mainly due to the fact that MeHg is able to bioaccumulate in different biota and can biomagnify along food chains (Gong et al., 2019).

iii. **Mercury in soil:**
In many artisanal gold mining communities in the world, especially in developing countries, mercury pollution has been found significantly (Gyamfi et al., 2021). When discussing mercury from an agricultural perspective, its transfer from agricultural soils to different crops depends not only on the build-up of mercury but on several soil properties as well. These include oxidation-reduction potential (E_h) (Randall et al., 2004), residue time, soil pH, and soil organic matters (Qian et al., 2009). It must be noted that this scenario is different from the mercury accumulation found in mining soils (Feng et al., 2010).

iv. **Anthropogenic sources:**
Global mercury contamination in the modern age has been predominantly by anthropogenic influences, such as fossil fuel combustion, electroplating, and smelting, among others (Sun et al., 2014). The effects

of these influences are more serious in developing countries because of their high amount of industrialization in the last few decades (Sun et al., 2014; UNEP, 2013).

Some of the ways that humans become the cause of their own mercury exposure are listed subsequently:

- Human contribution to atmospheric mercury: The ways in which humans contribute to atmospheric mercury, include mining and coal burning (gold and mercury in particular)
- Incorporation of mercury in dental amalgamations

4.4 GENERAL ROUTES OF EXPOSURE OF MERCURY IN HUMANS

Mercury is an element that vaporizes with ease at room temperature. It has been established that about 85% mercury enters human systems by inhalation, among which 80% of metallic mercury vapour is outgassed from amalgams (Kudsk, 2009). The gravity of this percentage absorbed through inhalation can be understood when we compare it to the fact that about 7 to 10% ingested metallic mercury is absorbed in the body (Fowler et al., 2007), and about 1% of metallic mercury is absorbed through skin contact (Fowler et al., 2007). Knowledge about the general routes of exposure remain one of the top priorities in designing studies and is one of the initial steps in framing effective targeted policies towards dealing with mercury.

Some of the main routes of exposure of mercury into the body are as follows:

i. **Dietary route (ingestion)**:

   When atmospheric elemental mercury settles in water, it has been found that microorganisms are able to convert it into organic (methyl or ethyl) mercury. Small fishes and other small creatures in the water body intake this organic form of mercury and go on to be eventually consumed by larger fish. Fish at the top of the food chain (like shark, swordfish, or, tuna) are able accumulate considerable amounts of mercury in their tissues. Methyl and dimethyl mercury are usually found to originate chiefly from fresh or saltwater fish (biological sources) (Bernhoft, 2012).

   Ingestion of mercury that is bonded to organic moieties (methyl, dimethyl, or ethyl mercury), with its primary source as seafood, is one of the major pathways through which mercury affects the human body. There are reportedly over 3,000 lakes in the United States that have been closed to fishing owing to mercury contamination in the respective areas (Fowler et al., 2007).

ii. **Inhalation**

   Some of the chief ways in which human mercury exposures occur are through the inhalation of elemental mercury vapour via occupational exposure and, interestingly, through something that has been permitted for use in medicine – dental amalgams. Most of the human metallic mercury exposure has been found to come from mercury vapour outgassing from amalgam fillings. An estimated rate of the outgassing has been found to be about 2 to 28 micrograms per facet surface per day, and it is a very concerning fact that out of this amount, about 80% is absorbed, according to the World Health Organization and Berglund et al. (1988).

A comparatively less popular source of mercury vapour is spilled mercury, and there has also been a report of idiopathic thrombocytopenic purpura caused due to vacuuming spilled mercury, which produced a serious case of acute exposure to mercury vapour) (Schwartz et al., 1992).

iii. Skin contact

Elemental mercury vapour has been found to be very slowly absorbed through the skin in high concentrations. It causes irritation of both the eyes and skin and may produce contact dermatitis (ASTDR, n.d.).

4.5 TOXICITY

The various toxic effects that are caused by different forms of mercury upon their entry into the body to produce external symptoms are discussed subsequently:

(a) Inorganic forms and their toxicity

i. Metallic mercury vapour

Mercury has been observed to poison cellular function in the body, where the disruption occurs when the mercury (in whichever form it is present) alters the tertiary and quaternary structure of proteins and/or binds with selenohydryl and sulfhydryl groups. Consequently, mercury can also potentially impair the function of any organ or any subcellular structure. Although the main target organ of mercury vapour is the brain, the renal function, the immune function, along with functioning of the peripheral nerves and muscles, are also affected. With massive acute exposure to mercury vapour, bronchiolitis and erosive bronchitis may result, potentially leading to respiratory failure which may be accompanied by symptoms indicating damage to the central nervous system, such as tremors or erethism (Garnier et al., 1981). Chronic exposure to doses of mercury vapour considered clinically significant has usually been found to produce neurological dysfunction. At low-level exposures, non-specific symptoms like fatigue, anorexia, weakness, gastrointestinal disturbance, and weight loss have been described (Friberg and Vostal, 1972). Higher exposure levels have been associated with mercurial tremor, in which symptoms include fine muscle fasciculations punctuated every few minutes by a coarse shaking (Berglund et al., 1988).

ii. Mercurous (Hg_2^{2+}) mercury

Even though the effects of mercury have been well documented, products such as calomel (Hg_2Cl_2) continue to be used still in parts of the world as a laxative. Although calomel is poorly absorbed, some of it is converted to mercuric mercury (Bernhoft, 2012) – a form of mercury that is considerably easier to be absorbed and which induces toxicity.

iii. Mercuric mercury

The general targets observed while discussing acute poisoning with mercuric salts (classically $HgCl_2$) are the gastrointestinal (GI) tract and the kidneys. An extensive precipitation of enterocyte proteins has been found to occur, in conjunction with abdominal pain, bloody diarrhoea, and vomiting, along with probable gut mucosa necrosis. This condition may lead to the death of an individual either from septic or hypovolemic shock or peritonitis. Surviving patients have been found to commonly develop renal tubular necrosis with anuria (Barnes et al., 1980). Chronic poisoning with mercury salts has rarely been observed. It usually also involves concomitant occupational exposure to mercury vapour (Bernhoft, 2012).

(b) Organic forms and their toxicity

Methyl mercury reacts with sulfhydryl (-SH) groups throughout the body and has been said to hold the potential to interfere with functions of any cellular or sub-cellular structures. Mercury is also believed to interfere with DNA transcription and protein synthesis (Gruenwedel and Lu, 1970), including protein synthesis in the developing brain, with disappearance of ribosomes and destruction of the endoplasmic reticulum (Herman et al., 1973).

There is considerable evidence that suggests that the disruption of numerous sub-cellular elements in the CNS, or central nervous system, and other organs and in mitochondria can be attributed to methyl mercury toxicity. Several adverse effects on heme synthesis have also been described on many bodily functions (Fowler and Woods, 1977), including free radical generation (Crespo-López et al., 2009; Olivieri et al., 2001), cell membrane integrity in many locations (Berlin et al., 2007), neurotransmitter disruption, and stimulation of neural excitotoxins (Berlin et al., 2007), leading to damage to many parts of the brain and peripheral nervous system (Berlin et al., 2007). Methyl mercury has also been associated with reduction in natural killer cell activity (Vimercati et al., 2001), as well as an imbalance in Th2:Th1 ratios favouring autoimmunity (Park et al., 2000; de Vos et al., 2007). Mercury is possibly associated with disruption of DNA repair (Crespo-López et al., 2009).

4.6 ACUTE AND CHRONIC EXPOSURE OF MERCURY WITH EFFECTS

Mercury poisoning is often found to be misdiagnosed because of the somewhat peculiar onset of symptoms. The clinical presentation of a person exposed to mercury depends upon the length of exposure, dose, and form of exposure. While chronic toxicity is linked with exposure to organic mercury commonly, acute toxicity is more commonly linked with elemental mercury inhalation or inorganic mercury ingestion (WHO, 2017). It is noteworthy that irrespective of the form of mercury that is detected in the body, the kidneys and the CNS have been found to be the two primary target organs of toxicity. Some of the effects of acute and chronic toxicity on different organs of the body have been shown in Figure 4.1.

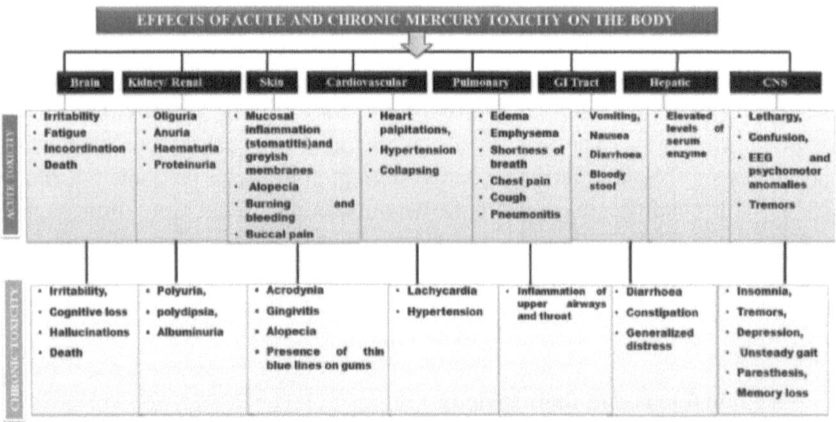

FIGURE 4.1 Effects of acute and chronic mercury toxicity on different parts of the body.

Source: Broussard et al., 2002; CDC, n.d.

4.7 PHARMACOKINETICS OF MERCURY EXPOSURE

Pharmacokinetics is a term that is used to express the effects of different drugs when they are administered into living systems. In order to estimate the effects of each form of mercury on the body as a whole, the fate of each form upon entry into the body's circulation, and, consequently, the entry and fate into different organs of the human system must be looked at. In this section, we will be discussing the pharmacokinetics of different types of inorganic and organic mercury in the circulation and in the brain, thyroid, gut, and kidney. We will also aim to understand the biotransformation processes related to the different forms of mercury, from entry of mercury into the body to excretion. The pharmacokinetics of elemental mercury in circulation and in the different organs discussed subsequently have been illustrated in Figure 4.2.

4.7.1 Circulation

Metallic mercury has been found to be quickly oxidized to mercuric mercury upon its entry to the bloodstream (Fowler et al., 2007). Similar to the action of metallic mercury in the bloodstream, mercuric mercury adheres to –SH groups or sulfhydryl groups present on erythrocytes, metallothionein, or glutathione or is suspended in plasma (Clarkson et al., 1961). Mercurous mercury salt is poorly soluble in water in the form of calomel or Hg_2Cl_2 and is poorly absorbed by the intestine. Some portion of the salt, however, is speculated to undergo oxidation to forms that can be absorbed with more ease (Goodman and Gilman, 1955). An incessant absorption of MeHgCI is observed to occur from the stomach into the systemic circulation, and this is where a swift exchange of CHsHg + to –SH-containing ligands takes place.

To date, there have been quite a few reservations about whether mercurous mercury survives in the body in any form other than as a transitional form (Fowler et al.,

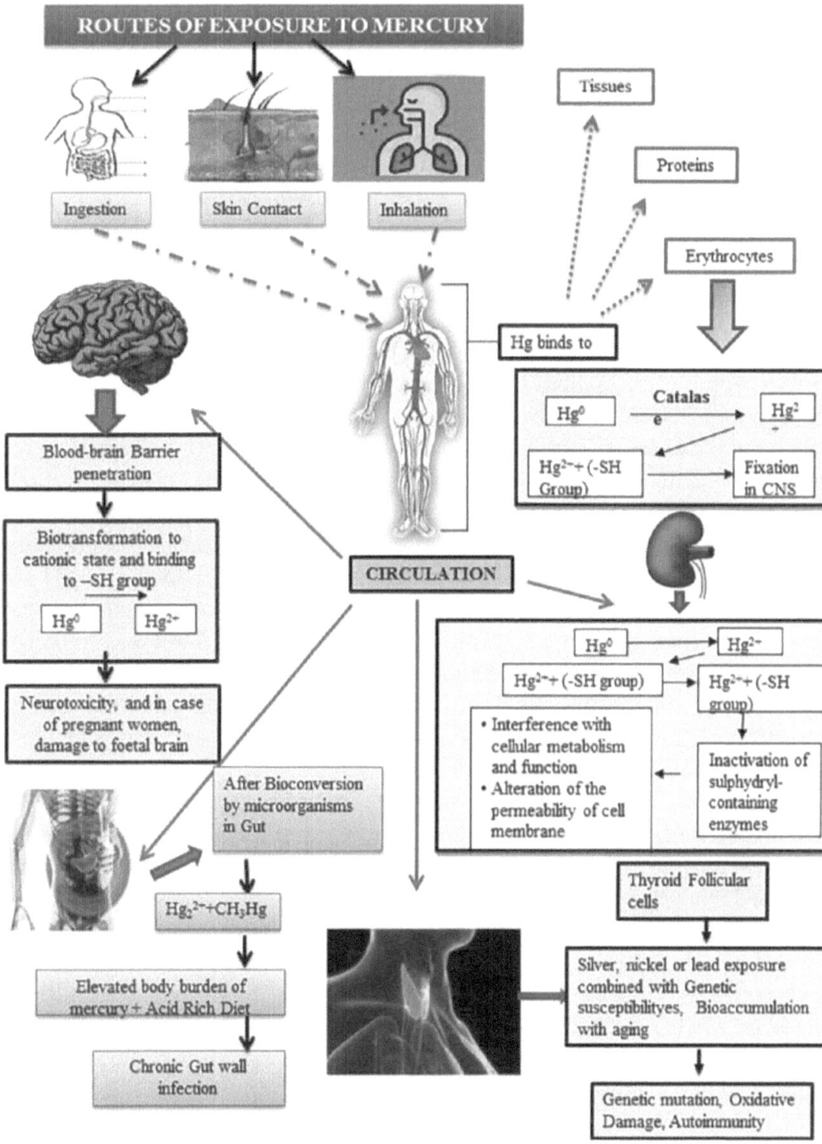

FIGURE 4.2 Exposure, distribution, and cellular targets of mercury.

2007). Some absorption unmistakably occurs, since calomel has often been associated with pink disease or acrodynia (a disease which has been attributed to chronic mercury poisoning). Methyl mercury enters into the body mainly through ingestion of fish containing the same, among other means. Upon entry into circulation, methyl mercury binds to –SH or sulfhydryl groups (particularly to those in cysteine). Its deposition has been found to be ubiquitous in the body, and approximately four days after exposure, equilibrium between the blood and body occurs. Distribution to peripheral tissues has been speculated to occur in literature through one or more transporters, especially the cysteine transporter, which has been found to be adherent to the sulfhydryl group in cysteine (Bernhoft, 2012). In case of mercuric mercury, modes of entry into the body have been through skin creams, among other products that contain compounds with mercuric chloride (Bernhoft, 2012). About 2% of ingested mercuric chloride has been found to be absorbed initially (Norseth and Clarkson, 1970) in the body. It is also believed that the corrosive effect of mercuric mercury on the intestine can increase permeability and, consequently, absorption, with exposure that is prolonged (Kostial et al., 1978).

4.7.2 Brain

Mercury vapour is transported to the brain in two forms – either adherent to red cell membranes or dissolved in serum. Metallic mercury has been found to pass easily through the blood brain barrier (Nordberg and Serenius, 2009). In case of mercury exposure in pregnant women, metallic mercury has been found to pass easily through the placenta and then the blood-brain barrier of the foetus, following which it lodges in the foetal brain (Clarkson et al., 1972). The blood-brain barrier is a particularly specialized structure that comprises of endothelial cells. The endothelial cells of the blood-brain barrier, unlike those in other organs of the body, form a continuous layer that is mainly characterized by tight junctions (Brightman and Reese, 1969). These junctions are the sites where adjoining cells physically merge—something that is responsible for not allowing most substances from entering the brain between cells and opposing osmotic and oncotic forces that usually control exchange of materials between the blood and tissues (Aschner and Aschner, 1990).

Compared to ingested elemental mercury, mercury vapour exposure leads to much higher concentrations of inorganic mercury in the central nervous system of humans (Magos, 1967), culminating in neurotoxicity. Because of its high lipophilicity, inhaled mercury vapour traverses the alveolar membrane fairly quickly (Magos, 1967). In the plasma, this monoatomic mercury gas is oxidized to its divalent ionic form (after erythrocytic diffusion). Despite this rapid oxidation by RBCs, some mercury vapour persists in the bloodstream sufficiently long to diffuse across the blood-brain barrier, where it is oxidized by the endogenous hydrogen peroxide catalase system to the divalent cation—Hg^{2+} or mercuric mercury (Magos, 1967). Although metallic mercury can pass through membranes in both directions, this last oxidation reaction results in fixation of mercury in the central nervous system by its binding to –SH-containing ligands (Aschner and Aschner, 1990).

Mercuric mercury in itself has not been found to cross the blood-brain barrier efficiently (as opposed to metallic mercury), but it does accumulate in significant quantity in the placenta, amniotic fluid, and foetal tissues (Neamtu et al., 2021). There is also evidence that shows that mercuric mercury transport occurs via one or more amino acid transporters (particularly that for cysteine), which may result in accumulation in the brain (Jansson, 2001). The vapour of mercury has exhibited affinity for sulfhydryl (–SH) groups and is found to bond to sulfur-containing amino acids throughout the body (Bernhoft, 2012). Methyl mercury also has a high affinity for sulfhydryl groups in tubulin due to which it inhibits the organization of microtubules that are important in CNS development (Ponce et al., 1994; Sager et al., 1984; Li et al., 2021). The binding to –SH groups also interferes with the intracellular signaling of multiple receptors (e.g., muscarinic, dopaminergic) and promotes the blockade of Ca^{++} channels in neurons (Castoldi et al., 2000). Inorganic mercury can also increase the permeability of chloride channels of GABA A receptors in the dorsal root ganglion (associated with neuronal hyperpolarization). In case of mercurous (Hg_2^{2+}) mercury, it is doubtful whether it survives in the body, other than as a transitional form between metallic mercury and mercury (Fowler et al., 2007).

4.7.3 Thyroid

Exposure to mercury can lead to different forms of mercury like elemental (Hg^0) mercury and methyl mercury (MeHg) entering the body, which go on to affect the thyroid gland. Upon entering into the body, these forms of mercury enter thyroid follicular cells. It has been established that mercury is commonly present in human adult thyroid follicular cells (Pamphlett et al., 2021). In case the human has been exposed to other metals along with mercury, like cadmium, lead, silver or nickel, we can expect these metals to react with mercury (Pamphlett et al., 2021). This reaction will produce effects that will depend upon the genetic susceptibilities of the individual, synergy between toxic metals, and other factors. The effects on the thyroid seen in the domains of genetic mutations, autoimmunity, and oxidative damage are thyroid cancer, autoimmune thyroiditis, and hyperthyroidism (Pamphlett et al., 2021).

Being a known promoter of autoimmune reactions (Pollard et al., 2019), autoimmune thyroiditis could result when the amount of mercury in thyroid follicular cells reaches a critical level in people with a genetic predisposition to autoimmunity (Saranac et al., 2011). Autoimmune thyroiditis and thyroid cancer have also been linked by several researchers, with the proposal that the cancer may arise from the chronic inflammation of the thyroiditis (Resende De Paiva et al., 2017). Recent findings regarding mercury in follicular cells suggest that mercury toxicity underlying both disorders could be another reason these two disorders often co-exist (Pamphlett et al., 2021).

4.7.4 Gut

Elemental mercury is found to be bio-converted through the action of microorganisms to mercurous (Hg_2^{2+}) and methylmercury ($CH3Hg^{+1}$) in the gut (Norseth and Clarkson, 1970). Only around 2% of ingested mercuric chloride is absorbed by the body initially. It has often been speculated that its corrosive effect on the intestine may increase permeability and, with prolonged exposure, the absorption capacity as well. (Norseth and Clarkson, 1970). Mercury toxicity is often associated with a serious infection of the gut called chronic gut wall infection. When one consumes an acid-forming diet and, on top of this consumption, possesses the stress of an elevated body burden of mercury from dental amalgams or other sources (fishes, varieties of paint, or industrial exposures), these two conditions lead to a transitory leaky gut syndrome (Bolan, 2021).

In a person with leaky gut syndrome, the lining of the gut develops microscopic holes. Microorganisms like yeast, anaerobic bacteria, and fungi then go on to leak in through the gut wall and start living permanently in the gut associated lymphoid tissue

(GALT). These are the bacteria that are known to produce acidic toxins and which have been seen to have a systemic effect on health in the long run. Normally, the gut lining is found to regulate intestinal flora to produce eubiosis by placing a secretory immunoglobulin A (IgA) on the surface, and this secretory IgA antibody is responsible for killing dysbiotic organisms on contact. Thus, when an adequate amount of secretory IgA is not present, the wrong organisms proliferate in the gut and produce gas, chronic persistent bloating, constipation, and/or diarrhoea (Smothermon, 2006).

4.7.5 Kidney

Mercury has been labelled to be a very harmful environmental nephrotoxicant (Christy et al., 2017). This is mainly due to its capability of concentrating and reabsorbing different divalent metals and ions. In the kidney, mercury is concentrated in the epithelial cells of the proximal convoluted tubules (Sutton and Tchounwou, 2007). The mercuric ion readily forms a covalent linkage with sulphur, replacing a hydrogen atom from a sulphydryl group, thus inactivating sulphydryl-containing enzymes and, as a result, interfering with cellular metabolism and function (Madsen, 1980).

There lies a special significance in mentioning inorganic mercurial salts and organicumbaias while discussing renal toxicity, including phenyl mercury and alkoxyalkyl mercury compounds, because they are degraded to inorganic forms in the body and consequently give rise to kidney damage. Inhalation of metallic mercury vapour also affects the kidney (Kazantzis, 1970). Metallic mercury vapour disseminates into the tissues followed by oxidation into the mercuric form, where some of the metallic mercury has been observed to be bound to plasma proteins and to haemoglobin at first. The kidney is thus a very important organ to study during toxicological studies related to mercury as it retains more mercury than any other organ in the body (Kazantzis, 1970).

4.7.6 Other Organs

Metallic mercury is deposited in the thyroid, liver, kidneys, myocardium, skin, adrenals, pancreas, enterocytes, lungs, salivary glands, testes, sweat glands, breast, muscles, and prostate, in addition to the brain. Mercury has also been said to be associated with organ dysfunction on numerous occasions.

4.7.7 Excretion

Metallic mercury is mainly excreted as mercuric mercury (Fowler et al., 2007). The excretory half-lives of mercuric mercury and metallic mercury vary extensively. The half-life duration simply depends on the organ in which different forms of mercury are being deposited along with the redox state, with values ranging from a few days to several months (Fowler et al., 2007) and some pools (like the CNS) having a half-life exceeding several years.

Measuring the amount of mercury in hair is also a method of mercury estimation, but it's not a good indicator of total body burden due to the different forms of accumulated mercury reflecting blood methyl mercury, but not elemental mercury, at the time of incorporation. To quantify metallic mercury in the body, analyzing urine and stool is a better option. Methyl mercury has a human excretory half-life of approximately 70 days, with most being excreted through stool. Some is shed through sweat, breast milk, and saliva, where about 20% of methyl mercury has been found to be excreted through breast milk, depending on exposure severity (Fowler et al., 2007). Half-lives of different forms of mercury appear to be multiphasic, as with metallic mercury, with human studies suggesting an effective half-life of 42 days for 80% of an oral tracer dose. The remaining 20% did not seem to have a measurable rate of excretion (Rahola et al., 1971). This may mirror demethylation to metallic mercury in the brain and other organs or mechanisms that are yet to be determined (Bernhoft, 2012).

4.8 TREATMENT OF MERCURY TOXICITY: CHELATION THERAPY

While treating mercury toxicity, the first step is to remove the source from which the body is receiving the mercury, followed by drawing out the toxic mercury from the body. Popular therapeutic methods of mercury toxicity treatment include chelation therapy, among other forms of therapy. Chelation therapy has been found to be a safe and effective method for drawing toxins and metabolic wastes from the bloodstream (Sanoviv Medical Institute, 2021). It involves the intravenous infusion of a prescription medicine called ethylenediaminetetraacetic acid (EDTA) (Sanoviv Medical Institute, 2021). EDTA has one of the most popular treatment options for removing undesirable metals from the body since, with prolonged treatment, evidences of decreased body burden have been inferred (Bernhoft, 2012).

Protocols published by the U.S. Department of Health and Human Services call for provocation with DMPS (2,3-dimercaptopropanol-sulfonic acid) in conjunction with or minus the EDTA. These protocols have been designed primarily for diagnostic breadth and for safety. DMPS has been found to comparatively have far better affinity for different forms of mercury than EDTA. Provocation with both DMPS and EDTA gives a better picture of the overall body burden of the metal. Patients with glutathione S-transferase (GST) enzyme abnormalities may also receive glutathione to accelerate the excretion of chelated metal. For a few unknown reasons, preferential excretion occurs during treatment, where patients with GST polymorphisms tend to excrete mercury later than other heavy metals (Bernhoft and Buttar, 2008). Due to the preferential excretion of lead and other metals earlier than mercury, we may often have an early-stage false-negative result for the metal. All effective chelation protocols have thus called for replacement of those beneficial minerals also removed by EDTA and DMPS (Bernhoft, 2012).

4.9 MITIGATION MEASURES

Given subsequently are some of the proposed methods for dealing with mercury pollution and the toxicity caused subsequently in various ecosystems (GreenFacts, n.d.):

- **Reduction at source**: One of the most effective measures to reduce mercury pollution and its consequent effects on humans would be its reduction at the source. The releases that are currently sourced from human activities may be limited by control and/or preventive measures. Several preventative measures such as reduction of the utilization of products containing mercury and raw materials containing unwanted mercury impurities can be taken. Other preventative measures include improving efficiency of processes in which mercury or mercury-based products are a necessity as a raw material or present in the fuel by using alternatives and low-mercury fuel, such as natural gas instead of coal, and using fuels with a composition that makes mercury easier to control.

- **Replacement with non-mercurous materials**: Replacing those products and processes that use or contain mercury with ones that do not may significantly reduce mercury in various households, in the waste stream, and in our environment. Such steps tend to be cost-effective, especially as demand grows, with a few exceptions and some conceivable trade-offs. Today, for example, we have commercially available low-energy fluorescent lamps that do have mercury content but may have a lower overall environmental impact than ordinary bulbs, because less mercury-containing fuel may be burnt to generate the essential electricity.
- **End-of-pipe techniques**: These include control methods at the point of emission. Techniques such as filtering of exhaust gases can be categorized under this. These techniques are useful when very small amounts of mercury are present in raw materials used in cement production, fossil-fuelled power plants, and metal mining and processing (GreenFacts, n.d.).
- **Effective waste management**: This is a control method which can reduce the mercury releases into the environment originating from spills or gradual leakage (e.g. from broken auto switches, broken thermometers, and dental amalgams). There are methods that propose to render mercury content of the waste inert before disposal so as to minimize the release in the future. One country where this is actively practiced in is Sweden, where waste with higher mercury concentrations is required to be deposited in landfills that are specially equipped in order to limit evaporation and leaching, or in 'final storage' deep underground.
- **Dental amalgams**: It is ideal that dentists prepare mercury amalgam fillings using other suitable alternatives for mercury. Dental amalgams should also be removed before cremation, and even if cremation is carried out with the same, the chimney gases may be filtered.

  Industrial alternatives: It is difficult to introduce an outright ban on mercury usage in industries. What we can do instead is develop policies of lesser use of the same and design methods for reducing mercury escaping into the atmosphere. Some proposed plans are as follows (GreenFacts, n.d.):
  - Cleaning up of the fuel before burning with respect to mercury, or the chimney gases after burning with respect to the same, can also help. However, we will again be left with the mercury that is removed – a form of mercury that then becomes waste which will need to be managed further.
  - Industries that use raw materials containing trace amounts of mercury, like cement, mining, and metal industries, may be prescribed to use a different raw material or to implement end-of-pipe control techniques, as mentioned earlier.

- The scrap steel industry can separate mercury-containing components, such as switches and lights, beforehand.
- For chlor-alkali producers, the regulating authorities may mandate strict mercury accounting procedures, along with the use of exhaust air filtration, implementation of proper leak detection protocols, and the proper management of waste at every facility.

Use of iron-based solvents: The aquatic environment is one of the biggest sources of exposure of humans to mercury. Remediation in the aquatic environment can be brought about by the use of iron sorbents like iron oxides (Richard et al., 2016), zerovalent iron (Zhou et al., 2013), and iron sulphides (Han et al., 2014). They are considered as effective sorbents for mercury, mainly due to their particle size being controllable, their high reactivity, their low cost, and their environmental compatibility (Gong et al., 2019).

4.10 CONCLUSION

A significant amount of literature has established the fact that mercury exposure can be through a variety of routes, and the effects that various forms of mercury produce throughout the body are varied in terms of manifestations. Studies have linked mercury toxicity to a variety of diseases, ranging from neurodegenerative diseases, diseases relating to the renal and cardiovascular system, among various others. Yet we find that mercury toxicity is not a very popular parameter that is included in the differential diagnosis of human subjects who display symptoms such as fatigue, weight loss, memory loss, anxiety, depression, and difficulty concentrating, even though these have been established to be symptoms of low-grade chronic mercury exposure (Bernhoft, 2012).

Mercury has tremendous potential to do damage to humans if not handled in the right way; events such as the Minamata tragedy, due to aquatic mercury toxicity, effects of which were seen long after the source of methylmercury was cut off, bears testament to this statement. Since diagnosis of mercury overload has often been found to be difficult, extra care must be taken to develop new strategies of detection for the same. The commonly used modalities of today to detect mercury in the body like blood, urine, and/or hair levels do not really correlate with total body burden and have been found to offer information that is of little use diagnostically. Provocation with DMPS has been found to produce a more accurate assessment of the total mercury body burden. Chelation therapies with EDTA and DMPS have also proven to be good methods of inducing a more efficient removal of mercury from the body. However, as discussed, mercury toxicity will continue to persist as a problem until proper policies are designed and amended from time to time for reduction of mercury usage and tackling mercury pollution in every sector of society, from industry to medicine to manufacturing.

REFERENCES

Aschner, M., & Aschner, J. L. (1990). Mercury neurotoxicity: Mechanisms of blood-brain barrier transport. *Neuroscience & Biobehavioral Reviews*, *14*(2), 169–176. https://doi.org/10.1016/s0149-7634(05)80217-9

ASTDR. (n.d.). *ATSDR MMG for Mercury*. astdr.cdc. Retrieved November 26, 2023, from https://www.atsdr.cdc.gov/mhmi/mmg46.pdf

Bakir, F., Damluji, S. F., Amin-Zaki, L., Murtadha, M., Khalidi, A., Al-Rawi, N. Y., Tikriti, S., Dhahir, H. I., Clarkson, T. W., Smith, J. C., & Doherty, R. A. (1973). Methylmercury poisoning in Iraq. *Science*, *181*(4096), 230–241. https://doi.org/10.1126/science.181.4096.230

Barnes, J. L., McDowell, E. M., McNeil, J. S., Flamenbaum, W., & Trump, B. F. (1980). Studies on the pathophysiology of acute renal failure. *Virchows Archiv B Cell Pathology Including Molecular Pathology*, *32*(1), 233–260. https://doi.org/10.1007/bf02889030

Berglund, A., Pohl, L., Olsson, S., & Bergman, M. (1988). Determination of the Rate of Release of Intra-oral Mercury Vapor from Amalgam. *Journal of Dental Research*, *67*(9), 1235–1242. https://doi.org/10.1177/00220345880670091701

Berlin M, Zalups RK, Fowler BA. Mercury. In: Nordberg GF, Fowler BA, Nordberg M, Friberg LT, editors. *Handbook on the Toxicology of Metals*. 3rd edition. chapter 33. New York, NY, USA: Elsevier; 2007.

Bernhoft, R. A. (2012). Mercury toxicity and treatment: A review of the literature. *Journal of Environmental and Public Health*, *2012*, 1–10. https://doi.org/10.1155/2012/460508

Bernhoft, R. A., & Buttar, R. A. (2008). Autism: A multi-system oxidative and inflammatory disorder. In *Townsend letter* (pp. 86–90).

Bolan, S., Seshadri, B., Keely, S., Kunhikrishnan, A., Bruce, J., Grainge, I., Talley, N. J., & Naidu, R. (2021). Bioavailability of arsenic, cadmium, lead and mercury as measured by intestinal permeability. *Scientific Reports*, *11*(1). https://doi.org/10.1038/s41598-021-94174-9

Bridges, C. C., & Zalups, R. K. (2017b). The aging kidney and the nephrotoxic effects of mercury. *Journal of Toxicology & Environmental Health Part B: Critical Reviews*, *20*(2), 55–80. https://doi.org/10.1080/10937404.2016.124350

Brightman, M. W., & Reese, T. S. (1969). Junctions between intimately apposed cell membranes in the vertebrate brain. *Journal of Cell Biology*, *40*(3), 648–677. https://doi.org/10.1083/jcb.40.3.648

Broussard, L. A., Hammett-Stabler, C. A., Winecker, R. E., & Ropero-Miller, J. D. (2002). The toxicology of mercury. *Laboratory Medicine*, *33*(8), 614–625. https://doi.org/10.1309/5hy1-v3ne-2lfl-p9mt

Castoldi, A. F., Barni, S., Turin, I., Gandini, C., & Manzo, L. (2000). Early acute necrosis, delayed apoptosis and cytoskeletal breakdown in cultured cerebellar granule neurons exposed to methylmercury. *Journal of Neuroscience Research*, *59*(6), 775–787.

CDC. (n.d.). *Mercury (elemental): Lung damaging agent | NIOSH | CDC*. CDC.gov. www.cdc.gov/niosh/ershdb/emergencyresponsecard_29750021.html

Choi, B. H., Cho, K. H., & Lapham, L. W. (1981). Effects of methylmercury on human fetal neurons and astrocytes in vitro: A time-lapse cinematographic, phase and electron microscopic study. *Environmental Research*, *24*(1), 61–74. https://doi.org/10.1016/0013-9351(81)90132-8

Clarkson, T. W., Gatzy, J., & Dalton, E. (1961). *UR-582*. Rochester, NY: Division of Radiation Chemistry and Toxicology, University of Rochester Atomic Energy Project.

Clarkson, T. W., Magos, L., Cox, C., Greenwood, M. A., Amin-Zaki, L., Majeed, M. A., et al. (1981). Tests of efficacy of antidotes for removal of methylmercury in human poisoning during the Iraq outbreak. *Journal of Pharmacology and Experimental Therapeutics*, *218*, 74–83.

Clarkson, T. W., Magos, L., & Greenwood, M. (1972). The transport of elemental mercury into fetal tissues. *Neonatology*, *21*(3–4), 239–244. https://doi.org/10.1159/000240512

Crespo-López, M. E., Macêdo, G. L., Pereira, S. I., Arrifano, G. P., Picanço-Diniz, D. L., Nascimento, J. L. M. D., & Herculano, A. M. (2009). Mercury and human genotoxicity: Critical considerations and possible molecular mechanisms. *Pharmacological Research*, *60*(4), 212–220. https://doi.org/10.1016/j.phrs.2009.02.011

de Vos, G., Abotaga, S., Liao, Z., Jerschow, E., & Rosenstreich, D. (2007). Selective effect of mercury on Th2-type cytokine production in humans. *Immunopharmacology and Immunotoxicology*, *29*(3–4), 537–548. https://doi.org/10.1080/08923970701690993

Diagnostic and Treatment Protocols for safer, effective mercury human biohazard management. (2003). In *Consensus Development Working Group of the International College of Integrative Medicine*.

Duntas, L. H. (2008). Environmental factors and autoimmune thyroiditis. *Nature Clinical Practice Endocrinology & Metabolism*, *4*(8), 454–460. https://doi.org/10.1038/ncpendmet0896

Feng, X., Foucher, D., Hintelmann, H., Yan, H., He, T., & Qiu, G. (2010). Tracing mercury contamination sources in sediments using mercury isotope compositions. *Environmental Science & Technology*, *44*(9), 3363–3368. https://doi.org/10.1021/es9039488

Fowler, B. A., Friberg, L. T., Nordberg, G. F., & Nordberg, M. (2007). *Handbook on the toxicology of metals*. Elsevier.

Fowler, B. A., & Woods, J. S. (1977). Ultrastructural and biochemical changes in renal mitochondria during chronic oral methyl mercury exposure. *Experimental and Molecular Pathology*, *27*(3), 403–412. https://doi.org/10.1016/0014-4800(77)90010-7

Friberg, L., & Vostal, J. (1972). *Mercury in the environment: An epidemiological and toxicological appraisal*.

Garnier, R., Fuster, J. M., Conso, F., Dautzenberg, B., Sors, C., & Fournier, E. (1981). Acute mercury vapour poisoning (author's transl). *Toxicological European research. Recherche EuropeenneenToxicologie*, *3*(2), 77–86.

Goodman, L. S., & Gilman, A. (1955). *The pharmacological basis of therapeutics*. The Macmillan.

Gong, Y., Huang, Y., Wang, M., et al. (2019). Application of iron-based materials for remediation of mercury in water and soil. *Bulletin of Environmental Contamination and Toxicology*, *102*, 721–729. https://doi.org/10.1007/s00128-019-02559-4

GreenFacts. (n.d.). Mercury. *Greenfacts.org*. www.greenfacts.org/en/mercury/l-2/mercury-6.htm/

Gruenwedel, D. W., & Lu, D. S. (1970). Changes in the sedimentation characteristics of DNA due to methylmercuration. *Biochemical and Biophysical Research Communications*, *40*(3), 542–548. https://doi.org/10.1016/0006-291x(70)90936-8

Gyamfi, O., Sørensen, P. B., Darko, G., Ansah, E., Vorkamp, K., & Bak, J. L. (2021). Contamination, exposure and risk assessment of mercury in the soils of an artisanal gold mining community in Ghana. *Chemosphere*, *267*, 128910. https://doi.org/10.1016/j.chemosphere.2020.128910

Han, D. S., Orillano, M., Khodary, A., Duan, Y., Batchelor, B., & Abdel-Wahab, A. (2014). Reactive iron sulfide (FeS)-supported ultrafiltration for removal of mercury (Hg(II)) from water. *Water Research*, *53*, 310–321. https://doi.org/10.1016/j.watres.2014.01.033

Herman, S. P., Klein, R., Talley, F. A., & Krigman, M. R. (1973). An ultrastructural study of methylmercury-induced primary sensory neuropathy in the rat. *Laboratory Investigation*, *28*(1), 104–118.

Jansson, T. (2001). Amino acid transporters in the human placenta. *Pediatric Research*, *49*(2), 141–147. https://doi.org/10.1203/00006450-200102000-00003

Kazantzis, G. (1970). Mercury and the kidney. *The Transactions of the Society of Occupational Medicine*, *20*(2). https://doi.org/10.1093/occmed/20.2.54

Kostial, K., Kello, D., Jugo, S., Rabar, I., & Maljković, T. (1978). Influence of age on metal metabolism and toxicity. *Environmental Health Perspectives*, *25*, 81–86. https://doi.org/10.1289/ehp.782581

Kudsk, F. N. (2009). The influence of ethyl alcohol on the absorption of mercury vapour from the lungs in man. *Acta Pharmacologica et Toxicologica*, *23*(2–3), 263–274. https://doi.org/10.1111/j.1600-0773.1965.tb03593.x

Li, X., Pan, J., Wei, Y., Ni, L., Xu, B., Deng, Y., Yang, T., & Liu, W. (2021). Mechanisms of oxidative stress in methylmercury-induced neurodevelopmental toxicity. *NeuroToxicology*, *85*, 33–46. https://doi.org/10.1016/j.neuro.2021.05.002

Madsen, K. M. (1980). Mercury accumulation in kidney lysosomes of proteinuric rats. *Kidney International*, *18*(4), 445–453. https://doi.org/10.1038/ki.1980.157

Magos, L. (1967). Mercury—Blood interaction and mercury uptake by the brain after vapor exposure. *Environmental Research*, *1*(4), 323–337.

The Merck Index (1$0^{th}$h ed., p. 842). (1983). Rahway, NJ: Merck Co., Inc.

Mercury, Medical Management Guidelines. (n.d.). *ASTDR toxic substances portal*. https://wwwn.cdc.gov/TSP/MMG/MMGDetails.aspx?mmgid=106&toxid=24

Mercury. (n.d.). National Library of Medicine. https://pubchem.ncbi.nlm.nih.gov/compound/23931

Methylmercury (EHC 101, 1990). (n.d.). InChem. https://inchem.org/documents/ehc/ehc/ehc101.htm

Michielan, A., & D'Incà, R. (2015). Intestinal permeability in inflammatory bowel disease: Pathogenesis, clinical evaluation, and therapy of leaky gut. *Mediators of Inflammation*, *2015*, 1–10. https://doi.org/10.1155/2015/628157

National Toxicology Program, Institute of Environmental Health Sciences, National Institutes of Health (NTP). (1992). *National Toxicology Program Chemical Repository Database*. Research Triangle Park, NC.

NCBI. (n.d.). *PubChem*. https://pubchem.ncbi.nlm.nih.gov/

Neamtu, R., Craina, M., Dahma, G., Popescu, A., Erimescu, A., Citu, I., Dobrescu, A., Horhat, F., Vulcanescu, D., Gorun, F., Bernad, E., Motoc, A., & Citu, I. (2021). Heavy metal ion concentration in the amniotic fluid of preterm and term pregnancies from two cities with different industrial output. *Experimental and Therapeutic Medicine*, *23*(2). https://doi.org/10.3892/etm.2021.11034

Nordberg, G. F., & Serenius, F. (2009). Distribution of inorganic mercury in the guinea pig brain. *Acta Pharmacologica et Toxicologica*, *27*(5), 269–283. https://doi.org/10.1111/j.1600-0773.1969.tb00513.x

Norseth, T., & Clarkson, T. W. (1970). Studies on, the biotransformation of ^{203}Hg-labeled methyl mercury chloride in rats. *Archives of Environmental Health: An International Journal*, *21*(6), 717–727. https://doi.org/10.1080/00039896.1970.10667325

Olivieri, G., Brack, C., Müller-Spahn, F., Stähelin, H. B., Herrmann, M., Renard, P., Brockhaus, M., & Hock, C. (2001). Mercury induces cell cytotoxicity and oxidative stress and increases β-amyloid secretion and tau phosphorylation in SHSY5Y neuroblastoma cells. *Journal of Neurochemistry*, *74*(1), 231–236. https://doi.org/10.1046/j.1471-4159.2000.0740231.x

Pamphlett, R., Doble, P. A., & Bishop, D. P. (2021). Mercury in the human thyroid gland: Potential implications for thyroid cancer, autoimmune thyroiditis, and hypothyroidism. *PLoS One*, *16*(2), e0246748. https://doi.org/10.1371/journal.pone.0246748

Park, S. H., Araki, S., Nakata, A., Kim, Y. H., Park, J. A., Tanigawa, T., Yokoyama, K., & Sato, H. (2000). Effects of occupational metallic mercury vapour exposure on suppressor-inducer (CD4+CD45RA+) T lymphocytes and CD57+CD16+ natural killer cells. *International Archives of Occupational and Environmental Health*, *73*(8), 537–542. https://doi.org/10.1007/s004200000173

Pollard, K. M., Cauvi, D. M., Toomey, C. B., Hultman, P., & Kono, D. H. (2019). Mercury-induced inflammation and autoimmunity. *Biochimica et Biophysica Acta (BBA)—General Subjects*, *1863*(12), 129299. https://doi.org/10.1016/j.bbagen.2019.02.001

Ponce, R., Kavanagh, T., Mottet, N., Whittaker, S., & Faustman, E. (1994). Effects of methyl mercury on the cell cycle of primary rat CNS cells in vitro. *Toxicology and Applied Pharmacology*, *127*(1), 83–90. https://doi.org/10.1006/taap.1994.1142

Qian, J., Zhang, L., Chen, H., Ming, H., Yun-Fei, N., Xu, Z., & Liu, H. (2009). Distribution of mercury pollution and its source in the soils and vegetables in Guilin area, China. *Bulletin of Environmental Contamination and Toxicology*, *83*(6), 920–925. https://doi.org/10.1007/s00128-009-9853-y

Rahola, T., Hattula, T., & Lorolainen, A. (1971). The elimination of ^{203}Hg-methylmercury in man. *Scandinavian Journal of Clinical and Laboratory Investigation*, *27*(Suppl. 116), 77.

Randall, P. M., & Chattopadhyay, S. (2004). Influence of pH and oxidation-reduction potential (Eh) on the dissolution of mercury-containing mine wastes from the Sulphur Bank Mercury Mine. *Mining, Metallurgy & Exploration*. https://doi.org/10.1007/bf03403309

Resende De Paiva, C., Grønhøj, C., Feldt-Rasmussen, U., & von Buchwald, C. (2017). Association between umbaiao's thyroiditis and thyroid cancer in 64,628 patients. *Frontiers in Oncology*, *7*. https://doi.org/10.3389/fonc.2017.00053

Richard, J.-H., Bischoff, C., Ahrens, C. G. M., & Biester, H. (2016). Mercury (II) reduction and co-precipitation of metallic mercury on hydrous ferric oxide in contaminated groundwater. *Science of the Total Environment*, *539*, 36–44. https://doi.org/10.1016/j.scitotenv.2015.08.116

Sager, P., Aschner, M., & Rodier, P. (1984). Persistent, differential alterations in developing cerebellar cortex of male and female mice after methylmercury exposure. *Developmental Brain Research*, *12*(1), 1–11. https://doi.org/10.1016/0165-3806(84)90170-6

Sanoviv Medical Institute. (2021, November 1). Reversing mercury toxicity through chelation therapy. *PlacidWay*. www.placidway.com/article/116/Reversing-Mercury-Toxicity-Through-Chelation-Therapy.

Saranac, L., Zivanovic, S., Bjelakovic, B., Stamenkovic, H., Novak, M., & Kamenov, B. (2011). Why is the thyroid so prone to autoimmune disease. *Hormone Research in Paediatrics*, *75*(3), 157–165. https://doi.org/10.1159/000324442

Schwartz, J. G., Snider, T. E., & Montiel, M. M. (1992). Toxicity of a family from vacuumed mercury. *The American Journal of Emergency Medicine*, *10*(3), 258–261.

Smothermon, R. K. (2006, October 6). Mercury toxicity, gut wall infection, and fatigue. *The Anti-Aging Medical Clinic*. https://medical-library.net/mercury-toxicity-gut-wall-infection-and-fatigue/

Sun, R., Sonke, J. E., Heimbürger, L. E., Belkin, H. E., Liu, G., Shome, D., Cukrowska, E., Liousse, C., Pokrovsky, O. S., & Streets, D. G. (2014). Mercury stable isotope signatures of world coal deposits and historical coal combustion emissions. *Environmental Science & Technology*, *48*(13), 7660–7668. https://doi.org/10.1021/es501208a

Sutton, D., & Tchounwou, P. (2007). Mercury induces the externalization of phosphatidyl-serine in human renal proximal tubule (HK-2) cells. *International Journal of Environmental Research and Public Health*, *4*(2), 138–144. https://doi.org/10.3390/ijerph2007040008

Suzuki, T., Takemoto, T. I., Shishido, S., & Kani, K. (1977). Mercury in human amniotic fluid. *Scandinavian Journal of Work, Environment & Health*, *3*(1), 32–35. https://doi.org/10.5271/sjweh.2793

United Nations Environmental Programme (UNEP). (2013). *Global mercury assessment*. Geneva: UNEP Chemicals.

U.S. Coast Guard. (1999). *Chemical Hazard Response Information System (CHRIS)—Hazardous chemical data* (Commandant Instruction 16465.12C). Washington, DC: U.S. Government Printing Office.

U.S. Department of Health and Human Services, Public Health Service. (n.d.). *Toxicological profile for mercury*. ASTDR. www.atsdr.cdc.gov/toxprofiles/tp46.pdf

Vimercati, L., Santarelli, L., Pesola, G., Drago, I., Lasorsa, G., Valentino, M., Vacca, A., & Soleo, L. (2001). Monocyte-macrophage system and polymorphonuclear leukocytes in workers exposed to low levels of metallic mercury. *Science of the Total Environment*, *270*(1–3), 157–163. https://doi.org/10.1016/s0048-9697(00)00780-4

World Health Organization (WHO). (2017, March 31). *Mercury and health*. https://www.who.int/news-room/fact-sheets/detail/mercury-and-health

Zhang, T., Kucharzyk, K. H., Kim, B., Deshusses, M. A., & Hsu-Kim, H. (2014). Net methylation of mercury in estuarine sediment microcosms amended with dissolved, nanoparticulate, and microparticulate mercuric sulfides. *Environmental Science & Technology*, *48*(16), 9133–9141. https://doi.org/10.1021/es500336j

Zhou, X., Zhang, J. Z., Qiu, X. K., & Wang, D. Y. (2013). *Huan jing ke xue = Huanjing kexue*, *34*(11), 4304–4310.

5 Toxic Effluent—Harmful Effects of Industries on Public Health

Sonali Paul, Anushka Ghosh, Arpan Banerjee, Sohini Roy, Subhabrata Paul, Susmita Mukherjee

5.1 INTRODUCTION

Toxicity of the aquatic system is a result of various kinds of industrial effluent discharged in the environment. The industrial discharge involves effluents containing toxic metals like arsenic, cadmium, chromium, lead, and other harmful substances. Other than this, the effluents released are high in BOD and COD levels, making it more problematic. The negative impact which these effluents have is the highest on aquatic life. In addition, through the food chain, other organisms are also harmed (Wang et al., 2004). These effluents harm organisms which are dependent on the watercourse around a specific area. The health and proper growth of humans, domestic animals, and aquatic animals are also affected by the awful effects of untreated industrial effluents, which is a major concern. The contaminants are largely the result of agriculture run-off, disposal of solid waste, sewage, especially in the urban areas, and discharge of untreated industrial waste.

5.2 TYPES OF INDUSTRIES

The disposal of industrial effluents has been a burning issue for much time. In India, a huge amount of wastewater generated from different industries is discharged on land or into the running water, causing a huge problem to the environment.

Some of the industries which produce large amounts of effluent are mentioned as follows:

5.2.1 Iron and Steel Industry

The industries where iron is produced from its ores involve powerful reduction reactions in blast furnaces. Cooling waters used during the process are contaminated with harmful chemicals like ammonia and cyanide. Production of coke from coal in plants also requires water for cooling and the use of water in by-products separation. Contamination of waste streams occurs due to gasification products such as benzene, naphthalene, anthracene, cyanide, ammonia, phenols, cresols, together with a range of more complex organic compounds known collectively as polycyclic aromatic hydrocarbons (PAH). Contaminants like hydraulic oils, tallow, and particulate solids are seen in particular. Wastewaters include acidic rinse waters, together with

 DOI: 10.1201/9781003297901-5

waste acid produced as a result of the final treatment. Although many plants do acid recovery, where the mineral acid is boiled away and separated from the iron salts, there remains a large volume of highly acidic ferrous sulfate or ferrous chloride which needs to be removed. Many steel industry wastewaters are contaminated by hydraulic oil, also known as soluble oil, which is considered quite harmful (Das et al., 2018).

5.2.2 Textile Industry

Textile industries such as carpet manufacturers generate wastewater due to a wide variety of processes, including wool cleaning and finishing, yarn manufacturing, and fabric finishing. Primarily, contaminates include BOD, SS (suspended solids), oil and grease, sulfide, phenols, and chromium. The presence of insecticide particles in fleeces is a particular problem in treating waters generated in the wool processing industry. Animal fat is also a common pollutant of this industry.

Textile dyeing industries generate wastewater that contains synthetic dyes and natural dyestuff, gum thickener, and various wetting agents, pH buffers, and dye retardants or accelerators. Other than these, several particles contaminate the effluent released (Imtiazuddin et al., 2012).

5.2.3 Pulp and Paper Industry

Effluents released from the pulp and paper industry are biodegradable and non-toxic. But the same problem is faced in this case as the food industry due to high BOD and the presence of suspended solids (Toczyłowska-Mamińska, 2017). Effluents released from this industry are high in contaminants like chloroform, dioxins, furans, and phenols.

5.3 TYPES OF INDUSTRIAL EFFLUENTS

Industries like textile mills, steel plants, and the petrochemical industry produce a large amount of effluent which is laden with harmful chemicals which on release cause too much harm to the environment (Kanu and Achi, 2011).

This wastewater entering the surface water bodies leaves a great impact on the environment, which is considered to be one of the main reasons for water pollution also. Industrial effluents are characterized on the basis of their conductivity, chemical oxygen demand, turbidity, and total suspended solids (Staicu et al., 2017). Depending on the industries and the water used by them, wastewater can contain both biodegradable and non-biodegradable suspended solids, organic, dissolved acids and alkalis, colouring agents, etc. (Chhonkar et al., 2000). These are coming from various types of plants as well. Breweries, coal plants, pulp, and paper industries, iron, and steel mills emit mainly highly suspended solids and high acids. Colouring compounds come from pulp and paper mills, palm oil mills, textile industries, and tanneries (Dwivedi and Tomar, 2018). Chemical plants, textiles, and tanneries are also sources of high dissolved solids and high alkalis. Oxygen-consuming wastes come from breweries, dairies, distilleries, packaging houses, pulp, and paper industries. Petroleum refineries, metal finishing, laundries, and oil mills discard oily and greasy materials (Carolin et al., 2017).

There are some types of industrial effluents mentioned subsequently:

5.3.1 Liquid Wastes

Huge amounts of water are needed by the industries which can come into contact with radioactive materials, inorganic mixed liquids, solid particles, etc. Then these wastewaters are released into natural water bodies near the industry (Singarea and Dhabardeb, 2014).

5.3.2 Solid Wastes

Organic materials, inorganic materials, heavy metals, metal dust, etc. are released from many industries like the cement industry, coal plants, steel plants, etc. (Manikandan et al., 2015). They are mixed with effluent water.

5.3.3 Chemical Wastes

Scrap metals, oils, cyanides, metal phosphides, organosilicon, fluorides, and chlorine solvents like chloroform, chlorobenzene, non-chlorinated solvents, namely toluene, methanol, acids, alkalis, are major types of chemical wastes which are highly harmful to the ecosystem (Manikandan et al., 2015).

5.3.4 Toxic Wastes

Kerosene, gasoline, diesel fuel, antifreeze, and brake fluids are some examples of hazardous wastes (Sujatha et al., 2013).

5.3.5 Heavy Metals

Industrial sources like metal refineries, coal industries, nuclear power plants, textiles, plastic and paper industries, and petroleum industries are some examples from which several heavy metals are released. Such heavy metal effluents are arsenic (As), molybdenum (Mo), selenium (Se), nickel (Ni), cadmium (Cd), lead (Pb), mercury (Hg), etc.

5.3.6 Effluent Discharge—Few Specific Industries

5.3.6.1 Steel Plant

Ample amount of water is needed for the formation of steel and iron. The main source of water is from the purifying process of blast furnace gas, slag crushing, and cooling processing (Yılmaz et al., 2018). The wastewater from the blast furnace gas has a pH of 7–9. It has 33–670 mg/l of total suspended solids, nearly 200 mg/l of volatile suspended solids, 40–150 mg/l permanganate, 800–4000 mg/l total dissolved solids. Now the composition of wastewater from slag crushing are 3–4 meq/l alkalinity, 500–600 mg/l TSS, 400–500 mg/l TDS, nearly 30 mg/l of volatile suspended solids, 10–11 mg/l sulphite. It has the same alkalinity of that of slag crushing water and nearly the same TDS value (Choy et al., 2019). It also has 1000–1500 mg/l TSS, 2–50 mg/l lubricating oil, same sulphate quantity like the slag crushing wastewater, and 10–150 mg/l volatile solid in the dry residue. Another source of wastewater is the cooling of pig iron. It is basically water that contains TDS, TSS, permanganate volatile solid in dry residue in common with the others, and also 30–300 mg/l chloride ion, 200–300 mg/l calcium oxide, magnesium oxide 15–30 mg/l, and sulphate 20–650 mg/l (Morrow et al., 2014). Lastly comes the pickling wastewater that has 80–600 mg/l iron, 0–50 mg/l aluminium, 200–2000 mg/l sulphate ion, and 50–200 mg/l calcium ion, with a pH of 1.5–4.5.

5.3.6.2 Textile Industry

Chemicals like sodium hydroxide, sodium sulfide, sodium carbonate, sodium chloride, sulphuric acid, sulfate, sulfide, oxychloride, bleaching powder, tannic acid, detergents, acetic acids, mordants, dyes, starch, etc. are required in this industry. The effluents are observed to be of high salinity, high alkalinity, and high BOD and COD levels.

5.3.6.3 Pulp and Paper Industry

In 1876, India's first paper mill was established in West Bengal. Sodium sulfate, sodium sulfide, sodium carbonate, calcium hypochlorite, and magnesium bisulfate are the main chemicals used by this industry. Lignins and their derivatives were also one of the major effluents. They cannot be degraded because of their carbon-carbon linkage, and they also have some mutagenic effects. Lignin and its derivative are the reason for the brown colour of the wastewater. The wastewater from both large and small mills was found to have some other chemicals such as 6 mg/L N, 10 mg/L K, 1.5 mg/L P, 75 mg/L Na, 200 mg/L Mg, etc. They also contain resins, acids, chlorinated lignin, chloro fatty acids, tri and pentachlorophenols, etc. and toxic materials.

5.4 IMPACT ON ECOSYSTEM AND HUMAN HEALTH

An ecosystem is a structural complex community where all living organisms and all non-living components interact with each other in a systematic manner. From the last decade, human activities and a huge expansion in urbanization and industrialization have changed the main structure of the ecosystem, which made a functional degradation of the ecosystem. It has a negative impact on human survival. In this study, there is a brief discussion on the different types of industries and the effluents and the mainstream impacts of industrial effluents on the environment and largely on human health.

5.4.1 Implications of Pulp and Paper Industry Waste

The pulp and paper industry, like all other manufacturing industries, consumes a huge amount of energy and emits various hazardous pollutants, and along with that, it also produces greenhouse gases (Sujatha et al., 2013). The waste-contaminated water hampers aquatic life and the ecosystem. In the terrestrial ecosystem, this waste also has numerous adverse effects. Nickel, copper, chromium, lead, trichlorophenol, tetrachloroguaiacol, dichlorophenol, dichloroguaiacol, and pentachlorophenol are the main harmful pollutants that are emitted by the pulp paper industry and have numerous adverse effects on the ecosystem and the food chain (Regaldo et al., 2018).

These pollutants reduce the survival rate and cause damage at the molecular level (cell, individual and population). Under relevant rules of respective State Pollution Control Boards (India), the industry has to build a well-developed waste management system (Dudani et al., 2017).

5.4.1.1 Impact on the Environment, Food Chain, and Humans

As this study discussed earlier, various kinds of environmental toxic compounds are present in the waste of the pulp and paper industry. The major part of these pollutants is released into the water and causes surface water body pollution. It has also been found that the chemical pollutants present in this effluent, like resin acids, are the cause of genotoxicity in the sea ecosystem (Dutta et al., 2018). The pulp and paper industries consume a ton of energy and drastically emit various sizes of very toxic solid pollutants. Solid waste causes a serious problem to the aquatic ecosystems and ruins the food chain. No proper treatment of solid and liquid waste of pulp and paper industries causes denaturation of water bodies and is toxic to aquatic life. Endocrine-disrupting compounds (EDCs) contaminations are also a serious class of contamination in the water bodies (Holon et al., 2018). The impact on the food chain system is a very serious situation where the pulp and paperboard industrial waste affects the growth of large fishes. A study in Assam, India, demonstrated that the effluents highly increase the cadmium and chromium levels (Ahmad et al., 2019).

5.4.2 Implications of Steel Industry Waste

Steel industries are the most pollution-causing industry. A huge amount of water and energy is required in this industry, and numerous toxic pollutants are emitted from it. These waste products have various adverse effects on both air and water ecosystems. Mainly most of the waste contains suspended solids like fly ash, coke, coal dust and sand, lubricant oils, hydraulic oils, chemical solvents, etc. The emitted chemical solvents are very dangerous for the ecosystem. As per the report of the Central Pollution Control Board of India, the effluents of the steel industry hampered the whole aquatic ecosystem. The steel industries have to maintain wastewater treatment and dangerous chemical dilution processes to reduce the toxicity rate under the limit. Dangerous zinc emission from this industry completely destroys soil fertility, and excess deposition of Zn causes crop death (Lyu et al., 2018).

5.4.2.1 Impact on the Environment, Food Chain, and Humans

Emitted chemicals ruin aquatic life and destroy the aquatic food chain. Chemicals like fly ash, coke, coal dust and sand, lubricant oils, and hydraulic oils denature the water bodies and reduce the oxygen concentration in water. Coal dust and sand deposition at the bottom of the water body can cause the death of aquatic plants (especially bottom dwellers). Through the food chain of the aquatic ecosystem, the toxic effect of chemicals hampers every living being (Brar et al., 2017). By the food chain, these pollutants affect each level of the pyramid. Intake of these chemicals through the food chain may produce several acute dysfunctions in the human body like sporadic fever, cramps, renal damage, and hypertension (Matta et al., 2018).

5.4.3 Impacts of Textile Industry Waste

Textile industrial effluents are extremely toxic and chemical-intensive and involve the use of high amounts of numerous environment-unfriendly, non-biodegradable chemicals. These waste treatment processes have to be there in the industry. Negligence in this process will degrade the environment. Reactants like sodium hydroxide, sodium sulfide, sodium carbonate, sodium chloride, sulphuric acid, oxychloride, bleaching powder, tannic acid, detergents, acetic acids, mordants, dyes, starch, etc. are required in this industry, so the waste product has an adverse effect on the environment. As the processing of the textile industry requires a huge amount of fresh water for wet processing, it generates a huge volume of wastewater. High BOD, COD, total dissolved solids, total suspended solids, and low dissolved oxygen (DO) are the most dangerous characteristics of textile industrial effluents (Matta et al., 2018).

5.4.3.1 Impact on the Environment, Food Chain, and Humans

The effluent from textile industry, like sodium hydroxide, sodium sulfide, and sodium carbonate, have dangerous effects on aquatic life. Sodium hydroxide reacts with CO2 and produces a huge amount of sodium carbonate in water (Singh et al., 2019). It reduces the water pH. Acidic water kills the fishes and other aquatic organisms in the sea ecosystem. By the food chain, humans are also affected. Various bone diseases and acute nerve diseases occur. Excessive deposition of chemicals can cause ulcers in the intestine. Tannic acid and detergent-like toxic chemical intake through the food chain causes various skin-related diseases and weakens bones (Aniyikaiye et al., 2019). Overall, the waste should go through treatment, otherwise it has various adverse effects on the environment and humans.

5.4.4 Impact of Heavy Metals on Human Health

Heavy metals have a toxic impact on human health, and the intensity depends on exposure. In this study, we discussed effluents of the pulp and paper industry, steel industry, and textile industry. The common eight heavy metals emitted by these industries are arsenic, barium, cadmium, chromium, lead, mercury, selenium, and silver. These metals are existing in environments at very low concentrations. But the high concentration can cause various adverse effects on human health. By ingestion or inhalation, heavy metals are exposed to the human body. Heavy metals and their effects on human health are briefly discussed subsequently:

5.4.4.1 Arsenic (As)

Arsenic is one of the deadliest odourless heavy metals. It is known as a carcinogen. Deposition of arsenic causes skin cancer, liver cancer, and, by inhalation, it can cause lung cancer (Gdara et al., 2018). According to the Environmental Protection Agency (EPA), the regulatory limit or the safe limit is around 0.01 parts per million (ppm) for drinking purposes, and according to the Occupational Safety and Health Administration (OSHA), the regulatory limit or the safe limit is around 10 $\mu g/m^3$ of workplace air for a maximum 40-hour workweek (Aniyikaiye et al., 2019). A low level of As in the bladder can cause bladder cancer. A little bit of exposure can inhibit the production of red and white blood cells. Arsenic has a negative impact on plant growth and biomass development, which can be managed with the application of Mg salt (Paul et al., 2022).

5.4.4.2 Barium (Ba)

Barium is the most abundant heavy metal, and it is very important for various industrial products. It is mostly non-carcinogenic. By ingestion, barium can cause diarrhoea-like symptoms (vomiting and abnormal cramps). Also, in some cases, breathing difficulties and muscle weakness occurs. Blood pressure fluctuation is the deadliest symptom of barium toxicity in the body. Due to high fluctuation of blood pressure, there is an increased risk of chronic chest pain, heart attack, and possibly death. An excess amount of barium intake in the body can cause numbness around the face. A large amount of barium disposition in the vein changes heart rhythm and, in some cases, paralysis also (Lyu et al., 2018). According to the Environmental Protection Agency (EPA), the regulatory limit or the safe limit is around 2.0 ppm in drinking water, and according to the Occupational Safety and Health Administration (OSHA), the regulatory limit or the safe limit is around 0.5 milligram of workplace air for a maximum 40-hour workweek (Aniyikaiye et al., 2019).

5.4.4.3 Cadmium (Cd)

Cadmium is the most toxic heavy metal. It is emitted mainly from the energy industries. The earlier-mentioned three industries also heavily emit this metal. Cadmium is known as a human carcinogenic heavy metal. Smokers get heavily affected by this metal. Cadmium severely damages the alveoli of the lungs. Cadmium deposition also destroys the flexibility of veins and causes heart attack due to uncontrolled blood pressure (Holon et al., 2018). Cadmium deposition in the kidney can damage it completely. In some cases, kidney cancer and kidney failure also occur. Besides kidney and lung damage, it also damages bones and can cause fragile bones. Excess deposition in the stomach can damage the cells and severely irritate them (Manikandan et al., 2015). According to the Environmental Protection Agency (EPA), the regulatory limit or the safe limit is around 5 ppb or 0.005 ppm in drinking water, and according to the Occupational Safety and Health Administration (OSHA), the regulatory limit or the safe limit is around 5 milligram of workplace air for a maximum 40-hour workweek. From the current report of the Food and Drug Administration (FDA), 5 ppb or 0.005 ppm is the limit range of cadmium levels in drinking water (Aniyikaiye et al., 2019).

5.4.4.4 Chromium (Cr)

Chromium (III) is an essential nutrient for the regulation of various natural processes in the human body, and, on the other hand, chromium (VI) is very dangerous to human health; it is a carcinogenic heavy metal. Inhaling a high amount of Cr can totally damage the respiratory system and respiratory tract, like burning and irritation of the nose, always having a runny nose, and sometimes nose ulcer. Excess inhalation can cause severe breathing problems and breathing diseases like asthma, cough, incomplete breath, or wheezing. Various skin problems can arise from large exposure to chromium. Skin burn, skin ulcer, severe skin allergy, swelling, and inflammation of the skin have been noted. An excess amount of Cr ingestion can damage liver cells and cause kidney failure. According to the Environmental Protection Agency (EPA), the regulatory limit or the safe limit is around 0.1 ppm in drinking water, and according to the Occupational Safety and Health Administration (OSHA), the regulatory limit or the safe limit is around 0.0005 and 1.0 milligram of workplace air for a maximum 40-hour workweek. From the current report of the Food and Drug Administration (FDA), 1 ppm is the limit range of cadmium levels in drinking water (Aniyikaiye et al., 2019).

5.4.4.5 Lead (Pb)

Lead is not as abundant in the environment as other heavy metals. But huge anthropological activity increases the concentration of lead in an environment that is very toxic and has various adverse effects on human health. According to the report of the Environmental Protection Agency (EPA), lead is the most dangerous carcinogenic heavy metal for humans. Lead has adverse effects on every human organ and organ system. A large amount of lead intake in the body can damage the whole nervous system, and paralysis also has been noted. It also hampers blood cell production and causes anaemia. The excess amount in blood circulation causes fragile bones, weak wrist joints and finger bones, and, ultimately, paralysis. Long-term exposure to Pb can damage the brain cells and line cells of the kidney and cause death. Miscarriage in pregnant women can happen due to heavy exposure to lead. It can also inhibit sperm production in men by damaging the reproductive organs. According to the Environmental Protection Agency (EPA), the regulatory limit or the safe limit is around 15 ppb in drinking water (Aniyikaiye et al., 2019).

5.4.4.6 Mercury (Hg)

Mercury itself is a very poisonous heavy metal for human health. Other than that, its derivatives like mercuric chloride and methylmercury also have adverse effects on human health. According to the report of the Environmental Protection Agency (EPA), mercuric chloride and methylmercury are the main human carcinogenic elements. These forms of mercury can heavily damage the brain cells and nervous systems. Extremely high levels of mercury can deeply damage foetuses and also cause kidney failure. It induces many brain diseases (shyness, memory problems, vision and hearing problems, etc) and blood pressure fluctuation (Dutta et al., 2018). The short-term inhalation of Hg can heavily damage the lungs and create severe lung-related diseases and damage the stomach and cause diarrhoea and repeated vomiting (Regaldo et al., 2018). Allergies like eye irritation and skin allergies are the frontline symptoms of mercury toxicity in the human body. According to the Environmental Protection Agency (EPA), the regulatory limit or the safe limit is around 2 ppb in drinking water, and according to the Occupational Safety and Health Administration (OSHA), the regulatory limit or the safe limit is around 0.1 milligram of workplace air for a maximum 40-hour workweek. From the current report of the Food and Drug Administration (FDA), 1 part of methylmercury is the limit of cadmium levels in drinking water (Lyu et al., 2018).

5.4.4.7 Selenium (Se)

Selenium is mainly obtained from rocks and soil naturally, but huge uses of it increase its toxicity level. In a very minute amount, selenium is non-toxic (400 micrograms per day) and essential for human cellular function, but exposure to higher quantities have various adverse impacts on the human body (Dutta et al., 2018). Excess deposition in the intestine causes ulcers, vomiting, and diarrhoea (Ahmad et al., 2019). It also causes sclerosis when ingested (neural abnormalities and loss of hair are the main symptoms). It also damages the respiratory tract and respiratory organs. According to the Environmental Protection Agency (EPA), the regulatory limit or the safe limit is around 50 ppb in drinking water, and according to the Occupational Safety and Health Administration (OSHA), the regulatory limit or the safe limit is around 0.2 milligram of workplace air for a maximum 40-hour workweek (Gdara et al., 2018).

5.4.4.8 Silver (Ag)

It is the least harmful among all heavy metals which are secreted by the industries. According to the Environmental Protection Agency (EPA), silver is not a carcinogenic heavy metal to the human body but has some adverse effects on humans. Argyria occurs due to excess deposition of silver in the body. A blue-grey-coloured coating on the skin and various tissues is the symptom of argyria (Ahmad et al., 2019). Thc excess level of silver deposition can cause breathing problems like asthma and irritation in the nose, throat, and respiratory tract, etc. (Ahmad et al., 2019). Skin exposure to silver in a heavy amount can cause allergies like skin swelling and rashes. In some cases, skin inflammation also occurs. According to the Environmental Protection Agency (EPA), the regulatory limit or the safe limit is around 0.10 ppm in drinking water, and according to the Occupational Safety and Health Administration (OSHA), the regulatory limit or the safe limit is around 0.01 mg/m3 of workplace air for a maximum 40-hour workweek (Aniyikaiye et al., 2019).

5.5 A COMPARATIVE STUDY

In this study, we have analyzed the adverse effects of industrial effluents on human health and ecosystem. There are many industries, and so many harmful elements are discharged into the surface water bodies, posing a threat on us. The levels of different contaminants in the effluent released from industries vary depending on the production processes. Table 5.1 shows a comparative study on the effect of different industrial effluents on the human system.

TABLE 5.1
Comparative study of various industrial effluents and their impact

Name of toxic elements	Average level in effluents	Industrial source	Effect on human health
Permanganate (mg/l)	40–150	Steel	Pulmonary oedema Shortness of breath Kidney damage
Cyanide (mg/l)	0.6–1.3	Steel	Blocks the pathway of oxygen to the cells and affects the brain and heart
Thiocyanate (mg/l)	17	Steel	

TABLE 5.1 (*Continued*)
Comparative study of various industrial effluents and their impact

Name of toxic elements	Average level in effluents	Industrial source	Effect on human health
Fe (mg/l)	140–1180	Steel	Accumulation in organs, fatal damage to the brain and liver
Sulfite (mg/l) Sulfate (mg/l)	10–11 600–1000	Steel, Textile	Cardiovascular damage, neurological damages Irritation in eyes, lungs, skin, etc.
Phenol (%)	1–2.5	Steel	Skin, eye, throat irritation, liver damage, skin damage, nervous system irritation, skin burns, tremors, etc.
Lubricating oil (mg/l)	2–50	Steel	Dermatitis, eczema, xeroderma, etc.
Magnesium oxide (mg/l)	15–30	Steel	Bloating, diarrhoea
Calcium oxide (mg/l)	200–300	Steel	Pulmonary oedema, shortness of breath
Chlorinated lignin (%) Bleaching powder (mg/l)	65–75 900–2000	Paper-pulp Textile	Cytotoxic behaviour towards breast cancer cells and affects the liver Skin and eye irritation
Aluminium ion (mg/l)	0–50	Steel	Interstitial pneumonia, toxic myocarditis, ischemic stroke, thrombosis, anaemia, Alzheimer's disease, etc.
Sodium hydroxide (mg/l) Sodium carbonate (mg/l) Sodium chloride (mg/l)	10 20 200–1000	Textile Textile, paper-pulp Textile, paper-pulp	Corrosive to body tissues, necrosis, stricture of oesophagus Oedema High blood pressure, kidney failure, heart disease, etc.
Potassium (ppm)	481.33	Paper-pulp	Affects heart's muscles
Heavy Metals (As, Cd, Ag, Cr, Hg etc.)	<10	Textile, paper-pulp	Disrupt the function of the central nervous system, damage blood components, cause lung, kidney, and liver damage.

Source: Compilation done by Author

It can be concluded from Table 5.1 that some specific industrial effluents are very toxic to human health, causing pulmonary oedema, dermatitis, eye damage, kidney damage, thrombosis, to name a few, but, of course, the level of harm depends on the concentration of the toxic element in the effluent and exposure time.

The industrial effluents, treated or untreated, are discharged into the surface water bodies, and these affect the quality of water. A particular pollutant released from different industrial units might vary in quantity and harm the ecosystem accordingly. Table 5.2 shows how different water quality parameters vary based on the industrial processes and their relative effects on the ecosystem.

TABLE 5.2
A comparison among three different types of industrial effluents based on different parameters

Parameters	Industries	Standard level	Average level in effluent	Effect on ecosystem
pH	Steel industry	5.5–9	7–9	Alkaline pH decreases the ability for ammonia removal in fish, reduces incoming light; acidic pH reduces aquatic life, egg hatching ability of fishes.
	Paper pulp industry	5.5–9	2.6–10.5	
	Textile industry	5.5–9	7–9	
Total dissolved solids	Steel industry	2000 ppm	800–4000 ppm	Water hardness, water discoloration, and depletion of aquatic plant's survival rate.
	Paper pulp industry	2000 ppm	3750–4010 ppm	
	Textile industry	2000 ppm	3210–5290 ppm	
Total suspended solids	Steel industry	600 ppm (public sewers), 200 ppm (irrigation land)	400–2000 ppm	Reduces photosynthetic ability of aquatic plants, increases water temperature, depletes oxygen levels.

TABLE 5.2 (*Continued*)
A comparison among three different types of industrial effluents based on different parameters

Parameters	Industries	Standard level	Average level in effluent	Effect on ecosystem
	Paper pulp industry	600 ppm (public sewers), 200 ppm (irrigation land)	399.2–1125 ppm	
	Textile industry	600 ppm (public sewers), 200 ppm (irrigation land)	830–1580 ppm	
Biological oxygen demand	Steel industry	350 ppm	80.1 ppm	Rapid oxygen depletion, low dissolved oxygen concentration.
	Paper pulp industry	350 ppm	183.5–670.8 ppm	
	Textile industry	350 ppm	500–1010 ppm	
Chemical oxygen demand	Steel industry	250 ppm	340.1 ppm	Rapid reduction in dissolved oxygen concentration levels, formation of anaerobic environment.
	Paper pulp industry	250 ppm	802–1566 ppm	
	Textile industry	250 ppm	1600–3200 ppm	

From Table 5.2, it can be concluded that the level of all the parameters in the effluent must be maintained within the permissible limit so that the negative impact on the ecosystem can be reduced for a better world.

5.6 CONCLUSION

From the discussion, it can be concluded that the industrial effluent contains toxicants with higher limits than those set by the regulatory agencies of India. Industries must treat their effluent prior discharge into the surface water bodies. Untreated effluent is posing immense threat to the life of all living organisms including human beings. There are different low-cost and eco-friendly treatment methods for wastewater treatment that have not been discussed here; industries must adopt those. There has to be stringent monitoring and penalties to be implemented for non-compliance, to maintain a safe living environment.

REFERENCES

Ahmad, Z., Khan, S.M., Ali, M.I., Fatima, N. and Ali, S., 2019. Pollution indicandum and marble waste polluted ecosystem; role of selected indicator plants in phytoremediation and determination of pollution zones. Journal of Cleaner Production, 236:117709.

Aniyikaiye, T.E., Oluseyi, T., Odiyo, J.O. and Edokpayi, J.N., 2019. Physico-chemical analysis of wastewater discharge from selected paint industries in Lagos, Nigeria. International Journal of Environmental Research and Public Health, 16(7):1235.

Brar, A., Kumar, M., Vivekanand, V. and Pareek, N., 2017. Photoautotrophic microorganisms and bioremediation of industrial effluents: Current status and future prospects. 3 Biotech, 7(1):1–8.

Carolin, C.F., Kumar, P.S., Saravanan, A., Joshiba, G.J. and Naushad, M., 2017. Efficient techniques for the removal of toxic heavy metals from aquatic environment: A review. Journal of Environmental Chemical Engineering, 5(3):2782–2799.

Chhonkar, P.K., Datta, S.P., Joshi, H.C. and Pathak, H., 2000. Impact of industrial effluents on soil health and agriculture-Indian experience: Part I-distillery and paper mill effluents. JSIR, 59(5):350–361.

Choy, C.A., Robison, B.H., Gagne, T.O., Erwin, B., Firl, E., Halden, R.U., Hamilton, J.A., Katija, K., Lisin, S.E., Rolsky, C. and Van Houtan, K., 2019. The vertical distribution and biological transport of marine microplastics across the epipelagic and mesopelagic water column. Scientific Reports, 9(1):1–9.

Das, P., Mondal, G.C., Singh, S., Singh, A.K., Prasad, B. and Singh, K.K., 2018. Effluent treatment technologies in the iron and steel industry-A state of the art review: Das et al. Water Environment Research, 90(5):395–408.

Dudani, S.N., Lakhmapurkar, J., Gavali, D. and Patel, T., 2017. Heavy metal accumulation in the mangrove ecosystem of south Gujarat coast, India. Turkish Journal of Fisheries and Aquatic Sciences, 17(4):755–766.

Dutta, J., Ahmad, A. and Singh, J., 2018. Study of industrial effluents induced genotoxicity on Allium cepa L. Caryologia, 71(2):139–145.

Dwivedi, P. and Tomar, R.S., 2018. Bioremediation of textile effluent for degradation and decolourization of synthetic dyes: A review. International Journal of Current Research in Life Sciences, 7(4):1948–1951.

Gdara, I., Zrafi, I., Balducci, C., Cecinato, A. and Ghrabi, A., 2018. Seasonal occurrence, source evaluation and ecological risk assessment of polycyclic aromatic hydrocarbons in industrial and agricultural effluents discharged in Wadi El Bey (Tunisia). Environmental Geochemistry and Health, 40:1609–1627.

Holon, F., Marre, G., Parravicini, V., Mouquet, N., Bockel, T., Descamp, P., Tribot, A.S., Boissery, P. and Deter, J., 2018. A predictive model based on multiple coastal anthropogenic pressures explains the degradation status of a marine ecosystem: Implications for management and conservation. Biological Conservation, 222:125–135.

Imtiazuddin, S.M., Mumtaz, M. and Mallick, K.A., 2012. Pollutants of wastewater characteristics in textile industries. Journal of Basic & Applied Sciences, 8:554–556.

Kanu, I. and Achi, O.K., 2011. Industrial effluents and their impact on water quality of receiving rivers in Nigeria. Journal of Applied Technology in Environmental Sanitation, 1(1):75–86.

Lyu, R., Zhang, J., Xu, M. and Li, J., 2018. Impacts of urbanization on ecosystem services and their temporal relations: A case study in Northern Ningxia, China. Land Use Policy, 77:163–173.

Manikandan, P., Palanisamy, P.N., Baskar, R., Sivakumar, P. and Sakthisharmila, P., 2015. Physico chemical analysis of textile industrial effluents from Tirupur city, TN, India. International Journal of Advanced Research in Science and Technology, 4(2):93–104.

Matta, G., Kumar, A., Naik, P.K., Tiwari, A.K. and Berndtsson, R., 2018. Ecological analysis of nutrient dynamics and phytoplankton assemblage in the Ganga River system, Uttarakhand. Taiwan Water Conservancy, 66(1):1–12.

Morrow III, W.R., Hasanbeigi, A., Sathaye, J. and Xu, T., 2014. Assessment of energy efficiency improvement and CO_2 emission reduction potentials in India's cement and iron & steel industries. Journal of Cleaner Production, 65:131–141.

Paul, S., De, R., Sinha, S., Dutta, A., Chakraborty, A. and Mukherjee, S., 2022. Sustainable management of arsenic translocation in the paddy plants (*Oryza sativa* L) cultivated in the alluvial soil of Gangetic West Bengal, India. Applied Biochemistry and Biotechnology, 194:3974–3983.

Regaldo, L., Gutierrez, M.F., Reno, U., Fernández, V., Gervasio, S., Repetti, M.R. and Gagneten, A.M., 2018. Water and sediment quality assessment in the Colastiné-Corralito stream system (Santa Fe, Argentina): Impact of industry and agriculture on aquatic ecosystems. Environmental Science and Pollution Research, 25(7):6951–6968.

Singarea, P. and Dhabardeb, S., 2014. Toxic metals pollution due to industrial effluents released along Dombivali Industrial Belt of Mumbai, India. European Journal of Environmental and Safety Sciences, 2(1):5–11.

Singh, U., Tiwari, R.K. and Pandey, R.S., 2019. Physicochemical characteristics of fertilizer industry effluent and its toxicological impact on the activity of acetylcholinesterase (aChE) in freshwater teleostsHeteropneustesfossilis and Labeorohita. Croatian Journal of Fisheries, 77(2):77–86.

Staicu, L.C., Morin-Crini, N. and Crini, G., 2017. Desulfurization: Critical step towards enhanced selenium removal from industrial effluents. Chemosphere, 172:111–119.

Sujatha, D., Mani, U., Durai, M.F., Saxena, P.N., Murthy, R.C., Rose, C. and Mandal, A.B., 2013. Contamination of soil and water by industrial effluents and metal accumulation in plant produce of Ranipet area of Tamilnadu, India. Journal of Applied Phytotechnology in Environmental Sanitation, 2(2).

Toczyłowska-Mamińska, R., 2017. Limits and perspectives of pulp and paper industry wastewater treatment–A review. Renewable and Sustainable Energy Reviews, 78:764–772.

Wang, H., Magesan, G.N. and Bolan, N.S., 2004. An overview of the environmental effects of land application of farm effluents. New Zealand Journal of Agricultural Research, 47(4):389–403.

Yılmaz, S., Gerek, E.E., Yavuz, Y. and Koparal, A.S., 2018. Treatment of vinegar industry wastewater by electrocoagulation with monopolar aluminum and iron electrodes and toxicity evaluation. Water Science and Technology, 78(12):2542–2552.

6 Toxicity of Surface Water Bodies and Public Health

Magapu Solomon Sudhakar and S. Senthil Kumar

6.1 INTRODUCTION

6.1.1 Clean Water Is Highly Indispensable for the Day-to-Day Life of Mankind

According to the factsheet of *www.epa.gov/cleanwaterrule*, to have healthy communities downstream, we must have a supply of clean water upstream. The welfare of streams, ponds, inlets, and seaside waters depend on the streams and wetlands where they begin. Streams and wetlands contribute a plethora of profits to societies through entrapping floodwaters, recharging groundwater supplies, filtering pollution, and imparting habitat for fish and wildlife. Many people hang on to fresh water supplies for their health: about 117 million Americans—one in three humans—get drinking water from streams that have been susceptible to pollution earlier than the Clean Water Rule. Our daily lives rely on having access to clean water: wholesome environments provide natural world habitat and locations to fish, paddle, surf, and swim. Our financial system relies upon pure water: manufacturing, farming, tourism, leisure, energy production, and various financial sectors need clean, fresh water to function well and thrive.

According to the Clean Water Act by the EPA from the government of the US, the definition of clean water has its own rules and stringent laws to protect pure fresh water, and the pure water rule states the following:

- **Clearly defines and protects tributaries that impact the health of downstream waters.** The Clean Water Act protects navigable waterways and their tributaries. The rule says that a tributary must show physical features of flowing water—a bed, bank, and ordinary high-water mark—to warrant protection. The rule provides protection for headwaters that have these features and science shows can have a significant connection to downstream waters.
- **Provides certainty in how far safeguards extend to nearby waters.** The rule protects waters that are next to rivers and lakes and their tributaries because science shows that they impact downstream waters. The rule sets boundaries on covering nearby waters for the first time that are physical and measurable.
- **Protects the nation's regional water treasures**. Science shows that specific water features can function like a system and impact the health of downstream waters. The rule protects prairie potholes, Carolina, and Delmarva bays, pocosins, western vernal pools in California, and Texas coastal prairie wetlands when they impact downstream waters.

DOI: 10.1201/9781003297901-6

- **Focuses on streams, not ditches**. The rule limits protection to ditches that are constructed out of streams or function like streams and can carry pollution downstream. So, ditches that are not constructed in streams and that flow only when it rains are not covered.
- **Maintains the status of waters within Municipal Separate Storm Sewer Systems.** The rule does not change how those waters are treated and encourages the use of green infrastructure.
- **Reduces the use of case-specific analysis of waters.** Previously, almost any water could be put through a lengthy case-specific analysis, even if it would not be subject to the Clean Water Act. The rule significantly limits the use of case-specific analysis by creating clarity and certainty on protected waters and limiting the number of similarly situated water features.

> **The rule protects clean water without getting in the way of farming, ranching, and forestry.** Farms across America depend on clean and reliable water for livestock, crops, and irrigation. Activities like planting, harvesting, and moving livestock have long been exempt from Clean Water Act regulation, and the Clean Water Rule doesn't change that. The Clean Water Rule provides greater clarity and certainty to farmers and does not add any new requirements or economic burden on agriculture.

Although these rules are implemented in the USA, still the clean water bodies contamination is a growing menace for public health sectors. Globally, this has become the most impeding issue for a healthy society. The Third World countries have a marginalized supply of clean water supplies which is overridden by frequent floods, droughts, famine, and tsunamis, and today we are witnessing how proxy wars and wars are devasting the clean water systems. Now this chapter will shed some light on the various water bodies getting contaminated through various sources and their health hazards.

6.2 WATER BODIES GETTING CONTAMINATED THROUGH MINING WASTE

The quality of water both in surface (Alonso et al., 2020) and ground (Ghezzi et al., 2019) gets contaminated through the unprecedented exploitation of mines as these are the rich source of mineral ores from the earth's crust. Nevertheless, mines are indispensable resources for the economic development of a nation, but improper implementation of environmental safeguards lead to the catastrophe of the ecosystems. The seamless exploitation of water resources for mining activities skews up the salinity and heavy metal content in groundwater resources (Gomo and Vermeulen, 2013). Perceiving the spring of dense metal is indispensable for enhanced execution of controlling guidelines and schemes throughout the functioning phase of the mine and solicitation of remedial technologies after the shutting of the mine. In regions where extensive mining is done, mine workings and waste (tailings, waste rocks, and slag) are a principal basis of dense metal adulteration.

Many mines are active and abandoned all over the world. Both active or abandoned mine workings are a predominate polluter of groundwater, and emphatically, it influences hydrogeochemistry and hydrogeochemical methods in the adjoining areas (Nordstrom, 2011: Gomo and Vermeulen, 2013; Bozau et al., 2017).

The skewed-up heavy metal contamination in soil, ground, and surface water was the major factor of health risks to mankind which could have been caused through direct consumption or dermal contact (Park and Choi, 2013).

In fact, World Health Organization (WHO) has framed permittable parameters for the concentrations of heavy metals for the utilization of drinking water by the public. But the mines increase the concentration of these heavy metals in domestic water supplies, which exceeds the threshold limits set by WHO (Abraham and Susan, 2017).

The enormity, occurrence, and extent of human acquaintance on consuming heavy metal–contaminated water is the index of threat assessment of dense metal adulteration. In fact, it was reported by Lu et al., 2019 and Rehman et al., 2018 that the impurities in the water resources at the mining sites, which is attributed to heavy metal infiltration, are the potential cause of noncarcinogenic and carcinogenic health problems, more importantly in children rather than in adults. In fact, the elevated concentrations of mercury and lead are the major cause of noncarcinogenic conditions causing major health risks.

To curtail the problem of dense metal contamination through the mines and mine waste, phytoremediation stands out to be the potential means which is an environmental, cost-effective, and worthwhile resolution. The water usage capacity of some trees like the eucalyptus and pinus species is very high. In fact, eucalyptus plantations, as a biodrainage source, decrease the water ingression into mine workings (Dennis et al., 2020). In fact, the high concentrations of heavy metals at the sites of mines become a potential impediment to grow these plants which have the capacity of biodrainage, since the piling up of the dense metal impurities at the sites of mines hinders the growth of the plants, which eventually leads to the death of these plants. During the process of phytoremediation, the extraction of dense metals from the polluted water is salvaged through the accumulation in the various parts of the plant like shoots, roots, and leaves. But this means of clearing heavy metal contamination from the mine sites does not give a permanent solution for total elimination of heavy metal contamination since decomposition of these plants could always pave the way for reentry of these heavy metals into the soil, which again causes environmental problems. The other means of curtailing the menace of heavy metal contamination at the mine sites is to implement reforestation of abandoned mines, which could drastically reduce the impact on the environment. In fact, in Poland this approach was successfully achieved by implanting replanting of trees at mining lands and waste heaps of coal, lignite, and sulfur, as reported by Pietrzykowski and Krzaklewski, 2018. Similarly, in eastern Kentucky, USA, such approach was used wherein the replanting of trees at mine land began in 1996–97, and the lands were converted into forest biomes.

In summary, it is now a well-known fact that the pollution of mines is a stressful ecosystem system. In fact, contamination of mine waters assessment is done by two processes; that is, pollution and mortification of herbal resources. Mines placed in exclusive places of the world are causing infection of water, and currently, pollution of water at greater elevations are additionally a primary anxiety. Surface water bodies placed at greater heights are susceptible to atmospheric changes and the increase in temperature, caused by global warming, leads to the movement of heavy metals, which then settle as sediment. Furthermore, mortification of herbal resources such as water resources, wooded area cover, and soil for mines results in land use and land cover (LULC) which disturbs the recharge/discharge capability of aquifers, degrading the quantity and quality of water. Heavy steel intensity surpasses recommended admissible limits in ingesting water at most locations adjacent the mines. Use of heavy metal–contaminated water for irrigation and drinking purposes leads to fitness hazards. The enactment of phytoremediation applied sciences is particularly recommended. Acid neutralization and restoration of a deserted mine to a natural ecosystem is expensive and needs manpower. The utility of locally reachable low-priced cloth, for instance, using hen eggshell remains, agricultural and kitchen waste for neutralizing acid is suggested. Therefore, it's crucial to urgently enforce effective waste management and treatment policies to preserve the quality and quantity of water resources in mining areas.

6.3 CONTAMINATION OF WATER RESOURCES LIKE GROUNDWATER AND LAKES THROUGH HEAVY METALS

Groundwater is one of the most essential and pure forms of readily available water used for various activities, for example drinking, irrigation, etc. Groundwater might also be defined as the water present in the subsurface in the geological formations regarded as aquifers and are pumped out for a variety of things to do such as agricultural, domestic, or industrial (Ahamad et al., 2018, 2020a; Madhav et al., 2020a).

Vertical size of water availability is marked as water table. Out of all the available water, only 3% is usable, and a small portion of that is our groundwater. It is established that groundwater is less polluted as compared to various accessible forms of water include groundwater (lakes, rivers, ponds, etc.). Due to overextraction and overdependence on groundwater, we are going through groundwater crisis (Ahamad et al., 2020b; Madhav et al., 2018).

In the current situation, several of the regions are struggling because of water scarcity and infection problems. Furthermore, its condition is further compromised by groundwater contamination (that may also be chemical or biological, etc.). People's overdependence on groundwater has additionally led to mistreatment of this resource. Climate trade and pollution are different elements that have led to savings in surface water. As a result, we have become more reliant on groundwater sources for our water supply (Coyte et al., 2019). A foremost bloc of India grieves from periphery water stress. India is one of the largest customers of groundwater with approximately 230 km3 of annual groundwater consumption (Chindarkar and Grafton, 2019). It has led to a reducing in the groundwater table in many areas of the country. Mistreatment of these water assets has exhausted and has led to drought-like situations in many components of the country. Pollution from natural and human activities has worsened this situation. Contamination can occur from natural substances or human-made pollutants (Madhav et al., 2018, 2020b), whilst human interference that leads to water depletion is recognized as an anthropogenic source of water depletion. One such endeavor is depositing waste in open land, which in turn leads to groundwater pollution.

Many factors contribute to groundwater contamination, like the following:

1. Tanks warehousing chemicals in liquid form and gasoline subjected to leakage can cause severe groundwater contamination.
2. Sewage waste, when carried to specially designed septic tanks, can leak microbial bodies and other contaminants into groundwater if the tanks are damaged.
3. A plethora of uncontrolled lethal waste becomes a major potential threat for groundwater contamination.
4. Many landfills where waste is piled up is also a potential source wherein the groundwater contamination occurs through the seepage of contaminates from the waste accumulated.
5. Many chemicals in the form of fertilizers sprayed at public parks and roadside plantations can lead to the percolation of chemicals through soil, eventually contaminating groundwater.
6. Atmospheric contaminants can easily pollute groundwater as it is part of the water cycle.

As reported by Saha et al. (2014), most parts of India are facing the problem of groundwater contamination which is predominated with arsenic and fluoride along with heavy metal contamination. Groundwater existing below the water table in soil is one of the vital valuable varieties of water that is utilized for a range of tenacities. Indeed, if one prefers to use it extensively, they should check and take measures to prevent the risk of groundwater contamination. Progressive activities such as industrialization, strength generation, agricultural activities, etc. need to be checked out to defend the contamination of groundwater. The wastes produced out of these are predominant sources of groundwater contamination. It's crucial to actively focus on implementing solutions that center around protecting these subsurface sources from pollution. One main stoppage that we can take to prevent the groundwater contamination is to attempt to stop it at the first site. Furthermore, we can choose in situ remedy processes for stopping and treating the contaminated groundwater bodies. One can adopt for ex situ practices and different applied sciences that can be a helping hand to fixing this important matter. In conclusion, public awareness, coupled with collaborative efforts, can bring about significant changes and present solutions to this global environmental issue.

Lakes are inland water bodies and account for 50.01% of the water current on the floor of the Earth (Bhateria and Jain, 2016). They do not have any direct connections with oceans in terms of shape however are often supported by rivers and precipitation and, hence, can also be transient or permanent, depending upon the supply and their geographical vicinity (Kohfahl et al., 2008). Due to their restricted size, lakes may also serve as extraordinary habitats to study dynamics between physical, chemical, and organic approaches (Bhateria and Jain, 2016). Based on hydrogeochemistry and salinity, lakes have been labeled into freshwater, saline, or brackish (Bhateria and Jain, 2016). Inland saline water lakes make up around 85,000 km^3 of the total volume of water on Earth, which is barely less than 105,000 km^3 contributed via freshwater lakes (Shiklomanov, 1990). Locally, saltwater lakes may additionally outnumber freshwater lakes, thereby dominating the landscape and serving as a harbor to several endemic species as well as migratory and breeding birds (Jellison et al., 2004), and its internally drained basin, developed due to dissolution strategies or tectonic activity, intersects the water desk (Duffy and Al-Hassan, 1988; Yechieli and Wood, 2002).

They are usually determined in sub-humid and arid (25–200 mm annual precipitation) to semi-arid (200–500 mm) environments (Jellison et al., 2004; Kohfahl et al., 2008). They might also differ in their size, salinity, flora and fauna, ionic composition, as well as age, giving upward push from ephemeral playas to ancient lakes (Hammer, 1986; Williams, 1996). Moreover, the saltiness and mass are induced by using the altering balance of hydrological inputs such as rainfall, groundwater influxes, and floor runoff and outputs such as vaporization, seepage, etc. These instabilities might also be a result of climatical or periodic variability and/or anthropogenic activities. Generally, salt lakes have a high salinity, pH (8–12) and low dissolved oxygen (above saturation near surface to depleted in hypolimnia region) (Melack, 1983; Jellison et al., 2004). The sediments of a salt lake contain soluble evaporite minerals such as sulfates and halites, sparingly soluble minerals such as dolomite and calcite, inorganic materials deposited through eolian processes and surface runoff, and natural material which may additionally be auto- or allochthonous (Last and Schweyen, 1983; Hammer, 1986). Salt lakes contribute significantly to ecological, economic, and cultural values. Saline structures have season-based totally momentary bays and usually scarcity in outlets. Williams (1972) said that, owing to their simplicity, they are comparatively less difficult to examine as adversarial to their freshwater counterparts. Moreover, being susceptible to exchange in climatic regimes, saline lakes serve as excellent sites to study the paleoclimate of the region and examine sedimentary records (Kohfahl et al., 2008). Salt lakes have been most importantly used as a source of food. Different food sources extracted from them encompass birds, fishes, shrimps, spirulina (blue-green algae), flamingo eggs, and ephydrids pupae (Hammer, 1986).

Owing to a range of microorganism and enzymes found in the mud and water of saline lakes, they are also used as fitness spas (Hammer, 1986). They are reservoirs of salt which has trading records dating back to 2000 years (Williams, 1981). Several economically essential minerals are extracted from salt lakes, which include borax, sodium sesquicarbonate (trona), epsomite, gypsum, lithium, and uranium (Williams, 1981). They also furnish habitation to various quintessential species containing migratory birds. Thus, many salt lakes are being covered through the International Treaty of Wetlands of International Importance (Ramsar Treaty) and efforts of its partner corporations (Wetlands International, the International Union for Conservation of Nature, Birdlife International, Worldwide Fund for Nature). Also, various non-governmental organizations are dedicated to the protection of individual salt lakes (Jellison et al., 2004)

More than 200 nations have joined the Ramsar Convention (Treaty) to shield 3870 wetland websites, of which 202 include permanent inland salt lakes and 405 coastal brackish to saline lagoons (RSIS, 2020). While this quantity can also mirror merely a small element of necessary saline lakes globally, the listing is being up to date per annum.

In comparison to their freshwater equivalents, even a slight change in salt lakes can be highly amplified due to the delicately balanced physical, chemical, and biological cycles. They are prone to illness by way of weathering of rocks, leaching of soils, wastewater disposal, runoff, as well as farm animals' activities (Bhateria and Jain, 2016). Furthermore, as the salt awareness increases, it may additionally go with the flow into clean groundwater instructed through bulk forces, thereby affecting the groundwater niche of the surrounding area. The cell flow, recognized as 'free convective flow,' may additionally proceed beneath the floor of the lake if the brine is reachable (Fan et al., 1997). As the need for suburbanization and transformation has elevated over the years, the quality and quantity of available water resources, both surface and groundwater, have become a critical issue with the growing population. Over the years, very few salt lakes, including the Caspian Sea and Issyk-Kul, have been able to keep their quantity as well as salinity regardless of water diversions (Jellison et al., 2004). The demand for water has also increased across various sectors, including agriculture and various industries like chemicals, textiles, etc. Moreover, since floor water is effortlessly inclined to contamination, it has led to overexploitation and great load on groundwater assets (Hussain et al., 2012). In the Indian setup, scarcity of a desirable framework for treatment of wastewater before its discarding is poor in small cities and rural areas. Central Pollution Control Board of India (2009) has said that out of around 38 billion liters of wastewater and sewage produced every day, just 12 billion liters can be dealt with in herbs. Thus, direct land utility or untreated discharge is the major reason of air pollution of water resources. Jellison et al. 2004 proposed that ultimately the financial prominence of freshwater outweighs that of salt lakes and the noneconomic values such as biodiversity and cultural values are rarely contemplated for the duration of administration evaluations. This has led to lack of firm laws for their safety and terrible management practices. In this evaluation, we studied saline lakes from around the world and, based totally on most common troubles confronted by using them, endorse measures for revival and conservation.

In summary, human intervention in environments is a widely acknowledged fact and has been the primary cause of their degradation. Since saline lakes comply with closed gadget dynamics, they are susceptible to even minute changes. Moreover, as it is believed that saline lakes do no longer possess tons of commercial value, there are inadequate legal guidelines for their protection. Thus, overuse, urbanization, mineral useful resource extraction, careless discharge of effluents, encroachment, illegal activities, local weather change, as well as lack of ideal management have jointly led to illness of saline environments around the world. This has now started affecting the biodiversity as properly as the health of the natives. After plenty of outcry and debates, the socioeconomic fee of saline lakes is being identified, and steps are being taken for better management of the troubles brought about due to overexploitation and negligence. However, the true impact and extent to which salt lakes are being affected globally need to be fully understood, despite the widespread acknowledgment of the impact of water harvesting structures, water diversions for irrigation, and urbanization on large lakes. Additionally, solar panels and coastal lagoons, which additionally represent saline settings of commercial, social, and ecological importance, are dealing with similar threats due to dearth of management measures. Thus, it is advised to involve stakeholders at each level inclusive of neighborhood community, administration associations, as well as governing bodies for a better appreciation of the issues, generating cooperation amongst themselves. Sterner laws and punishments for defaulters may also, in addition, help in curbing malpractices. Apart from conservation, revival of damaged areas ought to be highlighted, principally focusing on strategies which pose negligible intrusion to the natural placing of the area.

6.4 HEAVY METAL CONTAMINATION OF WATER RESOURCES

Besides survival of life, water is indispensable for healthful ecosystems, manufacturing of food, and energy (Gill et al., 2019; UN-WATER, 2020). It additionally serves as an essential hyperlink between the human life and environment as it is the most quintessential factor for local weather adaptation (Konapala et al., 2020). Primarily, it is immensely important, and, in fact, water is at the core of sustainable socioeconomic development (Tortajada, 2020; Sdiri et al., 2018; UN-WATER, 2020). Hence, enough providence of genuine good water is usually required for agricultural, industrial, and domestic practices. Nevertheless, in newest times, water pollution has come to be one of the best encounters for the international union. Water scarcity issues, worsened by water pollution, are becoming increasingly critical challenges for water quantity and quality in many parts of the world (Boretti and Rosa, 2019; Kummu et al., 2016; Mekonnen and Hoekstra, 2016). To a large extent, human activities such as overexploitation of water resources, using water systems for transportation, and the discharge of untreated wastewater from industries into water bodies often lead to water pollution. However, launch of large volumes of untreated industrial wastewater streams and home sewage are responsible for devastating and catastrophic water pollution and produce long-lasting grave consequences on aquatic environments and ecosystems (Singh et al., 2020; Schweitzer and Noblet, 2018; Speight, 2019). These water brooks of devastation, broadly speaking, have a complicated chemical composition owing to the presence of numerous types of contaminants ranging from artificial dyes, heavy metals, pigments, cyanides, nitrates to organic solvents such as xylene, toluene, crude oil, and other hydrocarbons. Among all the pollutants, heavy metals won the paramount appeal from environmentalists, chemists, and biologists due to their extremely hazardous nature (Bansal, 2020; Chen et al., 2015; Masindi and Muedi, 2018). Further, owing to nonbiodegradability of heavy metals, these contaminants impose a serious threat to human and aquatic existence via bioaccumulation and biomagnifications even when existing in very much less concentrations in water sources (Ahmed et al., 2019; Maurya et al., 2019). Heavy metals, in general, no longer have any particular or well-defined characterization; however, in literature, heavy metals are referred to as naturally occurring factors having high bulk density and atomic mass and are denser than water. Generally, they are described as the species having precise density larger than 5 g/cm3. Metals such as arsenic, lead, chromium, mercury, cobalt cadmium, nickel, and selenium come below the category of heavy metals (Alam et al., 2020; Fergusson, 1990; Mishra et al., 2018; Nriagu, 1989). It is true that some of the heavy metals are crucial micronutrients, especially zinc and selenium, and traces of prescribed quantities of such metals are fantastically essential for human health, animals, and plant life (Bhattacharya et al., 2016; Chen et al., 2019; Kieliszek, 2019; Rengel, 2004). Nonetheless, presence of heavy metals above tolerable limits in water sources and over ingestion both directly via consuming water or by means of food chains is extraordinarily hazardous and fatal. The discharge of inadequately-dealt-with wastewater streams of manufacturing and waste recycling industries are amongst the chief sources of heavy metallic infection in surface water as properly as in groundwater resources. The elevated presence of heavy metals in water can have severe adverse effects on the ecological health of aquatic animal species and are primarily accountable for declines in the populations of aquatic species of each animal and plant, especially the fish species. Owing to the robust neurotoxin nature of heavy metals for fish species, they interrupt the verbal exchange of fish species with the environment (Ahmed et al., 2016; Djedjibegovic et al., 2020; Huseen and Mohammed, 2019). Additionally, illness of heavy metal in water is regarded to reason irregularities in fishes, as examined in natural populations and in laboratories as well. Generally, such abnormalities serve as effectively reachable and exceptional biomarkers for identification of heavy metallic contamination in water (Javed and Usmani, 2019; Hemmadi, 2017; Sabullah et al., 2015). These fishes are amongst the primary sources of protein for human consumption also, the direct consumption of water contaminated with heavy metals can have harmful effects; food chains are other important pathways via which heavy metals migrate from contaminated floor water and groundwater to the human diet. Hence, the impact on human health and aquatic life is regularly assessed by various bodies, including the World Health Organization (WHO) (Jaishankar et al., 2014; Morais et al., 2012; Shen et al., 2019). For example, extra consumption of cadmium damages the kidney and bones. Mercury, which gets into the human system via fishes (food chains), causes neurological problems, specifically to the fetus in pregnant women. Lead is known for its neurotoxic effects and significantly reduces blood levels when consumed in excess of the recommended limits, especially in children. Arsenic comes from food chains (through rice and fishes) and causes most cancers and liver and kidney damage. The hazardous results of heavy steel illness in water are further elaborated in subsequent sections (Francesconi, 2010; Hong et al., 2014; Lavoie et al., 2018). Therefore, understanding the current state of heavy metal poisoning in water is of wonderful environmental and socioeconomic importance. This contributes significantly to the creation or development legislations, policies, and administration protocols for industries that create big streams of heavy-steel-contaminated effluents. Furthermore, it can provide guidance to researchers and engineers so that cutting-edge technologies, not only for treating wastewater before discharge but also for recovering valuable heavy metals from waste streams, can be designed and developed (Preisner et al., 2020; Schellenberg et al., 2020).

It enhances water remedy techniques in a synergic way. The treated water can be reused for various purposes in the same location, and impurities of heavy metals can be recollected as resources. The current chapter provides an all-inclusive overview of the current situation of heavy metal illness in floor water and groundwater. The cutting-edge technologies outlined in the chapter offers a state-of-the-art image of this subject for the readers who are working in one-of-a-kind arenas and spheres, such as ecologists, environmentalists, biologists, physicists, industrials, people related with water treatment plants, and environmental researchers and students. It includes some case studies with research perspectives which reveal the detrimental effects of heavy metallic contamination in water systems on local habitats. Moreover, several key elements of heavy metal contamination in water resources such as destruction of flora and fauna in waters, migration of heavy metals by using food chains and its implications on human fitness are discussed via a very intensive probing of literature reports. In the near future, this could culminate with a strategic plan and progress in management policies and the development of treatment technologies for the prevention and remediation of heavy metal pollution in water sources.

6.5 CONCLUSION

Heavy metals contamination due to exponential increase in industrialization and urbanization, along with a multifold increment in populace increase and agricultural practices, has imposed a foremost threat to all spheres of the environment, mainly to the water. Heavy metal illness adversely influences almost the whole thing inclusive of water, soil, humans, animals, and aquatic and plant life. The demand for remediation of heavy metallic illness in water and groundwater has intensified immensely in the past few years. Further, due to the entrance of hazardous heavy metals into the human system through food chains, this demand is greater in agricultural regions. Besides the conventional diseases, several neurological and cancer-related disorders are also observed in areas where irrigation involves the use of heavy metal-contaminated groundwater. Therefore, intensive examination of the agricultural soil and water is integral prior to initiation of agricultural practices, and particular sampling and traceable analyses must be finished and evaluated as per prescribed and established standards and protocols. In places where heavy metals in water are above the permissible limits, vital and well-timed actions ought to be taken, and remediation methods must be designed as per the availability of nearby sources and demands. However, from the perspective of heavy steel contamination in water, comprehensive data series and improvement of databases is immensely required through integration of all factors releasing heavy metals into water, specifically the anthropogenic factors. Without such databases, it is very challenging to execute heavy metallic remediation plans, and, importantly, floor degree influences of such strategies are no longer observed. Prior to implementation of administration tactics and treatment technologies, in-depth mapping of the sources of heavy steel pollution is necessary.

The mapping of these sources at the microlevel/local scale is a necessity before venturing to any water conservation programs. The sustainable improvement of water sources and management of heavy metallic contamination demand an adoption of a holistic approach. A recent development in wastewater remedy technologies, in adsorption, appears to be a wonderful methodology for removing heavy metals from water, as it is exceedingly superb not solely for disposing heavy metals from water but, additionally, for recollecting the adsorbed metals. Therefore, strengthen the waste-to-wealth approaches. Moreover, there are many other factors that should be taken into consideration in order reap an excessive overall functioning of remediation. The most fundamental elements encompass characterization of chemical composition and volume of heavy metal contaminated water, selection of adsorbent system, and regeneration and disposal of adsorbents. Importantly, enough attention must be given to protected drinking/transportable water as it is the most important requirement for survival, after oxygen. Consequently, monitoring drinking water sources for contamination with heavy/dense metals should be carried out in all regions of the world. However, accurate testing practices and methods must be employed. Additionally, adequate measurement should include both the sampling and analysis of contaminated water. Crucially, there is a need for a global initiative to ensure access to clean drinking and irrigation water. In addition to this, a variety of legal guidelines and guidelines for conservation of water sources, as properly as checking on industrial releases, need to be updated and revised or, if they no longer exist, need to be created, both in developed and underdeveloped economies. Educational programs are required for spreading facrs about the hazardous effects related to ingesting heavy-steel-contaminated water. Finally, the designing and improvement of commercially manageable and sustainable therapy strategies are extraordinarily vital for providing a sufficient supply of high-quality water, especially for rural or low-income households.

In fact, toxicity by chemicals is very hazardous for human fitness as it may instigate several sorts of extreme sickness subsequently with prolonged consumption. Their path of entrance in the system may be through food, water, or air. It furthermore depends on the exception of the environmental circumstance (e.g., features of air, water, etc.) alongside geographic reasons. Long-lasting ailments in human beings are due to their consumption which may additionally include cardiovascular disorders, carcinogenicity, neurological problems, brain disorders, etc. Heavy metals like cadmium, lead, mercury, copper, arsenic, fluoride, nitrate, chromium, and pesticides could be the major factors causing long-term ailments. These metals make their way into drinking water resources through the process of incomplete burning of fossil fuels as well as industrial waste discharge. To defend human health, WHO, USEPA, and BIS have given some up-to-date values for these chemical substances in water. In fact, approaches like bioremediation, more importantly plant-based bioremediation, adsorption, ion-exchange, photocatalysis, etc., were cited in the literature providing insight into the methods for eliminating dense or heavy metal impurities from water. Nevertheless, adsorption and phytoremediation were more predominant applications to remove the denser metals from soil and water. Further research is required to enhance efficient and less expensive strategies to supply safe water for all, which would be in accordance with one of the sustainable improvement goals. Microbial fuel cells (MCFs) are devices which could be of prime importance for the elimination of metals and pollution from the laboratory discharge. MFCs can be a modern way to employ a range of substrates with unnoticed, undesirable microorganism to attain bio-energy production. MFCs can be efficiently adapted as sustainable energy generators, offering potential benefits for improved safety and a more environmentally friendly approach. The production of bioelectricity from laboratory-generated wastewater by means of MFCs can be achieved with pencil graphite as electrodes (economical as well) in the absence of poisonous mediators and is additionally membraneless. The ability to convert organic material in wastewater into bioelectricity is a remarkable phenomenon; however, a deeper understanding of microbiological fundamentals and further advancements may be necessary. MFCs offer a capable prospect in the era of electrical energy from laboratory wastewater; however, similarly, enhancement and optimization are required to acquire better outcomes for generating electrical energy and putting off pollutants. Today many nanocomposites are being synthesized from different plant parts like the roots, leaves, stem, fruits, flowers, and other derivatives which have the practicable approach to act as multipurpose sellers such as decreasing, soothing, and covering, for the synthesis of nanocomposites. Nanocomposites manufactured from phytoplants are trivial, cost-effective, ecologically friendly, and a nontoxic route for the synthesis of phyto-nanocomposites. These nanocomposites display large functions in the fields of photodegradation of dyes, purification of water, antibacterial activity, antifungal activity, and antioxidants.

REFERENCES

Abraham, M.R., Susan, T.B., 2017. Water contamination with heavy metals and trace elements from Kilembe copper mine and tailing sites in Western Uganda: implications for domestic water quality. Chemosphere 169, 281–287.

Ahamad, A., Madhav, S., Singh, A.K., Kumar, A., Singh, P., 2020a. Types of water pollutants: conventional and emerging. In Sensors in Water Pollutants Monitoring: Role of Material. Springer, Singapore, pp. 21–41.

Ahamad, A., Madhav, S., Singh, P., Pandey, J., Khan, A.H., 2018. Assessment of groundwater quality with special emphasis on nitrate contamination in parts of Varanasi City, Uttar Pradesh, India. Applied Water Science 8 (4), 115.

Ahamad, A., Raju, N.J., Madhav, S., Khan, A.H., 2020b. Trace elements contamination in groundwater and associated human health risk in the industrial region of southern Sonbhadra, Uttar Pradesh, India. Environmental Geochemistry and Health 42, 3373–3391.

Ahmed, A.S., Sultana, S., Habib, A., Ullah, H., Musa, N., Hossain, M.B., et al., 2019. Bioaccumulation of heavy metals in some commercially important fishes from a tropical river estuary suggests higher potential health risk in children than adults. PLoS One 14, e0219336.

Ahmed, M.K., Baki, M.A., Kundu, G.K., Islam, M.S., Islam, M.M., Hossain, M.M., 2016. Human health risks from heavy metals in fish of Buriganga river, Bangladesh. SpringerPlus 5, 1–12.

Alam, R., Ahmed, Z., Howladar, M.F., 2020. Evaluation of heavy metal contamination in water, soil and plant around the open landfill site Mogla Bazar in Sylhet, Bangladesh. Groundwater for Sustainable Development 10, 100311.

Alonso, D.L., Pérez, R., Okio, C.K.Y.A., Castillo, E., 2020. Assessment of mining activity on arsenic contamination in surface water and sediments in southwestern area of Santurbán paramo, Colombia. Journal of Environmental Management 264, 110478.

Bansal, O.P., 2020. Health risks of potentially toxic metals contaminated water. In Nduka, J.K., Rashed, M.N. (Eds.), Heavy Metal Toxicity in Public Health. Intech Open. http://dx.doi.org/10.5772/intechopen.92141; https://www.intechopen.com/chapters/72026

Bhateria, R., Jain, D., 2016. Water quality assessment of lake water: a review. Sustainable Water Resources Management 2 (2), 161–173.

Bhattacharya, P.T., Misra, S.R., Hussain, M., 2016. Nutritional aspects of essential trace elements in oral health and disease: an extensive review. Scientifica 2016, 1–12. Article ID 5464373. http://dx.doi.org/10.1155/2016/5464373

Boretti, A., Rosa, L., 2019. Reassessing the projections of the world water development report. NPJ Clean Water 2, 1–6.

Bozau, E., Licha, T., Ließmann, W., 2017. Hydrogeochemical characteristics of mine water in the Harz Mountains, Germany. Chemie der Erde 77, 614–624. http://dx.doi.org/10.1016/j.chemer.2017.10.001

Central Pollution Control Board, 2009. Status of water supply, wastewater generation and treatment in Class-I cities and Class-II towns of India. Published by Central Pollution Control Board (Series: COPOCS/27/2002-03, CUPS/4/1978-79, and CUPS/6/30/1989-90). Ministry of Environment and Forests, Govt. of India, Delhi. www.cpcb.nic.in

Chen, G., Shi, H., Tao, J., Chen, L., Liu, Y., Lei, G., et al., 2015. Industrial arsenic contamination causes catastrophic changes in freshwater ecosystems. Scientific Reports 5, 17419.

Chen, H.-G., Chen, Y.-J., Chen, C., Tu, Z.-Z., Lu, Q., Wu, P., et al., 2019. Reproducibility of essential elements chromium, manganese, iron, zinc and selenium in spot samples, first-morning voids and 24-h collections from healthy adult men. British Journal of Nutrition 122, 343–351.

Chindarkar, N., Grafton, R.Q., 2019. India's depleting groundwater: when science meets policy. Asia & the Pacific Policy Studies 6 (1), 108–124. https://doi.org/10.1002/app5.269

Coyte, R., Singh, A., Furst, K.E., Mitch, W., Vengosh, A., 2019. Cooccurrence of geogenic and anthropogenic contaminants in groundwater from Rajasthan, India. Science of the Total Environment. https://doi.org/10.1016/j.scitotenv.2019.06.334

Dennis, R., Dennis, I., Mokadem, N., Smit, S., 2020. Investigate the possible reduction of mine water ingress by introducing tree plantations: case study of Cooke 4 mine (South Africa). Journal of African Earth Sciences 161, 103660.

Djedjibegovic, J., Marjanovic, A., Tahirovic, D., Caklovica, K., Turalic, A., Lugusic, A., et al., 2020. Heavy metals in commercial fish and seafood products and risk assessment in adult population in Bosnia and Herzegovina. Scientific Reports 10, 1–8.

Duffy, C.J., Al-Hassan, S., 1988. Groundwater circulation in a closed desert basin: topographic scaling and climatic forcing. Water Resources Research 24 (10), 1675–1688.

Fan, Y., Duffy, C.J., Oliver Jr, D.S., 1997. Density-driven groundwater flow in closed desert basins: field investigations and numerical experiments. Journal of Hydrology 196 (1–4), 139–184.

Fergusson, J.E., 1990. The Heavy Elements: Chemistry, Environmental Impact and Health Effects. Pergamon Press, Oxford, pp. 85–547.

Francesconi, K.A., 2010. Arsenic species in seafood: origin and human health implications. Pure and Applied Chemistry 82, 373–381.

Ghezzi, L., D'Orazio, M., Doveri, M., Lelli, M., Petrini, R., Giannecchini, R., 2019. Groundwater and potentially toxic elements in a dismissed mining area: thallium contamination of drinking spring water in the Apuan Alps (Tuscany, Italy). Journal of Geochemical Exploration 197, 84–92.

Gill, J.C., Mankelow, J., Mills, K., 2019. The role of earth and environmental science in addressing sustainable development priorities in Eastern Africa. Environmental Development 30, 3–20.

Gomo, M., Vermeulen, D., 2013. Investigation of hydrogeochemical processes in groundwater resources located in the vicinity of a mine process water dam. Journal of African Earth Sciences 86, 119–128.

Hammer, U.T., 1986. Saline Lake Ecosystems of the World (Vol. 59). Springer Science & Business Media, Dordrecht.

Hemmadi, V., 2017. A critical review on integrating multiple fish biomarkers aş indicator of heavy metals contamination in aquatic ecosystem. International Journal of Bioassays 6, 5494–5506.

Hong, Y.-S., Song, K.-H., Chung, J.-Y., 2014. Health effects of chronic arsenic exposure. Journal of Preventive Medicine and Public Health 47, 245.

Huseen, H.M., Mohammed, A.J., 2019. Heavy metals causing toxicity in fishes. Journal of Physics: Conference Series, 062028.

Hussain, H.M., Joshi, H., Singhal, D.C., Kumar, S., Rao, M.S., 2012. Development of an index of aquifer water quality within GIS environment. Iranian Journal of Earth Sciences 4 (1), 44–50.

Jaishankar, M., Tseten, T., Anbalagan, N., Mathew, B.B., 2014. Toxicity, mechanism and health effects of some heavy metals. Interdisciplinary Toxicology 7, 60–72. https://doi.org/10.2478/intox-2014-0009

Javed, M., Usmani, N., 2019. An overview of the adverse effects of heavy metal contamination on fish health. Proceedings of the National Academy of Sciences, India Section B: Biological Sciences 89, 389–403.

Jellison, R., Zadereev, Y.S., DasSarma, P.A., Melack, J.M., Rosen, M.R., Degermendzhy, A.G. et al., 2004. Conservation and management challenges of saline lakes: a review of five experience briefs. Lake Basin Management Initiative: Thematic Paper.

Kieliszek, M., 2019. Selenium fascinating microelement, properties and sources in food. Molecules 24, 1298.

Kohfahl, C., Rodriguez, M., Fenk, C., Menz, C., Benavente, J., Hubberten, H., et al., 2008. Characterising flow regime and interrelation between surface-water and ground-water in the Fuente de Piedra salt lake basin by means of stable isotopes, hydrogeochemical and hydraulic data. Journal of Hydrology 351 (1–2), 170–187.

Konapala, G., Mishra, A.K., Wada, Y., Mann, M.E., 2020. Climate change will affect global water availability through compounding changes in seasonal precipitation and evaporation. Nature Communications 11, 1–10.

Kummu, M., Guillaume, J., De, M.H., Eisener, S., Flörke, M., Porkka, M., et al., 2016. The world's road to water scarcity: shortage and stress in the 2^{0t}h century and pathways towards sustainability. Nature Science Reports 6, 38495.

Last, W.M., Schweyen, T.H., 1983. Sedimentology and geochemistry of saline lakes of the Great Plains. Hydrobiologia 105 (1), 245–263.

Lavoie, R.A., Bouffard, A., Maranger, R., Amyot, M., 2018. Mercury transport and human exposure from global marine fisheries. Scientific Reports 8, 1–9.

Lu, J., Lu, H., Lei, K., Wang, W., Guan, Y., 2019. Trace metal element pollution of soil and water resources caused by small-scale metallic ore mining activities: a case study from a sphalerite mine in North China. Environmental Science and Pollution Research 26, 24630–24644.

Madhav, S., Ahamad, A., Kumar, A., Kushawaha, J., Singh, P., Mishra, P.K., 2018. Geochemical assessment of groundwater quality for its suitability for drinking and irrigation purpose in rural areas of Sant Ravidas Nagar (Bhadohi), Uttar Pradesh. Geology, Ecology, and Landscapes 2 (2), 127–136.

Madhav, S., Ahamad, A., Singh, A.K., Kushawaha, J., Chauhan, J.S., Sharma, S., et al., 2020a. Water pollutants: sources and impact on the environment and human health. In Sensors in Water Pollutants Monitoring: Role of Material. Springer, Singapore, pp. 43–62.

Madhav, S., Kumar, A., Kushawaha, J., Ahamad, A., Singh, P., Dwivedi, S.B., 2020b. Geochemical assessment of groundwater quality in Keonjhar City, Odisha, India. Sustainable Water Resources Management 6, 46.

Masindi, V., Muedi, K.L., 2018. Environmental contamination by heavy metals. Heavy Metals 10, 115–132.

Maurya, P.K., Malik, D., Yadav, K.K., Kumar, A., Kumar, S., Kamyab, H., 2019. Bioaccumulation and potential sources of heavy metal contamination in fish species in River Ganga basin: possible human health risks evaluation. Toxicology Reports 6, 472–481.

Mekonnen, M., Hoekstra, A., 2016. Four billion people facing severe water scarcity. Science Advance 2, e1500323.

Melack, J.M., 1983. Large, deep salt lakes: a comparative limnological analysis. Hydrobiologia 105 (1), 223–230.

Mishra, V., Asoka, A., Vatta, K., Lall, U., 2018. Groundwater depletion and associated CO_2 emissions in India. Earth's Future 6, 1672–1681.

Morais, S., Costa, F.G., Pereira, M.D.L., 2012. Heavy metals and human health. Environmental Health Emerging Issues and Practice 10, 227–246.

Nordstrom, D.K., 2011. Hydrogeochemical processes governing the origin, transport and fate of major and trace elements from mine wastes and mineralized rock to surface waters. Applied Geochemistry 26, 1777–1791.

Nriagu, J.O., 1989. A global assessment of natural sources of atmospheric trace metals. Nature 338, 47–49.

Park, J.-H., Choi, K.-K., 2013. Risk assessment of soil, water and crops in abandoned Geumryeong mine in South Korea. Journal of Geochemical Exploration 128, 117–123.

Pietrzykowski, M., Krzaklewski, W., 2018. Reclamation of mine lands in Poland. Bio-Geotechnologies for Mine Site Rehabilitation 493–513.

Preisner, M., Neverova-Dziopak, E., Kowalewski, Z., 2020. An analytical review of different approaches to wastewater discharge standards with particular emphasis on nutrients. Environmental Management 66, 694–708.

Rehman, I., Ishaq, M., Ali, L., Khan, S., Ahmad, I., Din, I.U., et al., 2018. Enrichment, spatial distribution of potential ecological and human health risk assessment via toxic metals in soil and surface water ingestion in the vicinity of Sewakht mines, district Chitral, Northern Pakistan. Ecotoxicology and Environmental Safety 154, 127–136.

Rengel, Z., 2004. Heavy metals as essential nutrients. In Prasad, M.N.V. (Ed.), Heavy Metal Stress in Plants. Springer, Berlin, pp. 271–294. https://doi.org/10.1007/978-3-662-07743-6_11

RSIS, 2020. Ramsar sites information service. https://rsis.ramsar.org/ris-search/?f%5B0%5D5wetlandTypes_en_ss%3AInland%20wetlands

Sabullah, M., Ahmad, S., Shukor, M., Gansau, A., Syed, M., Sulaiman, M., et al., 2015. Heavy metal biomarker: fish behavior, cellular alteration, enzymatic reaction and proteomics approaches. International Food Research Journal 22.

Saha, D., Singh, B., Srivastava, S., Dwivedi, S., Mukherjee, R., 2014. Concept Note on Geogenic Contamination of Ground Water in India. Central Ground Water Board (CGWB), Bujal Bhawan,, Faridabad, Haryana.

Schellenberg, T., Subramanian, V., Ganeshan, G., Tompkins, D., Pradeep, R., 2020. Wastewater discharge standards in the evolving context of urban sustainability the case of India. Frontiers in Environmental Science 8, 30.

Schweitzer, L., Noblet, J., 2018. Water contamination and pollution. In Dransfield, T. (Ed.), Green Chemistry. Elsevier, USA. pp. 261–290. http://dx.doi.org/10.1016/b978-0-12-809270-5.00011-x

Sdiri, A., Pinho, J., Ratanatamskul, C., 2018. Water resource management for sustainable development. Arabian Journal of Geosciences 11, 124. https://doi.org/10.1007/s12517-018-3411-z

Shen, X., Chi, Y., Xiong, K., 2019. The effect of heavy metal contamination on humans and animals in the vicinity of a zinc smelting facility. PLoS One 14, e0207423.

Shiklomanov, I.A., 1990. Global water resources. Nature and Resources 26 (3), 34–43.

Singh, J., Yadav, P., Pal, A.K., Mishra, V., 2020. Water Pollutants: Origin and Status. In: Pooja, D., Kumar, P., Singh, P., Patil, S. (Eds), Sensors in Water Pollutants Monitoring: Role of Material. Springer, Singapore, pp. 5–20.

Speight, J.G., 2019. Natural Water Remediation: Chemistry and Technology. Butterworth-Heinemann, Oxford.

Tortajada, C., 2020. Contributions of recycled wastewater to clean water and sanitation sustainable development goals. NPJ Clean Water 3, 1–6.

UN-WATER. 2020. UN World Water Development Report 2020. United Nations. www.unwater.org/publications/un-world-water-development-report-2020

Williams, W.D., 1972. The uniqueness of salt lake ecosystems. In Kajak, S., Hillbricht-Illkowska, A. (Eds.), Productivity Problems of Freshwaters. Polish Academy of Science, Warsaw, pp. 349–361.

Williams, W.D., 1981. Inland salt lakes: an introduction. In Salt Lakes. Springer, Dordrecht, pp. 1–14.

Williams, W.D., 1996. The largest, highest and lowest lakes of the world: saline lakes. Internationale Vereinigungfu¨rtheoretische und angewandteLimnologie: Verhandlungen 26 (1), 61–79.

Yechieli, Y., Wood, W.W., 2002. Hydrogeologic processes in saline systems: playas, sabkhas, and saline lakes. Earth-Science Reviews 58 (3–4), 343–365.

7 Anthracene in Aquatic Ecosystems
A Review

Susri Nayak, Sthitaprajna Nath Sharma, Smruti Prajna Pradhan, Subhashree Nayak, Lipika Patnaik

7.1 INTRODUCTION

Anthracene ($C_{14}H_{10}$) is a polycyclic aromatic hydrocarbon, predominant in aquatic bodies due to several anthropogenic activities. Its various toxicities have been shown in various organisms including fish in aquatic medium. The means of anthracene toxicity in aquatic animals is primarily associated with its absorption through gills and skin, accumulation in gills, gastrointestinal tract and liver and excretion through kidneys. Various studies have reported the toxic effects of anthracene, affecting various biochemical, haematological and antioxidants parameters of various aquatic animals. Changes in the these parameters can aid in the biomonitoring of anthracene pollution in aquatic environment. The expansion and growth of aquatic species depends on the quality of the environment they are residing in. However, the survival ability and development of aquatic organisms become tough as aquatic bodies are known to be the final basin for the release of maximum pollutants, toxicants, xenobiotics and chemicals. Aquatic organisms are commonly used as biological indicators to assess the health status of any aquatic body, though fishes are always the first choice as test organisms as they are exceedingly responsive to any slight change in aquatic environment. Fishes have always been used as indicators for biomonitoring of various aquatic environmental pollutants including PAHs. In ecotoxicological studies, they are used as test organisms as they are very strong models to establish the numerous changes in various assessed biomarkers on exposure. Anthracene is known to affect the physiology of fishes like growth, enzyme activities, haematological parameters, gene expression, stress parameters and oxidants. Consumption of anthracene-contaminated fish and other aquatic animals can result in affecting the health conditions of humans. Anthracene targets skin, blood, stomach, intestines and the lymph system on bioaccumulation. Nausea, loss of appetite and swelling of stomach and intestine tissues is also noticed in humans consuming anthracene-contaminated food. Anthracene is known to be a skin irritant and causes dermatitis. However, anthracene is considered to be non-carcinogenic to humans as it is categorized under Category IV, as described by USEPA (2008). Persistent effects of pollutants and xenobiotics on various aquatic organisms' models may aid in setting up average criterion associated with assessment of risk and safety

DOI: 10.1201/9781003297901-7

measures for aquatic environment. Hence, various studied toxicological biomarkers together bestow with a complete understanding on the health status of both the aquatic organisms and environment and also act as indicators of environmental pollution stress.

The present review aims to provide an overview of the occurrence of anthracene in the environment, its chemical character, environmental fate and toxicological impact on aquatic organisms. Data on the biological action of anthracene will help to predict biotransformation and side effect of the toxicant. Analysis of various haematological, biochemical, enzymatic and histopathological studies of organs can be used to decide the toxicity, the mode of action of the toxicant and ecological risk associated with anthracene.

7.2 CHEMISTRY OF ANTHRACENE

Anthracene is a solid polycyclic aromatic hydrocarbon and is composed of three benzene rings fused together. It is a colourless organic tricyclic aromatic hydrocarbon with a chemical formula $C_{14}H_{10}$and a molecular mass of 178.23g/mol. The melting point of anthracene is 216°C, and its boiling point is 341.3°C. Its water solubility is 0.044 mgL^{-1} at 25°C. It is soluble in organic solvents like ethanol, methanol, hexane, toluene and carbon tetrachloride. The density of anthracene is 1.25 g/cm^3 (20° C). The other common names of anthracene are para naphthalene, green oil, anthracin, anthracene, tetra olive N2G and anthrazen. Anthracene is unsafe due to its high toxicity, persistent nature and low molecular weight (Dey et al. 2019). Anthracene is capable of undergoing oxidation and reduction reactions. Anthracene is more prone to oxidation-reduction reactions as compared to benzene because of its high resonance energy. Anthracene undergoes oxidation to yield anthraquinone, whereas it undergoes reduction with alkali metals to give deep-coloured radical anion salts. Anthracene is capable of undergoing photodimerization in the presence of UV light. Anthracene is commercially used on a large scale in textile and dye industries as it is converted to anthraquinone which is a precursor to many bright and vibrant dyes. It is used in the production of red dye alizarin. Anthracene is a semiconductor and used as scintillators for detection of high-energy particles like electrons, photons and alpha particles. Polyvinyl toluene, a plastic, is doped with anthracene for production of water-equivalent plastic scintillator which is used extensively for radiation therapy dosimetry. Anthracene is used on a large scale for manufacture of insecticides, wood preservatives and coating materials. It is also used in smoke screens and production of chemotherapeutic agent Amsacrine (Sunmonu et al. 2009). Anthracene is extensively used as a UV tracer in printed wiring board conformal coatings. The anthracene tracer allows for UV inspection of the conformal coating. Derivatives of anthracene like 1-hydroxyanthracene and 2-hydroxyanthracene, commonly known as anthrol and anthracenol, are used in pharmacological industries.

7.3 OCCURRENCE OF ANTHRACENE IN AQUATIC ENVIRONMENT

Anthracene is found everywhere as a product of incomplete production of fossil fuels (air), release of by-products from fabric and textile industries (water) and used as pesticides (soil), thus making its contamination ubiquitous in nature. Various authors have reported its occurrence in rivers, lakes, estuaries and wastewater (Table 7.1) (Yamei et al. 2009; Mohamad et al. 2011; Nagy et al. 2012; Zhao et al. 2014; Edokpayi et al. 2016; Adeniji et al. 2019a). Run-offs from coal-tar production and distillation units introduce anthracene in the environment. Accidental oil spills from ships, petroleum refineries and discharges from the ports also lead to anthracene pollution in and around coastal areas. Anthracene is also a component of petroleum and its by-products. In India various researchers have studied about anthracene contamination in the environment. Few notable works on PAHs contamination indicating anthracene contamination have been carried out in soils of Guwahati, Assam (Das et al. 2016) and in Bharalu tributary of the Brahmaputra River (Hussain et al. 2014). Anthracene was also detected in the groundwater of six districts of southern Jharkhand (Ambade et al. 2021). Anthracene was also detected in the work carried out by Panda et al. (2020), who detected the type and concentration of PAHs in Mahanadi Estuary at Paradeep, Odisha. The estuary is surrounded by industrial units, and they found anthracene concentration in the range of 1.2–4.1 ppb, which was much higher than the WHO prescribed guidelines i.e 50 ngL^{-1}. Anthracene was found concentrated in the muscle tissues of *Channa punctatus* collected from Gomti River, India (Malik et al. 2008). Anthracene is usually in combination with other PAHs as a mixture and is much perilous than its specific form (Irwin et al. 1997). Anthracene has been continuously polluting various water bodies and is considered to be equally hazardous as other PAHs. Low-molecular PAHs are considered more lethal as compared to high-molecular PAHs because they tend to be highly persistent in aquatic bodies and bioaccumulate in more concentrations in aquatic organisms (Deb et al. 2000; Dey and Ghosh 2019).

TABLE 7.1
Overview of the occurrence of anthracene compounds in the natural environment

Water bodies	Water (ng/L)	Sediments (ng/g)	Tissues (ng/g)
Johan River, Malaysia	7100 (dry season), 2200 (wet season) (Tan et al. 2020)		
Antwerp Port, Belgium, North Sea	2.67 (Teuchies et al. 2020)		
Mvudi River, South Africa	256000 (Edokpayi et al. 2016)	440 (Edokpayi et al. 2016)	
Nzhelele River, South Africa		540 (Edokpayi et al. 2016)	
Danube River, Hungary	<1–15 (Nagy et al. 2012)		
Buffalo River Estuary, South Africa	1.97 (Adeniji et al. 2019b)	230 (Adeniji et al. 2019b)	

(Continued)

TABLE 7.1 (*Continued*)
Overview of the occurrence of anthracene compounds in the natural environment

Water bodies	Water (ng/L)	Sediments (ng/g)	Tissues (ng/g)
Niger and Benue River confluence, Lokoja, Central Nigeria		50000–60000 (Ekere et al. 2017)	
Msunduzi River South Africa	14.12–228.30 (Munyengabe et al. 2017)	190–41130 (Munyengabe et al. 2017)	
Poyang Lake, China			*Aristichthysnobilis*—6.85–44 *Hypophthalmichthysmolitrix* –5.79–69.4 (Zhao et al. 2014)
Kor River, Iran	4.95–32.15 (Kafilzadeh et al. 2011)	7.91–11.38 (Kafilzadeh et al. 2011)	
Soltan Abad River, Iran	4.67–32.83 (Kafilzadeh 2015)	8.92–12.52 (Kafilzadeh 2015)	
Yangtze River Estuary, China.		1.7–20.6 (Yamei et al. 2009)	
Yellow River Estuary, China		0.3–10.4 (Yamei et al. 2009)	
AlkaberAljanuby River Estuary, Syria-Lebanon		3.8–107.8 (Mohamad et al. 2011)	
Atlantic Ocean, Pacific Ocean, Mediterranean Sea, Portugal			*Octopus vulgaris*: 0.06–0.15 *Octopus maya* 0.06 *Eledonecirrhosa* 0.14 (Oliveira et al. 2018)
Daya Bay, South China		0.8–2.4 (Yan et al. 2009)	
Gomti River, India	30–90 (Malik et al. 2011)	1.10–14.44 (Malik et al. 2011)	*Channapunctatus* 1.15 (Malik et al. 2008)
Clifton beach, Karachi coast, Pakistan			Fishes *Escualosa thoracatta*: 1550 *Apogos coccineues*: 190 *Johnius earutta*: 40 *Therapan jarbua*: 60 *Anchovies*: 10

TABLE 7.1 (*Continued*)
Overview of the occurrence of anthracene compounds in the natural environment

Water bodies	Water (ng/L)	Sediments (ng/g)	Tissues (ng/g)
			Srdinella sindensis: 110 Shellfishes *Naticavitellus*: 30 *Acanthopagrus latus*: 2310 *Portunuspelagicus*: 30 *Portunus sanguinolentis*: 640 (Siddiqi et al. 2009)
Bizerte lagoon, Tunisia			*Mytilusgalloprovincialis* 0.2–1.5 *Anguilla Anguilla* 0.2–1.0 (Barhoumi et al. 2016)
Bahía Blanca Estuary, Argentina			*Ramnogasterarcuata* 0.27–0.62 (Ronda et al. 2019)
Bahía Blanca Estuary, Argentina		9.2–23.2 (Recabarren-Villalon et al. 2019)	
Lake Chaohu, China	2.81 (Qin et al. 2020)	21.19 (Qin et al. 2020)	*Hemibarbus maculates*: 5.55 *Cyprinuscarpio*: 3.34 *Cultereryropterus*: 1.31 *Megalobrama amblycephala*: 3.19 *Hemisalanx prognathus*: 0.41 *Aristichthysnobilis*: 5.70 *Cipangopaludina chinensis*: 7.84 *Leander modestus*: 1.86 (Qin et al. 2020)
Mumbai harbour, India			Fishes *E. tetradactylum* 1.89 (Dhananjayan and Muralidharan 2012)
Hiroshima Bay Japan			Fishes (*Cepulaschlegeli, Paruplagusia japonica, Pleuronectesyokohama, Cociella crocodilla, Epinephelus tauvina, Apogennotatus, Peegrus major, Stephanolepis cirrhifer, Pterogobiouselapoides, Acropomajaponicum,umbaiap.*) 80–1850 (Deb et al. 2000)

(Continued)

TABLE 7.1 (*Continued*)
Overview of the occurrence of anthracene compounds in the natural environment

Water bodies	Water (ng/L)	Sediments (ng/g)	Tissues (ng/g)
Potter Cove, South Shetland Islands, Antarctica		2.1–20.8 (Curtosi et al. 2009)	*Notothenia coriiceps* 1.03–28.97 *Laternula elliptica* 0.99–1.32 *Nacella concinna* 0.76–2.64 (Curtosi et al. 2009)
East China Sea			Fishes 4.6–6.6 Molluscs 5.1–6.0 Crustaceans 6.1 (Wang et al. 2019)
Mochou Lake, Xuanwu Lake, Qianhu Lake, Yueya Lake, Pipa Lake, Yanque Lake, Wukesong Reservoir, Youyi River, Nan River Nanjing, China	205–744 (Wang et al. 2017)		
Weihe River, China	1.5–12.2 (Chen et al. 2021)	10.4–62.3 (Chen et al. 2021)	
Rufiji Estuary, Tanzania	0.20–15.6 (Shilla et al. 2018)	1.3–13.9 (Shilla et al. 2018)	*Terebralia sp.* 0.4–2.7 *Uca sp.* 0.80–2.60 *Panaeusmonodon* 1.9–11.8 *Hilsakelee* 1.3–10.2 *Trichiurus lepturus* 0.4–3.10 *Arius thalassinus* 0.7–1.8 (Shilla et al. 2018)
Mahanadi Estuary, Paradeep, Odisha, India	1200–4100 (Panda et al. 2020)		

TABLE 7.1 (*Continued*)
Overview of the occurrence of anthracene compounds in the natural environment

Water bodies	Water (ng/L)	Sediments (ng/g)	Tissues (ng/g)
Klang River Malaysia		0.04–0.23 (Bakhtiari et al. 2010)	
North Saskatchewan River, Alberta, Canada		0.50–2.43 (Stone et al. 2013)	
Yamuna River, Delhi, India		0.55 (Agarwal et al. 2006)	
Algoa Bay, South Africa	5610–6870 (Adeniji et al. 2019a)	334 (Adeniji et al. 2019a)	
Damodar River Basin	120 (Ambade et al. 2021)	5 (Ambade et al. 2021)	
Hooghly River, India	70 (Khuman et al. 2018)	5–8 (Khuman et al. 2018)	
Brahmaputra River India	30 (Khuman et al. 2018)	1–3 (Khuman et al. 2018)	
Mithi River, Mumbai India	3220–13700 (Singare 2016)		
Nansi Lake, China	0.61–2.13 (Zhang et al. 2015)	2.63–15.2 (Zhang et al. 2015)	*Carassiusauratus* 1.61–22.8 *Cyprinuscarpio* 3.88–17.9 *Channaargus* 3.52 *Macrobrachium nipponense* 3.34–19.6 (Zhang et al. 2015)
Mangrove Ecosystem, West Peninsula, Malaysia			Mangrove oyster *Crassostrea belcheri* 0.37–8.72 (Vaezzadeh et al. 2019)
Can Gio Coastal Wetland, Vietnam			*Crassostrea* sp. 0.24–3.19 *Cymatium sp.* 0.09–3.03 (Pham et al. 2019)

(Continued)

TABLE 7.1 (*Continued*)
Overview of the occurrence of anthracene compounds in the natural environment

Water bodies	Water (ng/L)	Sediments (ng/g)	Tissues (ng/g)
Soan River, Pakistan	8.7–28 (Aziz et al. 2014)		
White Nile, East Africa		17.2–129 (Abayi et al. 2021)	*Lates niloticus* 4.47–79.3 *Oreochromis niloticus* 21.1 *Clarias gariepinus* 5.98–16 (Abayi et al. 2021)
Langkawi Island, Malaysia	270–880 (Nasher et al. 2013)		
Woji creek, Sambreiro River, Nigeria	17000 (Ihunwo et al. 2021)		*Callinectes Amnicola* 958 (Ihunwo et al. 2021)
Catania Gulf Sicily, Italy			*Mullus surmuletus* 1.19 (Bua et al. 2021)
Roznow Dam Reservoir, Southern Poland		36700 (Baran et al. 2021)	Mussels *Anodonta anatine* 74700 (Baran et al. 2021)

7.4 BIOACCUMULATION

Many authors have stated in various studies that the highest accumulation of low-molecular PAHs like naphthalene and anthracene takes place in aquatic animals. Accumulation of PAHs is dependent on several factors like solubility of PAHs, its bioavailability, molecular weight, octanol-water partition coefficient (K08), exposure routes and behaviour of the inhabiting aquatic fauna (Connell and Miller 1984). Schrap and Opperhuizen (1990) had revealed that PAHs with low octanol-water partition coefficient (K08) do not bind much to the surrounding sediments and are thus quickly absorbed into the water, making the aquatic animals more prone to PAHs contamination. Anthracene compounds are widely reported in the aquatic environment, and constant exposure to high concentrations of anthracene in water has led to its bioaccumulation in fish and other aquatic species. Anthracene was found in the *Chanos chanos* fish tissues sampled from Gomti River, India, and the average concentration determined was 1.15 ngg^{-1} (Malik et al. 2008). Anthracene was found to be bioaccumulated in the muscle tissues of *Mustelus schmitti, Micropogonias furnieri, Cynoscion guatucupa and Ramnogaster arcuata* sampled from Bahía Blanca estuary, Argentina, and the amount varied from 0.27–0.62 ngg^{-1}. Various aquatic animals were sampled from Lake Chaohu, China, and naphthalene was found to be retained in the tissues of the sampled species. Naphthalene was found to concentrated in fishes like *Hemibarbus maculates* (5.5 ngg^{-1}), *Cyprinuscarpio* (3.34 ngg^{-1}), top mouth culter *Culter micropterus* (1.31 ngg^{-1}), bluntnose black bream *Megalobrama amblycephala* (3.19 ngg^{-1}), whitebait *Hemisalan prognathus Regan* (0.41 ngg^{-1}), Bighead carp *Aristichthysnobilis* (5.70 ngg^{-1}), snail *Cipangopaludina chinensis* (7.84 ngg^{-1}) and Chinese white prawn *Leander modestus* (1.86 ngg^{-1}) (Qin et al. 2020). Fishes like *Cepulaschlegeli, Paruplagusia japonica, Pleuronectes yokohama, Cociella crocodilla, Epinephelus tauvina, Apogennotatus, Peegrus major, Stephanolepis scirrhifer, Pterogobious elapoides, Acropomajaponicum, Nibeasp* sampled from Hiroshima Bay, Japan, also exhibited anthracene bioaccumulation ranging from 80–1850 ngg^{-1} (Deb et al. 2000) (Table 7.1). Big head carp and silver carp sampled from Poyang Lake, China, had bioaccumulated anthracene of concentration 5.79–69.4 ngg^{-1} (Zhao et al. 2014). Various authors have also mentioned the bioaccumulation of anthracene in a number of toxicological studies carried out in laboratories. Hall and Oris (1991) had reported the accumulation of anthracene in eggs and gonadal tissues (testis and ovary) of fathead minnows. The authors had administered anthracene of concentrations ranging from 6–12 µg/L initially, which was later increased to 12–20 µg/L for a total period of nine weeks, and bioaccumulation concentrations varied from 0.02–99.97 µg/L at the end of the exposure period. The authors had mentioned that anthracene had bioconcentrated in much higher amounts, and it depends on the lipid content of the tissues. Bioaccumulation of anthracene was also observed when *Chanoschanos* was exposed to concentrations of 0.011–0.176 mgL^{-1}. However, accumulation of anthracene varied from 8.20–94.9 mgL^{-1}. As the accumulation was higher than the concentrations administered, the authors had mentioned that anthracene was in bioavailable concentration in the test medium, which means the higher the concentration, the higher the accumulation (Palanikumar et al. 2012). Anthracene and its metabolites were also found to be bioaccumulated in different organs of *Oncorhynchuskisutch* (Roubal et al. 1977).

7.5 TOXICITY OF ANTHRACENE IN VARIOUS AQUATIC ANIMALS

Palanikumar et al. (2012) exposed the fingerlings of *Chanoschanos*to anthracene for a period of 96 hours and assessed its effects on various stress parameters. Levels of lipid peroxides increased in the tissues of brain, gills and muscles, which indicated that anthracene is capable of inducing oxidative stress. The effects of antioxidants were also studied, and it was noticed that the activity of GST decreased, whereas the activities of CAT and GSH increased. Similar work related to anthracene-induced changes in antioxidant levels had been carried out by Zhu et al. (2016) in *Ruditapes philippinarum. Ruditapes philippinarum* were exposed to anthracene concentrations of 0.007, 0.07 and 0.7 mgL^{-1}, and it was noticed that SOD levels increased in mantle, muscle and visceral mass tissues. CAT and GSH were first increased and then decreased with gradual time lapse and dose concentration. Combination of increase and decrease imply that these enzymes are scavengers of free radicals up to a certain extent due to high toxic stress. Vieira et al. 2008 had exposed common goby *Pomatoschistus microps* to anthracene concentrations of 0.25, 0.5, 1, 2 and 4 μgL^{-1} and studied the effects of various antioxidants to anthracene contamination. The activity of GST only decreased, whereas activities of other antioxidants like CAT, SOD, GR and

GPx increased, indicative of anthracene inducing oxidative stress in the animals. Vasanth et al. (2012) analyzed the effects of PAH anthracene, focusing on liver enzymes and antioxidant enzyme activity in *Labeo rohita*. The fishes were exposed to varying concentrations of anthracene (0.5, 1, 2, 3, 4 mgL^{-1}) for a period of five weeks for antioxidant enzymatic studies like CAT, SOD, GSH in liver, gills, muscles and kidneys. CAT, SOD and GSH increased, which is indicative of the role of antioxidants in scavenging oxyradicals generated as a result of anthracene contamination. Badreddine et al. (2017) exposed *Mytilus galloprovincialis* to 0.05, 0.15 and 0.4 μgL^{-1} concentrations of PAH anthracene for a period of two, four and eight days. GSH decreased in response to anthracene, thus supporting the uptake of anthracene in digestive gland. The decreased GSH response can be defined as a mechanism to remove hydrocarbons and counteract the lethal effects of ROS. MDA levels elevated indicating oxidative damage due to generation of ROS induced by anthracene. Yuan et al. (2017) determined the effects of both in-vivo and in-vitro effects of anthracene on *Mytilus edulis* by evaluating of various stress biomarkers and antioxidants. The laboratory experiments were carried out for 0, 3, 6, 9, 12, 15, 18 and 21 days. GST, GPx and GSH increased in general in digestive gland of mussels. GPx and GR increased in gill tissues of mussels. Stress markers like MDA and SOD also increased on exposure to anthracene. aChE activity was inhibited in the fingerlings of *Chanoschanos* on exposure to anthracene, indicating disturbance of cholinergic transmission (Palanikumar et al. 2012). Common goby *Pomatoschistus microps* also exhibited depletion in aChE activity on exposure to different concentrations of anthracene (0.25, 0.5, 1, 2 and 4 μgL^{-1}) for a period of 96 hours (Vieira et al. 2008). aChE activity also decreased in *Mytilus galloprovincialis* in response to anthracene presence, signifying neurotoxicity and change in the neurotransmission mechanism (Badreddine et al. 2017). Sunmonu et al. (2009) exposed African catfish, *Heterobranchus bidorsalis*, to anthracene of concentrations of 0.25, 0.50, 0.75, 1.0 and 1.25gL^{-1} for a period of 54 hours. Mortality percentage varied from 2.13 to 44.44%. Significant inhibition of enzymes like GGT, ALT and ALP were noticed in the liver and stomach tissues of experimental fishes. Highest inhibition of 89.74%, 79.96% and 46.26% was observed in case of ALT, GGT and ALP enzymes on exposure to anthracene. The decrease can be attributed to possible leakage of enzymes in the general blood circulation or dysfunction of organs. Similar enzymatic (LDH, AST, ALT, ALP) changes were observed in *Labeo rohita* on exposure to anthracene. LDH levels increased in all tissues, signifying anaerobic pathway of energy production due to anthracene stress. Similarly, ALP, ALT and AST levels also increased with increasing dose concentrations and time period. LDH activity had also increased in *Pomatoschistus microps* on anthracene contamination, suggesting an elevation of anaerobic pathway of energy production (Vieira et al. 2008). Dey et al. (2019) studied the toxicological responses of *Anabas testudineus* after exposing it to anthracene concentrations of 0.0075 and 0.015mgL^{-1} respectively. The test was carried out for a period of 1, 5, 10, 15 and 21 days. Haematological parameters like white blood cell (WBC) count, red blood cell (RBC) count, haemoglobin (Hb), percent lymphocyte (%LYM), packed cell volume (PCV), platelet (PLT), mean corpuscular volume (MCV), mean corpuscular haemoglobin (MCH) and mean corpuscular haemoglobin concentration (MCHC) were analyzed, and results revealed significant increase in WBC, %LYM, MCV, MCH and MCHC. RBC, Hb, PCV and PLT showed decreasing trends. Biochemical responses like cholesterol, protein, albumin, triglyceride, glucose, glutamic pyruvate transaminase, calcium, alkaline phosphatase, high density lipid, and low-density lipid were also analyzed. Cholesterol, protein, albumin, triglyceride values increased on the first day with subsequent reduction. Reverse results were observed in case of GLU, GPT, Ca, ALP, HDL, and LDL (Dey et al. 2019). Oris and Giesy (1985) also studied the toxicity of anthracene in the juveniles of sunfish. The study showed that anthracene-contaminated fish suffered from hypoxia, which was evident from increased opercular ventilation rates. The fishes also suffered from irritation. Damaged gills were evident from the histological studies which also supported the death of fishes due to asphyxiation. Irritation and hypoxia conditions were also observed in juveniles of bluegill sunfish (*Lepomis macrochirus Rafinesque*) (Oris and Giesy 1986). Peterson and Bain (2004) studied the toxicity of anthracene on mummichogs (*Fundulus heteroclitus*) and discovered that anthracene was capable of altering genes whose transcription was also altered. Sexually matured adult fathead minnows (*Pimephales promelas*) were exposed to 6 and 12 μgL^{-1} anthracene initially for a period of two weeks and then the concentration was increased to 12 and 20 μgL^{-1} for a period of three weeks. The authors observed that the number of eggs laid by anthracene-treated fish also decreased along with the survival rate. Various teratogenic effects like deformities in eyes and yolk, edema and internal haemorrhage were noticed on anthracene contamination. Anthracene was also seen to be maternally transferred to the embryos. The authors concluded that anthracene is capable of causing reproductive impairment (Hall and Oris 1991). Similar reproductive toxicity and decrease in estradiol levels were noticed in medaka (*Oryzias latipes*) (Cheek et al. 2001). Hatching and survival rates also decreased. Anthracene was found to destroy estrogen action. The number of neonates of *Daphnia magna* decreased when exposed to anthracene. Survival and fecundity rates also decreased (Holst and Giesy, 1989). Anthracene was also found to be lethal and affects the development of *Rana pipiens* in its late embryonic stage (Kagan et al. 1984). Various studies have also depicted the increased usage toxicity of anthracene in the presence of sunlight, leading to increased mortality of aquatic organisms (Bowling et al. 1983; Oris and Giesy 1985, 1986).

7.6 CONCLUSION

This review undoubtedly shows that there is substantial upsurge in anthracene contamination in the natural environment, especially in the aquatic environment. Several studies have shown the presence of anthracene in aquatic bodies, sediments and tissue across the globe, while data available from India is comparatively scanty. Anthracene is very high in China and Africa, as evident from the work reported in our review. Aquatic animals, on exposure to high concentrations of anthracene, can have changes in their biochemical parameters and at gene level, affecting transcription. Anthracene can accumulate and persist in aquatic sediments, which may be a risk factor to various organisms living at the bottom. Keeping in view the occurrence of anthracene in various water bodies, regular monitoring of fish and other aquatic organisms should be carried out to ensure that we do not consume fish, prawns, and crabs contaminated with PAH as it may pose several threats to human health. Various studies have revealed that anthracene is capable of altering the behaviour pattern, growth, development, coordination of the nervous system, reproduction and biochemical activities in aquatic animals. Reports have shown that anthracene disintegrates membrane structures and receptors, which, in turn, leads to oxidative stress. Data on the specific effect of anthracene toxicity on species is not very extensive. The present review is focussed on faunal component, but anthracene may also affect flora of aquatic bodies. Hence a detailed study on the uptake, accumulation and toxicity of anthracene compounds is essential for preventing PAH pollution in aquatic bodies and the impact of PAH pollutants on both fauna and flora.

REFERENCES

Abayi, Juma John Moses, Christopher Tombe Gore, Christine Nagawa, Benjamin A. Musa Bandowe, Henry Matovu, Edward Mubiru, Emily Chelangat Ngeno, Silver Odongo, Mika Sillanpää, and Patrick Ssebugere. "Polycyclic aromatic hydrocarbons in sediments and fish species from the White Nile, East Africa: Bioaccumulation potential, source apportionment, ecological and health risk assessment." *Environmental Pollution* 278 (2021): 116855. https://doi.org/10.1016/j.envpol.2021.116855

Adeniji, Abiodun Olagoke, O. O. Okoh, and A. I. Okoh. "Distribution pattern and health risk assessment of polycyclic aromatic hydrocarbons in the water and sediment of Algoa Bay, South Africa." *Environmental Geochemistry and Health* 41, no. 3 (2019a): 1303–1320. https://doi.org/10.1007/s10653-018-0213-x

Adeniji, A. O., O. O. Okoh, and A. I. Okoh. "Levels of polycyclic aromatic hydrocarbons in the water and sediment of Buffalo River Estuary, South Africa and their health risk assessment." *Archives of Environmental Contamination and Toxicology* 76, no. 4 (2019b): 657–669. https://doi.org/10.1007/s00244-019-00617-w

Agarwal, Tripti, P. S. Khillare, and Vijay Shridhar. "PAHs contamination in bank sediment of the Yamuna River, Delhi, India." *Environmental Monitoring and Assessment* 123, no. 1 (2006): 151–166. https://doi.org/10.1007/s10661-006-9189-6

Ambade, Balram, Shrikanta Shankar Sethi, Amit Kumar, Tapan Kumar Sankar, and Sudarshan Kurwadkar. "Health risk assessment, composition, and distribution of polycyclic aromatic hydrocarbons (PAHs) in drinking water of Southern Jharkhand, East India." *Archives of Environmental Contamination and Toxicology* 80, no. 1 (2021): 120–133. https://doi.org/10.1007/s00244-020-00779-y

Aziz, Faiqa, Jabir Husain Syed, Riffat Naseem Malik, Athanasios Katsoyiannis, Adeel Mahmood, Jun Li, Gan Zhang, and Kevin C. Jones. "Occurrence of polycyclic aromatic hydrocarbons in the Soan River, Pakistan: Insights into distribution, composition, sources and ecological risk assessment." *Ecotoxicology and Environmental Safety* 109 (2014): 77–84. http://dx.doi.org/10.1016/j.ecoenv.2014.07.022

Badreddine, Sellami, Khazri Abdelhafidh, Mouhamed Dellali, Ezzeddine Mahmoudi, David Sheehan, and Beyrem Hamouda. "The effects of anthracene on biochemical responses of Mediterranean mussels *Mytilus galloprovincialis*." *Chemistry and Ecology* 33, no. 4 (2017): 309–324.

Bakhtiari, Alireza Riyahi, Mohamad Pauzi Zakaria, Mohammad Ismail Yaziz, Mohamad Nordin HjLajis, Xinhui Bi, M. R. MohamadShafiee, and Mahyar Sakari. "Distribution of PAHs and n-alkanes in Klang River surface sediments, Malaysia." *Pertanika Journal of Science & Technology* 18, no. 1 (2010): 167–179.

Baran, Agnieszka, Agnieszka Klimkowicz-Pawlas, Aleksandra Ukalska-Jaruga, Monika Mierzwa-Hersztek, Krzysztof Gondek, Magdalena Szara-Bąk, Marek Tarnawski, and Iwona Spałek. "Distribution of polycyclic aromatic hydrocarbons (PAHs) in the bottom sediments of a dam reservoir, their interaction with organic matter and risk to benthic fauna." *Journal of Soils and Sediments* 21, no. 6 (2021): 2418–2431. http://dx.doi.org/10.1007/s11368-021-02968-1

Barhoumi, Badreddine, Yassine El Megdiche, Christelle Clérandeau, Walid Ben Ameur, Sabrine Mekni, Sondes Bouabdallah, Abdelkader Derouiche, Soufiane Touil, Jérôme Cachot, and Mohamed Ridha Driss. "Occurrence of polycyclic aromatic hydrocarbons (PAHs) in mussel (*Mytilus galloprovincialis*) and eel (*Anguilla anguilla*) from Bizerte lagoon, Tunisia, and associated human health risk assessment." *Continental Shelf Research* 124 (2016): 104–116. http://dx.doi.org/10.1016%2Fj.csr.2016.05.012

Bowling, J. W., G. J. Leversee, P. F. Landrum, and J. P. Giesy. "Acute mortality of anthracene-contaminated fish exposed to sunlight." *Aquatic Toxicology* 3, no. 1 (1983): 79–90.

Bua, Rosaria Ornella, Annalinda Contino, and Alessandro Giuffrida. "Polycyclic aromatic hydrocarbons in *Mullussurmuletus* from the Catania Gulf (Sicily, Italy): Distribution and potential health risks." *Environmental Science and Pollution Research* 28, no. 7 (2021): 7756–7765. https://doi.org/10.1007/s11356-020-11052-z

Cheek, A., T. H. Brouwer, S. Carroll, S. Manning, M. Brouwer, and J. A. McLachlan. "Developmental exposure to anthracene and estradiol alters reproductive success in medaka (*Oryzias latipes*)." *Environmental Science* 8 (2001): 31–45.

Chen, Yuyun, Yanxia Dong, Ting Duan, Junqin Zhang, and Yiqiang Zhou. "Polycyclic aromatic hydrocarbons in Weihe River, a typical river in arid and semi-arid region of Northwest China: Trends of pollution and risk." *Polish Journal of Environmental Studies* 30, no. 2 (2021). https://doi.org/10.15244/pjoes/124896

Connell, Des W., and Gregory J. Miller. *Chemistry and Ecotoxicology of Pollution*. Vol. 65. John Wiley & Sons, 1984.

Curtosi, Antonio, Emilien Pelletier, Cristian L. Vodopivez, and Walter P. Mac Cormack. "Distribution of PAHs in the water column, sediments and biota of Potter Cove, South Shetland Islands, Antarctica." *Antarctic Science* 21, no. 4 (2009): 329–339. http://dx.doi.org/10.1017/S0954102009002004

Das, Momita, Devendra Kumar Patel, Arup Kumar Sarma, Binode Kumar Baruah, Sofia Banu, and Jibon Kotoky. "Assessment of polycyclic aromatic hydrocarbons and heavy metals pollution in soils of Guwahati city, Assam, India." *Current Science* (2016): 2285–2292.

Deb, Swapan C., Taro Araki, and Takehiko Fukushima. "Polycyclic aromatic hydrocarbons in fish organs." *Marine Pollution Bulletin* 40 (2000): 882–885. http://dx.doi.org/10.1016/S0025-326X(00)00090-4

Dey, Sukhendu, and Apurba Ratan Ghosh. "Dose-specific biochemical and erythrocytic alterations of anthracene exposure on blood of *Anabas testudineus* (Bloch)." *Environmental Toxicology and Pharmacology* 72 (2019): 103247. https://doi.org/10.1016/j.etap.2019.103247

Dey, Sukhendu, Palas Samanta, Niladri Sekhar Mondal, Debraj Kole, Arghya Mandal, Atanu Patra, and Apurba Ratan Ghosh. "Dose specific responses of *Anabas testudineus* (Bloch) to anthracene (PAH): Haematological and biochemical manifestation." *Emerging Contaminants* 5 (2019): 232–239.

Dhananjayan, V., and S. Muralidharan. "Polycyclic aromatic hydrocarbons in various species of fishes from mumbai harbour, India, and their dietary intake concentration to human." *International Journal of Oceanography* 2012 (2012). http://dx.doi.org/10.1155/2012/645178

Edokpayi, Joshua N., John O. Odiyo, Oluwaseun E. Popoola, and Titus A. M. Msagati. "Determination and distribution of polycyclic aromatic hydrocarbons in rivers, sediments and wastewater effluents in Vhembe District, South Africa." *International Journal of Environmental Research and Public Health* 13, no. 4 (2016): 387. http://dx.doi.org/10.3390/ijerph13040387

Ekere, Nwachukwu, Newman Yakubu, and Janefrances Ihedioha. "Ecological risk assessment of heavy metals and polycyclic aromatic hydrocarbons in sediments of rivers Niger and Benue confluence, Lokoja, Central Nigeria." *Environmental Science and Pollution Research* 24, no. 23 (2017): 18966–18978. https://doi.org/10.1007/s11356-017-9511-6

EPA. *Polycyclic Aromatic Hydrocarbons (PAHs)-EPA Fact Sheet*. National Center for Environmental Assessment, Office of Research and Development, 2008.

Hall, A. Tilghman, and James T. Oris. "Anthracene reduces reproductive potential is maternally transferred during long-term exposure in fathead minnows." *Aquatic Toxicology* 19, no. 3 (1991): 249–264.

Holst, Linda L., and John P. Giesy. "Chronic effects of the photoenhanced toxicity of anthracene on *Daphnia magna* reproduction." *Environmental Toxicology and Chemistry: An International Journal* 8, no. 10 (1989): 933–942.

Hussain, Karishma, Rajesh Rajbangshi, and Raza Rafiqul Hoque. "Understanding levels and sources of PAHs in water of Bharalu tributary of the Brahmaputra River." *Asian Journal of Water, Environment and Pollution* 11, no. 2 (2014): 89–98.

Ihunwo, Owhonda Chikeru, and Millicent Uzoamaka Ibezim-Ezeani. "Distribution, Source Appropriation, and human health risk assessment of polycyclic aromatic hydrocarbons due to consumption of callinectesamnicola from Woji Creek in Sambreiro River." *Turkish Journal of Fisheries and Aquatic Sciences* 21, no. 5 (2021): 245–253. http://dx.doi.org/10.4194/1303-27142-v21_05_04

Irwin, R. J., Specialists, S. A. C., Van Mouwerik, M. A. R. K., Stevens, L., Seese, M. D., & Basham, W. (1997). Enviromental Contaminants Encyclopedia Coal tar Entry. Colorado State University.

Kafilzadeh, Farshid. "Distribution and sources of polycyclic aromatic hydrocarbons in water and sediments of the Soltan Abad River, Iran." *The Egyptian Journal of Aquatic Research* 41, no. 3 (2015): 227–231. https://doi.org/10.1016/j.ejar.2015.06.004

Kafilzadeh, Farshid, Amir Houshang Shiva, and Rokhsareh Malekpour. "Determination of polycyclic aromatic hydrocarbons (PAHs) in water and sediments of the Kor River, Iran." *Middle-East Journal of Scientific Research* 10, no. 1 (2011): 1–7.

Kagan, Jacques, Peggy A. Kagan, and Howard E. Buhse. "Light-dependent toxicity of α-terthienyl and anthracene toward late embryonic stages of *Rana pipiens*." *Journal of Chemical Ecology* 10, no. 7 (1984): 1115–1122.

Khuman, Sanjenbam Nirmala, Paromita Chakraborty, Alessandra Cincinelli, Daniel Snow, and Bhupander Kumar. "Polycyclic aromatic hydrocarbons in surface waters and riverine sediments of the Hooghly and Brahmaputra Rivers in the Eastern and Northeastern India." *Science of the Total Environment* 636 (2018): 751–760. http://dx.doi.org/10.1016/j.scitotenv.2018.04.109

Malik, Amrita, Priyanka Ojha, and Kunwar P. Singh. "Distribution of polycyclic aromatic hydrocarbons in edible fish from Gomti River, India." *Bulletin of Environmental Contamination and Toxicology* 80, no. 2 (2008): 134–138. https://doi.org/10.1007/s00128-007-9331-3

Malik, Amrita, Priyanka Verma, Arun K. Singh, and Kunwar P. Singh. "Distribution of polycyclic aromatic hydrocarbons in water and bed sediments of the Gomti River, India." *Environmental Monitoring and Assessment* 172, no. 1 (2011): 529–545. https://doi.org/10.1007/s10661-010-1352-4

Mohamad, Issam, Imad Hwejeh, and Mohammad Nasser. "Distribution of Polycyclic Aromatic Hydrocarbons (Pahs) in Marine Shore Sediments of Alkaber Aljanuby River Estuary, Boundary River (Syria-Lebanon)." *Fresenius Environmental Bulletin* 20, no. 10 (2011): 2624–2631.

Munyengabe, A., A. Mambanda, and B. Moodley. "Polycyclic aromatic hydrocarbons in water, soils and surface sediments of the Msunduzi River." *Journal of Environmental Analytical Chemistry* 4, no. 4 (2017): 1–13. http://dx.doi.org/10.4172/2380-2391.1000227

Nagy, Andrea Szabó, Gábor Simon, and István Vass. "Monitoring of polycyclic aromatic hydrocarbons (PAHs) in surface water of the Hungarian upper section of the Danube River." *Nova Biotech et Chim* 11 (2012): 27–35. http://dx.doi.org/10.2478/v10296-012-0003-2

Nasher, Essam, Lee Yook Heng, Zuriati Zakaria, and Salmijah Surif. "Concentrations and sources of polycyclic aromatic hydrocarbons in the seawater around Langkawi Island, Malaysia." *Journal of Chemistry* 2013 (2013). http://dx.doi.org/10.1155/2013/975781

Oliveira, Marta, Filipa Gomes, Álvaro Torrinha, Maria João Ramalhosa, Cristina Delerue-Matos, and Simone Morais. "Commercial octopus species from different geographical origins: Levels of polycyclic aromatic hydrocarbons and potential health risks for consumers." *Food and Chemical Toxicology* 121 (2018): 272–282. http://dx.doi.org/10.1016/j.fct.2018.09.012

Oris, James T., and John P. Giesy Jr. "The photoenhanced toxicity of anthracene to juvenile sunfish (Lepomis spp.)." *Aquatic Toxicology* 6, no. 2 (1985): 133–146. https://doi.org/10.1016/0166-445X%2885%2990012-8

Oris, James T., and John P. Giesy Jr. "Photoinduced toxicity of anthracene to juvenile bluegill sunfish (*Lepomis macrochirus Rafinesque*): Photoperiod effects and predictive hazard evaluation. *Environmental Toxicology and Chemistry* 5, no. 8 (1986): 761–768. https://doi.org/10.1002/ETC.5620050807

Palanikumar, L., A. K. Kumaraguru, C. M. Ramakritinan, and M. Anand. "Biochemical response of anthracene and benzo [a] pyrene in milkfish *Chanos chanos*." *Ecotoxicology and Environmental Safety* 75 (2012): 187–197. https://doi.org/10.1016/j.ecoenv.2011.08.028

Panda, Bhabani Shankar, and S. K. Mahanta. "Quantification of poly aromatic hydrocarbons in waters of Mahanadi estuary at Paradeep, Odisha." *International Research Journal of Engineering and Technology* 7, no. 6 (2020): 6971–6981.

Peterson, Janis S. K., and Lisa J. Bain. "Differential gene expression in anthracene-exposed mummichogs (*Fundulus heteroclitus*)." *Aquatic Toxicology* 66, no. 4 (2004): 345–355.

Pham, Luu Thanh, Thuy Thanh Thi Hoang, Loan Cam ThiTu, Yen Hoang Thi Tran, Bao Duy Le, Dong Van Nguyen, Huy Xuan Do, and Nam Van Thai. "Bioaccumulation and health risk assessment of polycyclic aromatic hydrocarbons in oyster (*Crassostrea* sp.) and gastropod (*Cymatium* sp.) species from the Can Gio Coastal Wetland in Vietnam." *Marine and Freshwater Research* 71, no. 6 (2019): 617–626. http://dx.doi.org/10.1071/MF19055

Qin, Ning, Wei He, Wenxiu Liu, Xiangzhen Kong, Fuliu Xu, and John P. Giesy. "Tissue distribution, bioaccumulation, and carcinogenic risk of polycyclic aromatic hydrocarbons in aquatic organisms from Lake Chaohu, China." *Science of the Total Environment* 749 (2020): 141577. http://dx.doi.org/10.1016/j.scitotenv.2020.141577

Recabarren-Villalon, Tatiana, Ana C. Ronda, and Andres H. Arias. "Polycyclic aromatic hydrocarbons levels and potential biomarkers in a native South American marine fish." *Regional Studies in Marine Science* 29 (2019): 100695. http://dx.doi.org/10.1016/j.rsma.2019.100695

Ronda, Ana Carolina, Ana Laura Oliva, Andrés Hugo Arias, Melina Mirta Orazi, and Jorge Eduardo Marcovecchio. "Biomarker responses to polycyclic aromatic hydrocarbons in the native fish *Ramnogaster arcuata*, South America." *International Journal of Environmental Research* 13, no. 1 (2019): 77–89. https://doi.org/10.1007/s41742-018-0155-2

Roubal, William T., Tracy K. Collier, and Donald C. Malins. "Accumulation and metabolism of carbon-14 labeled benzene, naphthalene, and anthracene by young coho salmon (*Oncorhynchus kisutch*)." *Archives of Environmental Contamination and Toxicology* 5, no. 1 (1977): 513–529.

Schrap, S. Marca, and Antoon Opperhuizen. "Relationship between bioavailability and hydrophobicity: Reduction of the uptake of organic chemicals by fish due to the sorption on particles." *Environmental Toxicology and Chemistry: An International Journal* 9, no. 6 (1990): 715–724.

Shilla, Dativa J., and Joyanto Routh. "Distribution, behavior, and sources of polycyclic aromatic hydrocarbon in the water column, sediments and biota of the Rufiji Estuary, Tanzania." *Frontiers in Earth Science* 6 (2018): 70. https://doi.org/10.3389/feart.2018.00070

Siddiqi, Hina A., Fayyaz A. Ansari, and Alia B. Munshi. "Assessment of hydrocarbons concentration in marine fauna due to Tasman Spirit oil spill along the Clifton beach at Karachi coast." *Environmental Monitoring and Assessment* 148, no. 1 (2009): 139–148. http://dx.doi.org/10.1007/s10661-007-0145-x

Singare, Pravin U. "Carcinogenic and endocrine-disrupting PAHs in the aquatic ecosystem of India." *Environmental Monitoring and Assessment* 188, no. 10 (2016): 1–25. https://doi.org/10.1007/s10661-016-5597-4

Stone, Mike, Adrian L. Collins, Curtis Brock, and Darcy McDonald. "Spatial and temporal distribution of pahs in the north Saskatchewan River, Alberta, Canada." *IAHS Proceedings and Reports* 361 (2013): 294–301. https://doi.org/10.26434/chemrxiv.14055398.v1

Sunmonu, T. O., O. D. Owolabi, and O. B. Oloyede. "Anthracene-induced enzymatic changes as stress indicators in African catfish, *Heterobranchus bidorsalis* Geoffroy Saint Hilaire, 1809." *Research Journal of Environmental Sciences* 3, no. 6 (2009): 677–686.

Tan Pei Jian, Brenda, Muhammad RazaUl Mustafa, Mohamed Hasnain Isa, AsimYaqub, and Yeek Chia Yeek Ho. "Study of the water quality index and polycyclic aromatic hydrocarbon for a river receiving treated landfill leachate." *Water* 12, no. 10 (2020): 2877. http://dx.doi.org/10.3390/w12102877

Teuchies, Johannes, Tom J. S. Cox, Katrien Van Itterbeeck, Filip J. R. Meysman, and Ronny Blust. "The impact of scrubber discharge on the water quality in estuaries and ports." *Environmental Sciences Europe* 32, no. 1 (2020): 1–11. https://enveurope.springeropen.com/articles/10.1186/s12302-020-00380-z

Vaezzadeh, Vahab, Mohamad Pauzi Zakaria, Chui Wei Bong, Najat Masood, Sami Mohsen Magam, and Sadeq Alkhadher. "Mangrove oyster (*Crassostreabelcheri*) as a biomonitor species for bioavailability of polycyclic aromatic hydrocarbons (PAHs) from sediment of the west coast of peninsular Malaysia." *Polycyclic Aromatic Compounds* 39, no. 5 (2019): 470–485. https://doi.org/10.1080/10406638.2017.1348366

Vasanth, S., Arul Ganesh, T. Siva Vijayakumar, S. Karthikeyeni, M. Manimegalai, and P. Subramanian. "Assessment of anthracene on hepatic and antioxidant enzyme activities in *Labeo rohita* (Hamilton, 1822)." *International Journal of Pharmacy & Life Sciences* 3, no. 5 (2012).

Vieira, L. R., A. Sousa, M. F. Frasco, I. Lima, F. Morgado, and L. Guilhermino. "Acute effects of Benzo [a] pyrene, anthracene and a fuel oil on biomarkers of the common goby *Pomatoschistus microps* (Teleostei, Gobiidae)." *Science of the Total Environment* 395, no. 2–3 (2008): 87–100. https://doi.org/10.1016/j.scitotenv.2008.01.052

Wang, Chunhui, Shenglu Zhou, Shaohua Wu, Jing Song, Yaxing Shi, Baojie Li, and Hao Chen. "Surface water polycyclic aromatic hydrocarbons (PAH) in urban areas of Nanjing, China." *Water Science and Technology* 76, no. 8 (2017): 2150–2157. https://doi.org/10.2166/wst.2017.387

Wang, Qian, Fei Peng, Yiqin Chen, Ling Jin, Jun Lin, Xu Zhao, Jie Yin, and Juan-Ying Li. "Heavy metals and PAHs in an open fishing area of the East China Sea: Multimedia distribution, source diagnosis, and dietary risk assessment." *Environmental Science and Pollution Research* 26, no. 21 (2019): 21140–21150. https://doi.org/10.1007/s11356-019-05355-z

Yamei, Hui, Minghui Zheng, Liu Zhengtao, and Gao Lirong. "Distribution of polycyclic aromatic hydrocarbons in sediments from Yellow River Estuary and Yangtze River Estuary, China." *Journal of Environmental Sciences* 21, no. 12 (2009): 1625–1631. https://doi.org/10.1016/S1001-0742(08)62465-1

Yan, Wen, Jisong Chi, Zhiyuan Wang, Weixia Huang, and Gan Zhang. "Spatial and temporal distribution of polycyclic aromatic hydrocarbons (PAHs) in sediments from Daya Bay, South China." *Environmental Pollution* 157, no. 6 (2009): 1823–1830. https://doi.org/10.1016/j.envpol.2009.01.023

Yuan, Mengqi, You Wang, Bin Zhou, Xiaoyang Jian, Wenlong Dong, and Xuexi Tang. "An integrated biomarker response index for the mussel *Mytilus edulis* based on laboratory exposure to anthracene and field transplantation experiments." *Chinese Journal of Oceanology and Limnology* 35, no. 5 (2017): 1165–1178.

Zhang, Guizhai, Zhaoke Pan, Xiaoming Wang, Xiaojie Mo, and Xiaoming Li. "Distribution and accumulation of polycyclic aromatic hydrocarbons (PAHs) in the food web of Nansi Lake, China." *Environmental Monitoring and Assessment* 187, no. 4 (2015): 1–12. https://doi.org/10.1007/s10661-015-4362-4

Zhao, Zhonghua, Lu Zhang, Yongjiu Cai, and Yuwei Chen. "Distribution of polycyclic aromatic hydrocarbon (PAH) residues in several tissues of edible fishes from the largest freshwater lake in China, Poyang Lake, and associated human health risk assessment." *Ecotoxicology and Environmental Safety* 104 (2014): 323–331. *h*ttps://doi.org/10.1016/j.ecoenv.2014.01.037

Zhu, Lin, Xuexi Tang, Ying Wang, Yadong Sui, and Hui Xiao. "Use of antioxidant enzymes of clam *Ruditapes philippinarum* as biomarker to polycyclic aromatic hydrocarbon pollution." *Chinese Journal of Oceanology and Limnology* 34, no. 2 (2016): 416–421.

8 Role of Allelochemicals in Alleviation of Stress in the Terrestrial and Aquatic Ecosystem

Pratik Talukder, Sounak Chanda, Meghna Mishra, Baishakhi Sinha, Anushka Dutta, Ayan Adhikary

8.1 INTRODUCTION

Plant metabolism can be defined as a complex pathway of physical and chemical events that helps in breaking down complex organic molecules and converting them into their simpler forms. These simpler forms can then be absorbed by the body to carry out other functions, or secreted outside, which may or may not have a specific function. The process of metabolism is carried on by pathways of respiration, photosynthesis as well as the synthesis of other organic compounds. The process of metabolism in organisms is highly specific, compartmented and is utilized in a myriad of actions and stimulations which carry out further pathways. These productions have both positive and negative impacts on the producing plant. Example of such pathways includes the production of volatile cues in order to attract the herbivore to detect the plant; this negatively affects the plant (War et al., 2012). Again, the production of toxins protects it from the same herbivore predators and other higher organisms. Broadly, plant metabolism can be classified into two major types: primary metabolism and secondary metabolism.

Primary metabolism is the production of the essential components of plants that are crucial for survival. These crucial components thus produced are called primary metabolites. They are directly involved in the growth and development of a plant. The produced by-products are often very simple chemical substances. They include carbohydrates, amino acids, organic acids, vitamins, hormones, flavonoids, phenolics, and glucosinolates. These are essential for plant growth, development, stress adaptation, and defence. The primary metabolites are known to determine the nutritional quality of food, colour, taste and smell. These compounds also have antioxidative, anticarcinogenic, antihypertensive, anti-inflammatory, antimicrobial, immuno-stimulating and cholesterol-lowering properties (Panche et al., 2016).

On the other hand, secondary metabolism results in the production of secondary metabolites. They are typically organic compounds produced through the modification of primary metabolite synthases. These are organic molecules that aren't directly engaged in an organism's regular growth and development. While primary

DOI: 10.1201/9781003297901-8

metabolites are important for species survival, as they play a role in photosynthesis and respiration, the absence of secondary metabolites does not result in immediate death, but rather in long-term impairment of the organism's survivability as they are often involved in plant defence. Sustenance of plants is achieved through their innate defence mechanism, most of which is achieved through these secondary metabolites. They are also used in signal transduction.

8.2 PLANT SECONDARY METABOLITES (PSMs)

Plants, being sessile, cannot defend themselves against herbivores and other infectious organisms by migration, nor do they have a superior immune system to protect them against bacterial, viral or other infectious attacks (Wink, 2003). Hence the PSMs have evolved over time to play a vital role in plant defence mechanisms. It has been experimentally proved that PSMs prevent the plant from snails and other herbivores. A noticeable difference is present in the concentration of the PSMs in various organs. Mostly, the organs related with defence and reproduction have the most potent PSM (Wink, 2003). Plant hormones are a noticeable example of secondary metabolites which help mainly in the regulation of metabolic activities within the cell. The sweet odour of flowers to attract the pollinators is due to the production of terpenoids, amines and phenylpropanoids. They exhibit a distinctive smell recognizable by specific organisms and thus act as pollinator attractants (Wink, 1992). But these secondary metabolites are often useful to mankind as well and are being widely used in industries. These have a huge range of applications: Atropine, derived from various plants, acts as a competitive antagonist for acetylcholine receptors, specifically those of the muscarinic type; hence this can be used in the treatment of bradycardia (McLendon et al., 2021). Antibiotics such as erythromycin and bacitracin are also considered to be secondary metabolites. Erythromycin, a widely used antibiotic produced from *Saccharopolyspora erythraea*, has a broad antibacterial range. It is usually manufactured in big amounts and taken orally (Pham et al., 2019). Some substances become hazardous only after being consumed; for example, glycol cyanide in cassava root releases cyanide only after the herbivore consumes it (Cereda et al., 1996). Foxgloves (*Digitalis purpurea*) yield a number of poisonous compounds, including cardiac and steroidal glycosides. Nausea, vomiting, hallucinations, convulsions and death can all result from its ingestion (Pillay et al., 2019). Codeine, also an alkaloid derived from the opium poppy, is considered the most widely used drug in the world according to the World Health Organization. Primarily codeine is used to treat mild pain and relief coughing, although in some cases it is used to treat diarrhoea and some forms of irritable bowel syndrome (Rosenblum et al., 2008).

8.3 ALLELOPATHY

Allelopathy is the process of production and secretion of biochemicals which influence and help in the survival, growth development and reproduction of the producing plant as well as the neighbouring plants. It can also be described as a plant-plant, plant-microbe, microbe-microbe, plant-insect or plant-herbivore interaction as microbes can also carry out allelopathy to overcome stress factors (Jilani et al., 2008;

Li et al., 2010). The biochemicals thus produced are termed as allelochemicals. The types of allelochemical produced in individual species vary, even of those under the same genus. In fact, this has enabled the taxonomists to successfully carry out subgroup classification via metabolite profiling (Muzell et al., 2016). These allelochemicals can have both positive (helpful for the plant) and negative effects (detrimental for the plant) on the secreting plant. Although in most cases, it acts as defence against neighbouring plants to help the secreting plant survive competition. They play a significant role in both interspecific and intraspecific competition in plants. Allelochemical interactions are a key component in determining the distribution and abundance of species within plant communities. This is the key factor that gives the surety of the success of many characteristically invasive plants. In walnuts (*Juglans nigra*), the roots, inner bark, nut husks and leaves contain a non-toxic chemical called hydrojuglone. When exposed to air or soil, it is oxidized into juglone that is biologically active and acts as a respiratory inhibitor to some plants (Rietveld, 1983). Ailanthus (*Ailanthus altissima*) produces an allelopathic chemical called ailanthone, which inhibits the growth of other plants. The inhibitors are strongest in the bark and roots but are also present in the leaves, wood and seeds of the plant (Heisy et al., 1996).

8.3.1 Root Exudation

Ions, inorganic acids, free oxygen, water and other carbon-based substances such as amino acids, polysaccharides and others are all exuded by roots into the surrounding soil rhizosphere due to stimulation through various biotic and abiotic stresses (Bais et al., 2006). These exudates play the role of an important mediator in the plant-plant or plant-microbe interaction (Bais et al., 2004). It has been noticed that rice (*Oryza sativa*) and sorghum (*Sorghum bicolor*), on sensing the presence of weeds, alter their root exudation and release increased amounts of allelochemicals such as momilactone and sorgoleone respectively, in order to suppress the growth of weeds (Zhao and Li, 2021; Dayan, 2006). The chemical response of allelopathic rice seedling changes under varying environmental conditions. Quite similarly, roots of *Fagopyrum esculentum* alters its exudates after it recognizes the weed *Amaranthus retroflexus* and hence suppresses its growth (Gfeller et al., 2018).

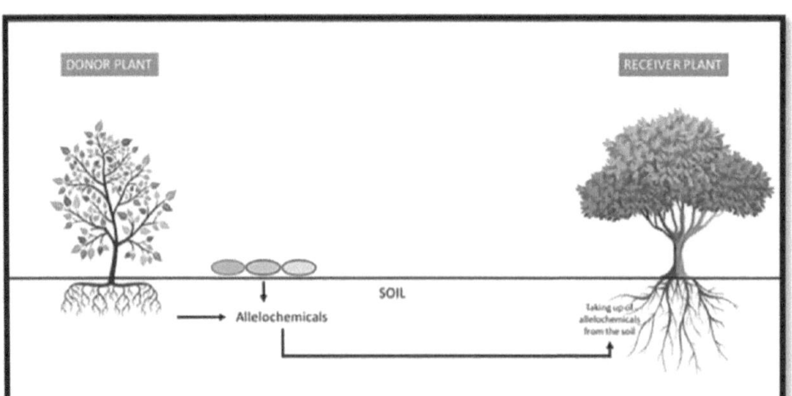

FIGURE 8.1 Secretion and uptake of allelochemicals by donor and receiver plant respectively.

8.3.2 Volatilization

Mostly plants of tropical and temperate regions use the method of volatilization for the release of allelochemicals. The produced chemicals are volatilized, and they either emerge through the stomata of the leaves and are taken up by other plants' stomata, or are solubilized in water and fall to the ground, mixing up with soil. Another indirect route of release of the volatile compounds is the decomposing parts of the producing plant (Singh et al., 2021). It was noticed in a closed experiment that the VOCs released by *Adenophora litter* have increased the survivability rate in the Indian and Chinese variants, but such significant results weren't present in the case of other variants. This concludes that allelopathic effect depends on the geographical location of the plant variant also (Baldwin et al., 2006).

8.3.3 Leaching

Leaching can be defined as a run-off by any liquid. In the case of plants, the surfaced allelochemicals in leaves and flowers can be leached and deposited in soil, thereby causing plant-plant, plant-microbe interactions. However, due to the uncertainty of the concentration of allelochemical present in the leachate, it is often referred to as an indirect process. The leachate from *Leucaena leucocephala* (Lam.) showed high levels of phenolic compounds (>1 mM) and mimosine, a non-protein amino acid. These chemicals increased membrane permeability in the water hyacinth (*Hyacinthus orientalis L.*) leaf tissue, disrupting the plasma membrane and resulting in cell death, hence restricting the further growth of the aquatic weed. *Leucaena leucocephala* leachate also reduces membrane integrity, rate of respiration and antioxidative defence of water hyacinth leaf tissues (Chai et al., 2013).

8.3.4 Decomposition

This can be considered majorly as an indirect process of release of allelochemicals. The plant residues, when acted on by soil microorganisms, release allelochemicals.

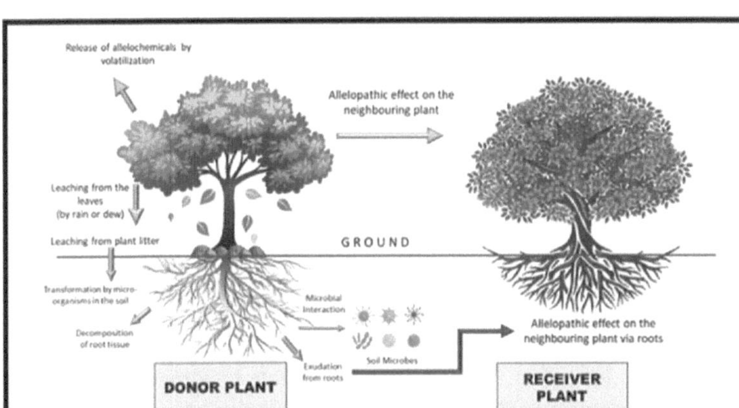

FIGURE 8.2 Routes for the release of allelochemicals from plants.

Depending on the type of interaction and subsequent transformation of the parent chemical, the bacteria stimulate or neutralize or even deactivate the allelopathic substances (Zhang et al., 2021).

8.3.5 Insect Repellent from Naturally Derived Plant Products

In the course of evolution, various types of insects have evolved in such a way that they share the same habitat with plants and animals. Around the 1980s, a major mosquito larvicide came into the limelight. Oil extracted from *Tagetes minuta* (Mexican marigold) had the presence of alpha-terthieynyl, which was extremely phototoxic to mosquitoes. The oil was highly efficient as a mosquito larvicide, which led to its commercialization. Various studies have also shown that to defend itself against the harmful invader Japanese beetles (*Popillia japonica*), geranium secrets a special compound called quisquatic, which possesses the capability to render the insect paralyzed (Koul and Walia, 2009). Many plant-based essential oils also have immense insecticidal properties. To name a few, citronella (*Ctmbopogan nardus*), tea tree (*Melaleuca alternifolia*), lemongrass (*Cymbopogan winterianus*), eucalyptus (*Eucalyptus globulus*), neem (*Azadiracta indica*), lavender (*Lavandula angustifolia*), clove (*Eugenia caryophyllus*) and aloe vera (*Aloe barbaensis miller*) are common and natural insect repellents that are very effective in warding off ants, flies, moths, mosquitoes, lice, ticks and fleas (Koul and Walia, 2009). Idioblasts are cells that help to protect the plants against invasion by insect and pests. They contain oil, latex, gum, resin, tannin and pigments. Some of them contain mineral crystals and poisonous calcium oxalate or carbonate or silica (Gajger et al., 2021). When the integrity of the cell is tampered due to water pressure, calcium oxalates are released, resulting in the stiffening of the tissue structure, following which sharp oxalate crystals tear the mouth parts of the insects during the feeding activity. The plant-based allelochemicals can also act as repellents, deterrents, growth inhibitors or a direct cause of mortality. In due course, insects have also maneuvered new strategies to overcome the resistance posed by these plant toxins. This coevolution based on competition between insects and plants finally leads to speciation.

8.3.6 Pest Control from Naturally Derived Plant Products

The usage of plant-based products such as neem, tulsi and citronella and their uses as a pesticide and insecticide has been known since time immemorial. As humans evolved, so did their lifestyle, and with that bloomed agriculture. With the onset of the agricultural form of livelihood, development of pest control and management also came into existence.

Plants' inbuilt defence mechanism ranges from having thick and waxy cuticle to a thick cell wall which acts as a mechanical blockage to pathogenic activity. Phytoanticipins are compounds which are well known for having antimicrobial and antifungal activities that provide resistance to pests and associated pathogens. Through extensive researches and recent studies, we have come to know about some of the major allelochemicals.

8.3.6.1 Alkaloids

Some of the major plant alkaloids are nicotine, caffeine, morphine, colchicine, quinine, etc. Alkaloids are capable of disintegrating insect cell membranes, nerve transmission and leakage of cells. Veratrin is a mixture of alkaloids separated from *Veratrum album* L. These alkaloids have a neurotoxin named pyrethrin which is responsible for tampering membrane depolarization (Koul and Walia, 2009). Another well-known rodenticide Ryanodine, extracted from *Liliaceae ryania* speciosa, are known to block calcium channels and subsequently causes muscle contractions and paralysis.

8.3.6.2 Limonoids and Quassinoids

The best known limonoid is Azadirachtin. Neem (*Azadiractin indica*), which in fact is a mixture of more than 100 limonoid compounds, namely salannin, Azadirachtin, nimbin and many others. Quassinoids are produced only in the family Simaroubaceae. Quassinoids are almost structurally similar to limonoids. Quassinoids are known to have antifeeding actions on *Tobacco budworms* (Koul and Walia, 2009). Limonin, a limonoid compound, is known for its bitterness, found in many citrus plants. Some other limonoid compounds known for bitterness are nimolin, nomilinic acid, ichagin and obacunoic acid. Out of these, limolin and nomilin are widely used for rendering resistance to lepidopterans and coleopterans.

8.3.6.3 Monoterpenoids and Sesquiterpenoids

Monoterpenes are made of terpenes such as terpinene, limonene and terpenoids such as geraniol, linalool, monocyclic alcohols, aliphatic aldehyde, aromatic phenols, bicyclic alcohol, monocyclic ketones, bicyclic ketones and esters. Monoterpenoids present in hemlock (*Tsuga canadensis L.*) mediates resistance to woolly adelgid infestations. The Colorado potato beetle was inhibited by the action of sesquiterpene 11 beta-acetoxy-5 alpha-(angelyloxy)-silphenen-3-one. Alpha-cyperone, a sesquiterpene isolated from the *Cyperes rotundus* tubers, has insecticidal properties against diamond black moth *P. xylostella* (Zhang et al., 2021).

8.3.6.4 Naphthoquinone

Naphthoquinones belong to the category of plant secondary metabolites that have well-developed cytotoxic properties which often interfere with cellular life processes. Naphthoquinones are best known for their highly efficient pesticidal, insecticidal and nematocidal activities. Naphthoquinones work by initiating and enhancing the production of reactive oxygen species (ROS), thereby resulting in cell death by causing oxidative stress.

8.3.6.5 Rocaglamide

Rocaglamides are cyclo penta tetrahydro benzofuran aromatic compounds that are mostly found in the genus *Aglaia*. Almost all the rocaglamide derivatives are known to exhibit insecticidal properties against the larvae of polyphagous pest insect *Spodoptera littoralis*. Rocaglamide isolated from the twigs of Chinese rice flower bush, *Aglaia orodata*, was found to inhibit larval growth and has an insecticidal effect on variegated cutworms *Peridroma saucia* and Asian armyworms, *Spodoptera litura* (Koul and Walia, 2009).

8.4 ALLELOCHEMICALS AS BIOHERBICIDES

Allelochemicals are collectively known as bio-communicators. The wonders of allelopathy and phytotoxic cooperation between plants are of utmost importance. These have no unwanted side effects on human health or on plants as their composition is different from that of synthetic or artificial herbicides (Soltys et al., 2013). For instance, we use soap water on some plants which are targeted by mites, spider, aphids, etc., which helps in the dehydration of the organisms, resulting in their death. Now many allelochemicals are being compared with artificial herbicides, and their mode of actions are used in controlling weeds as bioherbicides.
For example:

- Consider that we have two plants, plant A consisting of allelochemicals and plant B which is an unwanted weed.
- Plant A releases its allelochemicals which directly affects the unwanted weed, resulting in the withering of plant B.
- Plant A releases its allelochemicals which in turn get modified by some microorganisms and then act on the unwanted weed, plant B.
- Plant A releases its allelochemicals which, as a result, stimulate or activate some microorganisms, which in turn affects the unwanted weed, resulting in the death of plant B.

(Soltys et al., 2013)

8.5 STRAW MULCHING

The traditional practice of agriculture involving the use of herbicides in the field of weed management has harmful effects on the environment. Straw mulching is an

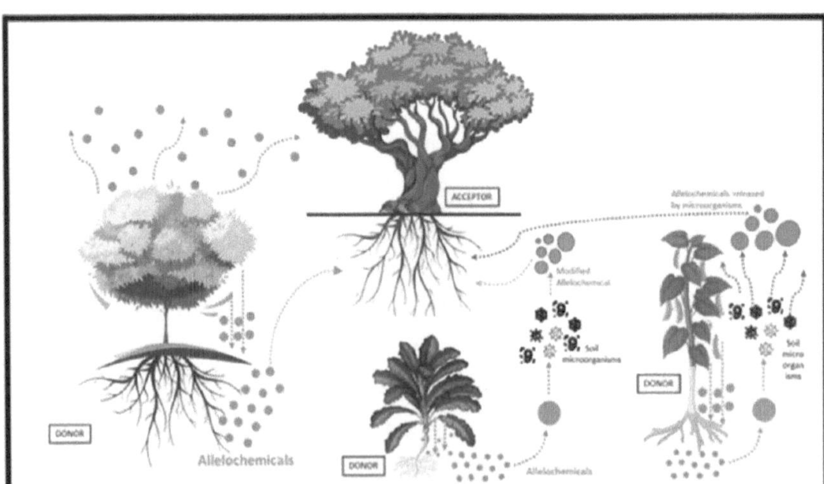

FIGURE 8.3 Effect of allelochemicals on acceptor plant via different modes.

allelopathic application which ensures feasible weed management and also brings down the harmful impact on the ecosystem. Allelochemicals extracted from decayed straw have potential use in weed management due their role as suppressants and also restraining the occurrence of diseases and pests. Straw mulching plays an important role in increasing soil fertility and ameliorating the soil organic matter. Studies have shown that rye mulch have decreased the growth and sprout of various troublesome agronomic plants and broad leaf weeds and green wheat straw obstructs the growth of Ipomoea weeds in corn and soybean fields, henceforth restraining the application of herbicides (Cheng and Cheng, 2015).

8.6 ROLE OF ALLELOCHEMICALS IN ALLEVIATION OF STRESS IN THE AQUATIC ECOSYSTEM

Harmful algal blooms (HABs) is a damaging phenomenon characterized by exponential algal growth induced by anthropogenic activities like eutrophication in water bodies. It includes all kinds of species that produce toxins and harmful secretions that affect the aquatic population (Kurashov et al., 2021). Effects are more prominent in shadow and stagnant water bodies (Hilt et al., 2006; Kaya et al., 2011; Mulderij et al., 2006; Švanys et al., 2014). Their deleterious effects include creation of dead zones in the water, increasing the biological oxygen demand (BOD) of the concerned aquatic system, interference of light penetration, thereby affecting the aquatic ecosystem (Scanferlato and Cairns, 1990). Thereby, it is imperative to control this phenomenon of algal growth (Saxena et al., 2021). Research has shown that allelopathy in aquatic plants has served to be favorable in controlling HABs.

Allelochemical-associated microalgal growth control is carried out in two major ways:

8.6.1 Planting Macrophytes

Macrophytes have shown constructive results in curbing the growth of algae by secreting allelochemicals which arrest the algal biomass. *M. spicatum* showed only short-term inhibitory effects on *Microcystis aeruginosa* by the secretion of catechin, caffeic acid, ellagic acid, gallic acid, nonanoic acid, pyrogallol and tellimagrandin II (Leu et al., 2002; Nakai and Hosomi, 2002; Nakai et al., 2005). Even negative effects on cyanobacteria biomass were also found (Švanys et al., 2014). Another notable in situ experimental study conducted in Lake Krumme in Berlin, Germany, exhibited positive results, where *Myriophyllum verticillatum* was found to show reduction in algal growth by secreting α-Asarone, phenylpropanoid glucosides and hydrolysable tannins (Aliotta et al., 1992; Hilt et al., 2006). In recent years, HABs have drawn attention to another worldwide phenomenon known as cyanobacterial bloom. Cyanobacteria release a biotoxin, microcystin, into the aquatic ecosystem. It has a damaging effect on the marine food chain and finally reaches higher organisms via biomagnification. Accumulation of nitrates and phosphates increases the amount of microcystins, thereby causing adverse changes in the aquatic system such as elevation in BOD levels (eutrophication). In several species of plants, the cyanobacterial toxins were seen to reduce the chlorophyll levels considerably, thereby affecting the rate of photosynthesis in them. Macrophytes like *M. spicatum* and *P. australis* were found to have inhibited photosynthetic rates under the effect of microalgae *Cladophora sp.* (Pflugmacher et al., 2002). Due to the aforesaid harmful effects of microcystins, it is necessary to suppress the growth of these algae. This is primarily mediated by allelochemicals, or more appropriately called algaecides, secreted by macrophytes. Studies have shown that macrophytes and cyanobacteria exhibit antagonistic relationships. Incorporation of lysine in cyanobacteria-rich water slows down the rate of biomass growth (Kaya et al., 2011).

8.6.2 Direct Addition of the Extracted Allelochemicals

This mechanism proceeds by isolation, purification, identification or synthetic manufacturing of allelochemicals such as gramine, pyrogallol, gallic acid, ellagic acid and catechin (Zhu et al., 2010). This approach is more practical as it involves less energy, labour investments and storage of extracted algaecides for further usage. Although there is a drawback of short-term algal inhibition, this mechanism comes with the advantage of rapid response (Men et al., 2007).

8.6.3 Algal Inhibition by Physiological Methods

Allelochemicals secreted by macrophytes often employ different mechanisms to disrupt important physiological processes in order to inhibit algal growth. One such notable method is changes in chlorophyll level of competing primary producers. It was noticed that polyphenolic allelochemicals secreted from macrophyte *Myriophyllum spicatum* considerably inhibited PSII due to the direct effects on the photosynthetic apparatus of algae (Leu et al., 2002). Root exudation is one of the major ways of release of allelochemicals in terrestrial as well as aquatic plants. In such cases, plant root is the first point of contact with the secreted allelochemical. Thus it can be concluded that allelochemicals affect the rate of ion water uptake via the root (Soltys et al., 2013, Reigosa et al., 2006).

8.7 EFFECT OF ALLELOCHEMICALS ON THE ENVIRONMENT

Studies have portrayed that the production of allelochemical as eco-friendly pesticides and controlling the growth of crops has showed the efficiency of allelochemicals in the field of agriculture (Cheng and Cheng, 2015). It is seen that there is a connection between the growth of the crop and the microorganisms present in the soil in the presence of allelochemicals. The microbe rhizosphere helps in the allopathic reaction of the plant positively. Microorganisms have the capability to change the constituents of allelochemicals which are liberated into the environment (Soltys et al., 2013).

8.8 CURRENT STUDIES AND FUTURE PROSPECT

Allelochemicals have been extensively used in agriculture, forestry and in the horticulture industry for many years. Numerous methods have been explored recently. They are classified into physical, chemical, biological and crop-growing approaches. The allelochemicals consists of a wide diversity of substances according to their chemical nature. It has positive and negative effects; for example, in agricultural management, weed control, crop protection and crop establishment are the positive effects, whereas autotoxicity, soil sickness and biological invasion are the negative impacts of allelochemicals. Allelochemicals are classified into 14 categories based on their chemical analogy (Soltys et al., 2013). Since the ancient times, allelopathy is used in agricultural practices, and till now it has developed vigorously. Allelochemicals can stimulate or inhibit plant germination and growth and permit the development of crops with low phytotoxic residue amounts in water and soil, thus it facilitates water treatment and recycling. It also helps in removing artificial herbicides (Awadallah, 2020). Earlier it was thought that allelochemicals neither show any toxic effects nor harm any other plants, but now it is proved that the efficacy and specificity of many allelochemicals are restricted and might arise as a harmful content to many weeds. Allelochemicals are also widely used in crop rotation and intercropping. Recently, a technique known as straw mulching was developed in South Africa. In traditional agricultural technique, weed control using herbicides was an expensive practice, along with its negative impact on the environment. By altering the methods and using allelopathic applications such as straw mulching, the problem can be minimized to some extent in an affordable cost. It is also eco-friendly and causes no harm to the environment. Moreover, straw mulching also improves the soil organic matter content and increases soil fertility. Both conventional breeding methods and those developed using transgenic technology can be applied in the breeding of allelopathic cultivars. Successful cultivars must also combine weed suppression ability with high yield potential and disease resistance. Recent research in allelopathy is striving towards the future for sustainable agriculture. This giant challenge has made allelopathy very important in agricultural production (Cheng and Cheng, 2015). Currently, agricultural production is characterized by heavy use of synthetic herbicides. Allelopathy can definitely reduce the amount of herbicide use. Progress towards using allelopathy-related techniques is an important part of crop and weed management in several agro-ecosystems. Significant results are leading to the development of successful ecological pest control.

8.9 CONCLUSION

Allelopathic communications have potential use in plant growth and development and suppression of spread of weed due to their dual role as stimulants and suppressants. Bioherbicides extracted from allelochemicals can be used in plant protection. Problems and possible proposals of utilizing allelochemicals in alleviation of stress in terrestrial and aquatic ecosystems are being widely discussed.

REFERENCES

Aliotta G, Molinaro A, Monaco P, Pinto G, Previtera L. Three biologically active phenylpropanoid glucosides from Myriophyllum verticillatum. 1992. *Phytochemistry* 31(1), 109–111.

Awadallah D. Phytotoxic effects of *Calotropis procera (Aiton)* W.T. leaves aqueous extract on seed germination of some cereal crops using probit analysis. 2020. *Scientific Journal of King Faisal University (Basic and Applied Sciences)* 21(1), 163–169.

Bais HP, Park SW, Weir TL, Callaway RM, Vivanco JM. How plants communicate using the underground information superhighway. 2004. *Trends in Plant Science* 9, 26–32. doi: 10.1016/j.tplants.2003.11.008

Bais HP, Weir TL, Perry LG, Gilroy S, Vivanco, JM. The role of root exudates in rhizosphere interactions with plants and other organisms. 2006. *Annual Review of Plant Biology* 57, 233–266. doi: 10.1146/annurev.arplant.57.032905.105159

Baldwin IT, Halitschke R, Paschold A, von Dahl CC, Preston CA. Volatile signaling in plant-plant interactions: "talking trees" in the genomics era. 2006. *Science* 311(5762), 812–815. doi: 10.1126/science.1118446

Cereda MP, Mattos MCY. Linamarin: The toxic compound of cassava. 1996. *Journal of Venomous Animals and Toxins* 2(1).

Chai TT, Ooh KF, Ooi PW, Chue PS, Wong FC. *Leucaena leucocephala* leachate compromised membrane integrity, respiration and antioxidative defence of water hyacinth leaf tissues. 2013. *Botanical Studies* 54(1), 8. doi: 10.1186/1999-3110-54-8

Cheng F, Cheng Z. Research progress on the use of plant allelopathy in agriculture and the physiological and ecological mechanisms of allelopathy. 2015. *Frontiers in Plant Science* 15(6), 1020.

Dayan FE. Factors modulating the levels of the allelochemical sorgoleone in Sorghum bicolor. 2006. *Planta* 224, 339–346. doi: 10.1007/s00425-005-0217-5

Gajger T, Dar I, Dar SA. Plant allelochemicals as sources of insecticide. 2021. *Insects* 12(3), 189. doi: 10.3390/insects12030189

Gfeller A, Glauser G, Etter C, Signarbieux C, Wirth J. *Fagopyrum esculentum* alters its root exudation after amaranthus retroflexus recognition and suppresses weed growth. 2018. *Frontiers in Plant Science* 9, 1–13. doi: 10.0033389/fpls.2018.00050

Heisy, Rod M. Identification of an allelopathic compound from *Ailanthus altissima* (Simaroubaceae) and characterization of its herbicidal activity. 1996. *American Journal of Botany* 83(2), 192–200. doi: 10.2307/2445938

Hilt S, Ghobrial MG, Gross EM. In situ allelopathic potential of myriophyllum verticillatum (haloragaceae) against selected phytoplankton species 1. 2006. *Journal of Phycology* 42(6), 1189–1198.

Jilani G, Mahmood, Chaudhry AN. et al. Allelochemicals: Sources, toxicity and microbial transformation in soil—a review. 2008. *Annals of Microbiology* 58, 351–357. doi: 10.1007/BF03175528

Kaya K, Shiraishi F, Uchida H, Sano, T. A novel retinoic acid analogue, 7-hydroxy retinoic acid, isolated from cyanobacteria. 2011. *Biochimica et Biophysica Acta (BBA)-General Subjects* 1810(4), 414–419.

Koul O, Walia S. Comparing impacts of plant extracts and pure allelochemicals and implications for pest control. 2009. *CAB Reviews: Perspectives in Agriculture, Veterinary Science, Nutrition and Natural Resources* 4, 9.

Kurashov E, Krylova J, Protopopova E. The use of allelochemicals of aquatic macrophytes to suppress the development of cyanobacterial "blooms". In: Pereira L, Gonçalves AM, editors. Plankton Communities. London: IntechOpen; 2021.

Leu E, Krieger-Liszkay A, Goussias C, Gross EM, Polyphenolic allelochemicals from the aquatic angiosperm myriophyllum spicatum inhibit photosystem II. 2002. *Plant Physiology* 130(4), 2011–2018. doi: 10.1104/pp.011593

Li ZH, Wang Q, Ruan X, Pan CD, Jiang DA. Phenolics and plant allelopathy. 2010. *Molecules* 15(12), 8933–8952. doi: 10.3390/molecules15128933

McLendon K, Preuss CV. Atropine. In: StatPearls. Treasure Island (FL): StatPearls Publishing; 2021.

Men YJ, Hu HY, Li FM. Effects of the novel allelochemical ethyl 2-methylacetoacetate from the reed (Phragmitis australis Trin) on the growth of several common species of green algae. 2007. *Journal of Applied Phycology* 19, 521–527.

Mulderij G, Smolders AJP, van Donk E. The allelopathic effect of Stratiotes aloides on phytoplankton under natural conditions. 2006. *Freshwater Biology* 51, 554–562.

Muzell TM, Vidal RA, Balbinot Junior AA, von Hertwig Bittencourt H, da Silva Souza Filho AP. Allelopathy: Driving mechanisms governing its activity in agriculture. 2016. *Journal of Plant Interactions* 11(1), 53–60. doi: 10.1080/17429145.2016.1159342

Nakai S, Hosomi M. Allelopathic inhibitory effects of polyphenols released by Myriophyllum spicatum on algal growth. 2002. *Allelopathy Journal* 10, 123–131.

Nakai S, Yamada S., Hosomi, M. Anti-cyanobacterial fatty acids released from Myriophyllum spicatum. 2005. *Hydrobiologia* 543, 71–78. doi: 10.1007/s10750-004-6822-7

Panche AN, Diwan AD, Chandra SR. Flavonoids: An overview. 2016. *Journal of Nutritional Science* 5, e47. doi: 10.1017/jns.2016.41

Pflugmacher S. Possible allelopathic effects of cyanotoxins, with reference to microcystin-LR, in aquatic ecosystems. 2002. *Environmental Toxicology: An International Journal* 17(4), 407–413.

Pham JV, Yilma MA, Feliz A, Majid MT, Nicholas M, Walker JR, Eunji K, Cho HJ, Reynolds JM, Song MC, Park SR, Yoon YJ. A review of the microbial production of bioactive natural products and biologics. 2019. *Frontiers in Microbiology* 10, 1404. doi: 10.3389/fmicb.2019.01404

Pillay VV, Sasidharan A. Oleander and datura poisoning: An update. 2019. *Indian Journal of Critical Care Medicine* 23(4), 250–S255.

Reigosa MJ, Pedrol N, González L. (Eds.). Allelopathy: A Physiological Process with Ecological Implications. Germany: Springer Science & Business Media; 2006.

Rietveld WJ. Allelopathic effects of juglone on germination and growth of several herbaceous and woody species. 1983. *Journal of Chemical Ecology* 9(2), 295–308. doi: 10.1007/BF00988047

Rosenblum A, Marsch LA, Joseph H, Portenoy RK. Opioids and the treatment of chronic pain: Controversies, current status, and future directions. 2008. *Experimental and Clinical Psychopharmacology* 16(5), 405–416. doi: 10.1037/a0013628

Saxena A, Marella TK, Singh PK, Tiwari, A. Indoor mass cultivation of marine diatoms for biodiesel production using induction plasma synthesized nanosilica. 2021. *Bioresource Technology* 332, 125098.

Scanferlato VS, Cairns JJ. Effect of sediment-associated copper on ecological structure and function of aquatic microcosms. 1990. *Aquatic Toxicology* 18(1), 23–34.

Singh AA, Gunasekaran R, Nirmal LA, Jacob S. Synthesis and extraction routes of allelochemicals from plants and microbes: A review. 2021. *Reviews in Analytical Chemistry* 40(1), 293–311. doi: 10.1515/revac-2021-0139

Soltys D, Krasuska U, Bogatek R, Gniazdowska A. Allelochemicals as bioherbicides—present and perspectives. 2013. *Herbicides—Current Research and Case Studies in Use.* doi: 10.5772/56185

Švanys A, Paškauskas R, Hilt S. Effects of the allelopathically active macrophyte Myriophyllum spicatum on a natural phytoplankton community: A mesocosm study. 2014. *Hydrobiologia* 737, 57–66. doi: 10.1007/s10750-013-1782-4

War AR, Paulraj MG, Ahmad T, Buhroo AA, Hussain B, Ignacimuthu S, Sharma HC. Mechanisms of plant defense against insect herbivores. 2012. *Plant Signaling & Behaviour* 7(10), 1306–1320. doi: 10.4161/psb.21663

Wink M. The role of quinolizidine alkaloids in plant insect interactions. In: E. A. Bernays, editor. Insect-Plant in-Teractions. Vol. IV. Boca Raton: CRC Press, 1992. pp. 133–169.

Wink M. Evolution of secondary metabolites from an ecological and molecular phylogenetic perspective. 2003. *Phytochemistry* 64(1), 3–19. doi: 10.1016/s0031-9422(03)00300-5

Zhang Z, Liu Y, Yuan L, Weber E, van-Kleunen M. Effect of allelopathy on plant performance: A meta-analysis. 2021. *Ecology Letters* 24(2), 348–362. doi: 10.1111/ele.13627

Zhao S, Li Y. Current understanding of the interplays between host hormones and plant viral infections. 2021. *PLoS Pathogens* 17(2), e1009242.

Zhu XG, Long SP, Ort DR. Improving photosynthetic efficiency for greater yield. 2010. *Annual Review of Plant Biology* 61, 235–261. doi: 10.1146/annurev-arplant-042809-112206

9 Comparative Assessment of Physico-Chemical and Biological Methods for the Treatment of Sanitary Landfill Leachate

Rachna Jain, Rita Mondal, Jhimli Bhattacharya, Akito I Sema, S. Saravanadevi, Dipanjali Majumdar

9.1 INTRODUCTION

Sanitary landfilling is globally recognized as the most efficient and cost-effective waste management methodology (Rajoo et al., 2020). Both gaseous emission and liquid discharge of landfill known as leachate are environmental concerns. Leachate is an indicator of the active metabolic potential of waste sites which generates due to the percolation of water alongside the heap of the dump. It extracts all the toxicants into a liquid phase and hence is considered as utmost toxic and a potential environmental threat. Assorted dumping, diverse sources of waste, and landfill age make leachate a frightening source of pollution for both surface and groundwater. It drastically reduces the dissolved oxygen content of surface water, and its prolonged exposure harms vegetation, aquatic organisms, and humans by polluting groundwater, impacting the quality of drinking water and irrigation water for agriculture (Arunbabu et al., 2017)

Landfill leachate containing high heavy metal concentrations can also cause genotoxicity and cytotoxicity to human and marine biota (Liu et al., 2019). Therefore, to minimize the negative environmental impact, the leachate needs to be effectively treated up to standard discharge limit before its recycling (Pajoudoro et al., 2020). The choice of treatment technology depends upon the leachate composition, and BOD/COD ratio is the primary determining factor for selecting an effective line of treatment. A >0.4 BOD/COD ratio signifies that the leachate is rich in organics. Hence, biological treatment may be effective, but a <0.01 ratio indicates the presence of a high concentration of chemical toxins, which needs some physico-chemical or advanced treatment technology for effective results.

Various physico-chemical and biological treatment technologies treat leachate up to safe discharge standards. The present chapter evaluates the efficacy of four most commonly used methods: aeration, advanced oxidation (fenton and photo-fenton), adsorption (activated charcoal), and microbiological (indigenous microbes) for the treatment of sanitary landfill leachate.

DOI: 10.1201/9781003297901-9

9.1.1 Aeration

Intensive aeration technology is an effective technique for converting conventional landfills to a more biologically active state under sustainable conditions, reducing the emission of hazardous and foul gases. This process is most suitable for organic matter decomposition and is ideal for young landfills with a high BOD/COD ratio. The major drawback of the whole process lies in the high energy consumption due to constant aeration, which increases the cost of the entire process and makes it unaffordable for low-budget treatment (Rich et al., 2008).

9.1.2 Advanced Oxidation Processes (AOPs): Fenton and Photo-Fenton

AOPs use peroxide as an oxidizing reagent, which oxidizes the chemical pollutants, and iron (II) (Fe^{2+}) acts as a catalyst that increases the rate of hydroxyl radical (OH) generation (Moradi and Farshid, 2014). Minimum energy consumption and easy alignment with other treatment methods make fenton a globally preferred choice for pilot-scale wastewater treatment.

Photo-fenton is another variant technique that involves the photo-de-carboxylation of ferric carboxylates by using UV radiation, apart from all the common chemicals used in the conventional fenton treatment process. Studies suggest the photo-fenton process as an efficient alternative process for treating landfill leachate.

9.1.3 Bioaugmentation

Re-inoculation of the active indigenous microbial culture into wastewater/leachate matrix for accelerating/augmenting the ongoing metabolic activities results in pollutant load reduction. It is the most efficient and cost-effective leachate treatment methodology.

9.1.4 Adsorption

Adsorption of different pollutants, viz. ammonia, phosphate, and heavy metals, from different influents using zeolite (clinoptilolite), pumice, and oyster shells have been previously reported (Wang et al., 2020). Remediation or treatment of wastewaters using activated carbon has been gaining more attention in recent years due to its simple operation, non-toxic nature, and low production cost (Alves et al., 2021). Studies have shown high sorption affinity toward leachate treatment from wastewater using a chemically modified biomass source (Mahdavi et al., 2018). The pros and cons of different leachate treatment techniques have been summarized in Table 9.1.

9.1.5 Materials and Methods

9.1.5.1 Leachate Sampling and Analysis

Samples for the present study have been collected from sanitary landfill (SLF) leachate located in Baidyabati Municipality and preserved as per standard protocol (Figure 9.1). The physico-chemical parameters for raw and treated leachate were analyzed following APHA, 2017 method (Jain et al., 2021; APHA, 2017).

TABLE 9.1
Comparison of commonly practiced leachate treatment techniques

Treatment method	Advantages	Disadvantages
Fenton	Applicable for low as well as high biodegradable leachate Can effectively reduce organic, inorganic, and heavy metal contaminants	The high amount of sludge generation, along with the release of toxic chemicals into the environments High cost Not suitable for large-scale application
Photo-fenton	Along with reducing the overall pollutant load, it also helps in decreasing the pathogenic load of the leachate	The high amount of sludge generation, along with the release of toxic chemicals into the environments High cost Not suitable for large-scale application
Activated charcoal adsorption	Adsorbs different toxic and recalcitrant chemicals at a higher rate	Effective for reducing the chemical load High operational cost Wider application is restricted
Bio-augmentation	Microbial isolates can be used for the reduction of different organic, inorganic, heavy metals, and other industrial recalcitrant chemicals from leachate Low operational cost Wider applicability	Isolation and characterization of efficient indigenous microbial isolates is a tedious task Elimination of pathogens through disinfection is critical before releasing the treated leachate in the environment

9.1.6 Leachate Treatment

9.1.6.1 Aeration

The SLF was equipped with two consecutive aeration ponds where the raw leachate undergoes primary treatment, that is, aeration. Afterward, it is directed to the sedimentation tank for settling for 24–36 hours. Settled effluent was collected and analyzed for different pollutant parameters and compared for the reduction in the pollutant load as compared to the raw sample.

9.1.6.2 Advanced Oxidation Methods

9.1.6.2.1 Fenton Treatment

The experiment was conducted in a batch reactor consisting of a 250 ml borosilicate glass beaker. A 100 ml leachate sample was taken, and pH was optimized and maintained at 3.5 throughout the experiment. The ratio of 2000:1000 mg/L of $FeSO_4$ and H_2O_2 was added, and the temperature was maintained at 50°C in a water bath for 30 minutes. Following, 5 ml samples were withdrawn after 15, 30, 45, and 60 minutes from the water bath for COD estimation (APHA, 2017).

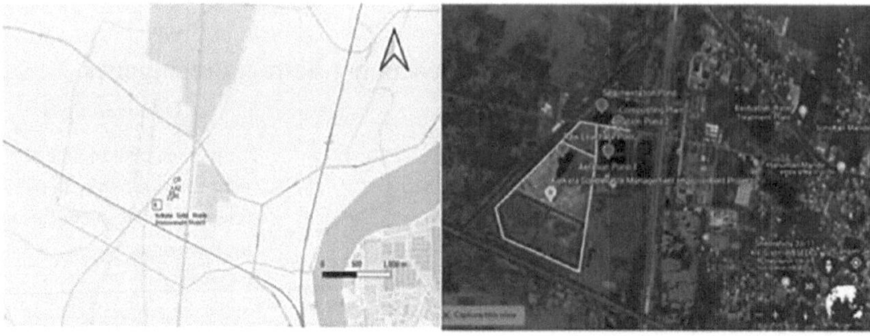

(a) (b)

FIGURE 9.1 (a) GIS-Based open street map (credit: QGIS Software) (b) Satellite view of the sampling site (credit: Google Earth), sampling site, Kolkata Solid Waste Improvement Project. Roads are mapped in orange and grey colors in the GIS map. Inside and main roads are mapped in yellow and magenta color, respectively. Samples for the present study were taken from the raw and sedimentation pond, respectively.

9.1.6.2.2 Photo-Fenton Treatment

In addition to the methodology mentioned earlier, the samples were kept under UV in the Laminar Air Flow Cabinet. Then 5 ml leachate samples were taken at regular intervals of 15, 30, 45, and 60 minutes to estimate COD.

9.1.6.3 Bioaugmentation

An indigenous microbial consortium was used as an inoculum for the present study to test its COD reduction potential under in situ conditions.

Microbial consortia was prepared by standard enrichment culture method; 1 ml of raw leachate was inoculated into 125 ml minimal salt medium (MSM), pH 7.0, and incubated for 24 hours at 37 °C and 100 rpm (repeated the cycle three times). Microbial consortia were mixed in landfill leachate in the ratio of 1:9 and incubated under aerobic culture conditions for 21 days. Residual COD concentration was observed at different time intervals: zero, seven, and 21 days, respectively.

Degradation kinetics was observed by plotting residual COD concentration (ln C_t/C_0) Vs incubation time as per first-order equation; where C_t is COD after t time incubation and C_0 is the initial COD. Kinetic constants and half-life were calculated using a linear regression equation.

9.1.6.4 Activated Charcoal Adsorption

For the experimental study, activated carbon dose of 0.5–3.0 g and 100 ml of varying concentration of COD concentration for both untreated and treated leachate was used. The original COD concentration for both treated and untreated was diluted to obtain the desirable concentration (treated leachate: 1105.39–1712.75 mg/L; untreated leachate: 1409.07–2550.91 mg/L) for further experimental analysis. After the successful adsorption process, the removal percentage and adsorption capacity were calculated using the following equation:

$$\text{Removal efficiency } (\%) = \frac{C_i - C_o}{C_i} \times 100 \qquad \text{Eq. (1)}$$

$$\text{Adsorption capacity } (q_e) = \frac{(C_i - C_o)}{M} V \qquad \text{Eq. (2)}$$

Where C_i is the initial concentration of leachate (mg/L) before adsorption; C_o represents the final concentration of leachate (mg/L) after adsorption; q_e is the amount of adsorbate adsorbed on the AC (mg/g); V is the volume of the leachate concentration; M is the amount of activated carbon dose

9.2 RESULTS AND DISCUSSION

9.2.1 Physico-Chemical Characteristics of Raw Leachate

The results of the physico-chemical analysis, as depicted in Table 9.2, revealed that all major pollutant variables, viz. pH, BOD, COD, TDS, TKN, chlorides, and phenols

TABLE 9.2
Physico-chemical analysis of raw and primary aerated sample

Physico-chemical parameter	Standard discharge limit*	Leachate characteristics	
		Raw	Primary treated
pH	5.5–9	7.6±0.1	7.9±0.1
Conductivity (µs/cm)	-	3250± 5	1758± 23.1
BOD (mg/l)	30	285±5	214.3±7.5
COD (mg/l)	250	6084.3±36.7	1274.7±16.7
TOC (mg/l)	-	784.7±17.5	325.2±2.9
Chloride (mg/l)	1000	637±40.7	554.0±7
Alkanity (mg/l)	-	4540.±50.5	2764.3±10
TSS (mg/l)	-	146.3±7.2	40.7±3.9
TDS (mg/l)	2100	8098±5.1	1492.7±36
Phosphate (mg/l)	-	45.4±0.5	27.4±1.7
Total kjeldahl nitrogen (mg/l)	100	877.7±18.8	573±16.5
Flouride (mg/l)	-	0.2±0	0.5±0
Total coliform (cells/100ml sample)	-	109.7±5.5	59.7±4.7
Total phenols (mg/l)	1.0	10.7±0.4	7.8±0.2

Source: *MOEF (2000)

were significantly high as compared to the standard discharge limit in both the raw as well as primary treated sample (Table 9.2). A high concentration of all the evaluated parameters indicates that the sanitary landfill is hazardous, and hence its discharge without treatment will affect the environment.

9.2.2 Application of Treatment Technologies

9.2.2.1 Aeration

Primary aeration treatment decreases the pollutant concentration by 13–81% of their initial value (raw leachate), as depicted in Figure 9.2. It showed a minor effect on the pH and chloride content of the sample, whereas the highest impact was observed on COD (79%) and TDS (81%). The primary treatment process reduced BOD, TKN, and phenols by 24, 34, and 27%. Reduction in all these parameters can be explained by the formation of colloids and higher molecular weight compounds, which can easily settle down in the sedimentation tank, and hence significant reduction was observed. But even after the primary treatment process, the pollutant concentration was significantly high from the standard discharge limit (Figure 9.2).

BOD and COD of the samples were 50–70% higher than the standard limit. TKN and total phenols were increased by 70–90%. Chloride and TDS were found lower than the standard discharge limit.

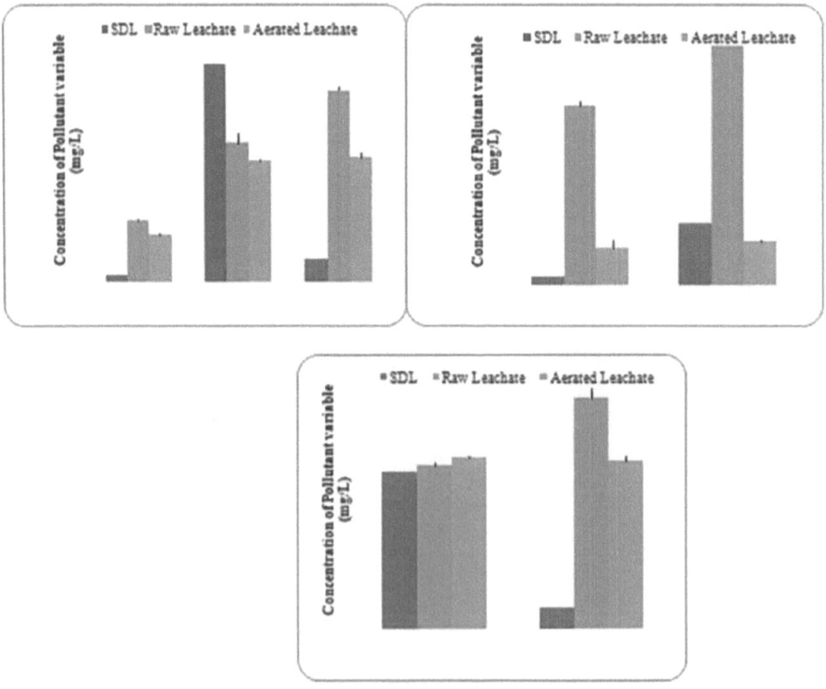

FIGURE 9.2 Comparative analysis of physico-chemical parameters of sanitary landfill leachate with respect to standard discharge limit (SDL) after primary aeration treatment.

9.2.2.2 Advanced Oxidation Methods

Advanced oxidation processes (AOPs) are a choice of interest for treating mature landfill leachate, as depicted in the present study (BOD/COD < 0.5). Hence, these AOPs can work well to decrease the pollutant load within the desirable limit. An optimal ratio of H_2O_2 and Fe^{2+} is necessary to avoid scavenging reactions and maximum COD removal. Hence this study has been designed under the previously optimized experimental conditions, H_2O_2: $FeSO_4$ ratio 2000:1000 (mg/L) and pH 3.5, to treat collected sanitary/mature landfill leachate. Fenton and photo-fenton treatment reduce COD by 83% and 95.3%, respectively as shown in Figure 9.3.

As reported by many researchers, COD removal efficiency is much superior to the photo-fenton process, which corroborates our findings. Hence it can be used as a secondary treatment method for the biologically treated mature landfill leachate. Incorporating UV radiation in the classical fenton method increases the organic carbon reduction rate almost three times. UV catalyzes the hydroxyl ion production rate, and it also helps in Fe^{2+} recycling, which ultimately results in speedy degradation. The method is also cost-effective as compared to the classical fenton method due to a significant reduction in Fe^{2+} consumption and reduced sludge formation.

9.2.2.3 Bioaugmentation

Microbiological screening by enrichment culture method resulted in a consortium comprising four distinct bacterial isolates. Primary morphological characterization of isolates was done by gram staining. Among the four screened cultures, three were gram-negative, and only one strain was found gram-positive. Their growth compatibility was checked by the antagonism test, which revealed that all the four isolates share a symbiotic relationship, and hence the growth rate exponentially increased upon mixing. Wang et al., 2018 have used an activated sludge process for landfill leachate treatment and observed its total nitrogen reduction potential. Therefore, a

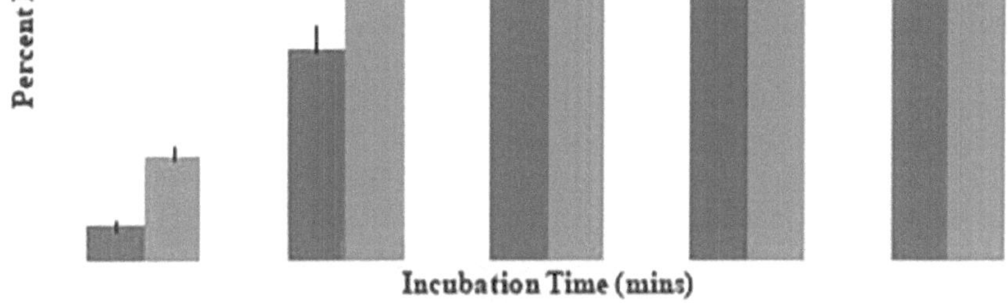

FIGURE 9.3 Timeline analysis of COD reduction by AOPs (fenton and photo-fenton methods).

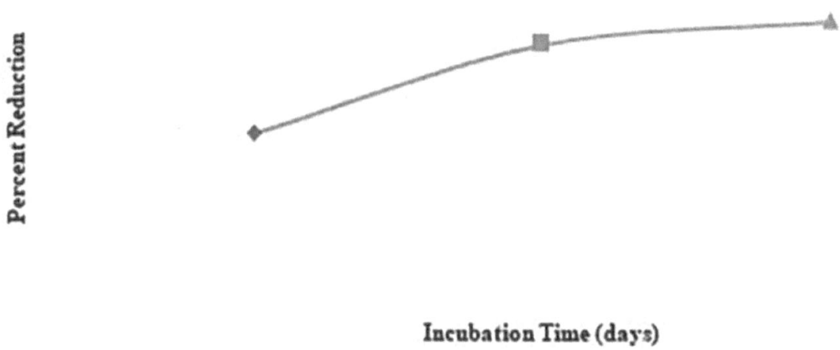

FIGURE 9.4 Timeline view for COD reduction by microbial consortia.

mix consortium of these four isolates was used to study the COD reduction potential of *in situ* landfill leachate. Poddar et al., 2019 have used microbial consortia containing *Klebsiella* spp., *Enterobacter* sp., and *Pantoea* sp. and obtained 16–20% degradation of mixed hydrocarbons within four days of treatment. Cyclic aliphatic 1-(1,5-Dimethylhexyl)-4-(4-Methylpentyl)-Cyclohexane was completely remediated, whereas naphthalene derivatives, aliphatic acids, and benzene derivatives were reduced by 36.72%, 78.13%, and 24.16%, respectively, within the same time frame.

In the present study, mixed microbial consortia reduced 58.5% of the initial COD load within seven days of incubation. Further extension of the incubation period increased the degradation by 96.6% within 21 days (Figure 9.4). Microbial reduction rate followed a standard pattern; percent reduction was low in the initial seven days, known as the adaptation time/lag phase. Afterward, it increased exponentially till 14 days, and then there was a slight static phase till 21 days of treatment.

9.2.2.4 Activated Carbon Adsorption

Activated carbon, also known as activated charcoal (AC), is a carbonaceous material used efficiently as an adsorbent material for wastewater treatment, catalysis, gas storage, etc. (Sema and Bhattacharyya, 2021). Though this carbon material has been used since Ancient Greece, due to its low-cost, well-developed porosity, simple operation, and non-toxic nature, it has attracted many researchers for its robustness (Wang et al., 2019). Usually, the ACs are synthesized either by physical or chemical route, using waste biomass as the precursor material. Some of these include pomelo peels (Li et al., 2016), orange peels, rice husk (Chuah et al., 2005), pine cone (Bhomick et al., 2018), Ravenna grass (Supong et al., 2019), coconut shell (Bernard et al., 2013), palm kernel shells (Zainal et al., 2018), etc. Additionally, the synthesized ACs provide well-developed porosity, large surface area, structural and thermal stability, and desirable functional groups (Peng et al., 2013). Therefore, commercial ACs were used for this study to eradicate the contaminants in leachate water. Herein, COD as the measure of source contaminant was used for the adsorption analysis. Results from the experimental analysis indicate the percentage removal increases with the increase in the adsorbent dose. The maximum removal efficiency for COD concentration (1105.39–1712.75 mg/L) for treated leachate (TL1 — TL4) was 98% and

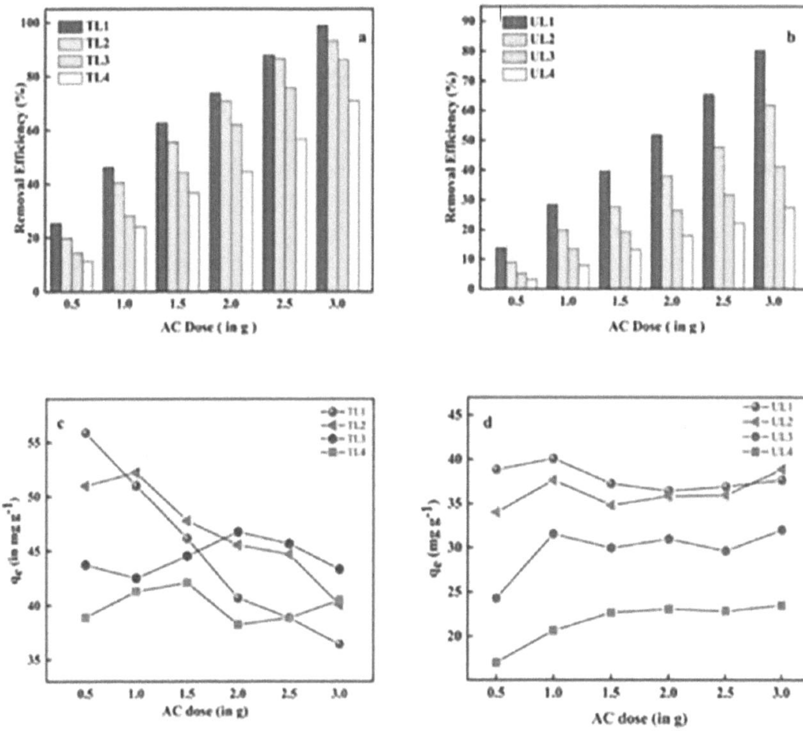

FIGURE 9.5 Removal efficiency (a and b) and adosption capacity (c and d) of treated (TL) and untreated (UL) leachate.

a minimum of 11% (Figure 9.5a) at different carbon doses ranging between 0.5 to 3g. Likewise, for the untreated leachate (UL1—UL4), the maximum and minimum removal percentage of COD (1409.07–2550.91 mg/L) by activated carbon was found to be 80% and 3% (Figure 9.5b). The pattern in the removal efficiency, with the increase in the adsorbent dose, is true for both treated and untreated leachate. Furthermore, an increase in the COD concentration of leachate (treated and untreated leachate) showed a gradual decrease in removal efficiency (Figure 9.5a and 9.5b).

The unusual decrease can be explained due to the presence of other organic or inorganic moieties present in the leachate water sample. The competition between different moieties for the surface-active sites or pores of ACs can ultimately result in poorer removal efficiency. Similar studies have also suggested that other moieties affect the removal efficiency of ACs. Adsorption capacity (q_e) gives valuable information about the amount of adsorbate that can get adsorbed on the adsorbent. Figure 9.5c for treated leachate and Figure 9.5d for untreated leachate show that the adsorption capacity decreases with the increase in adsorbate concentration. The decreasing trend in the adsorption capacity could be explained by the presence of more interfering species, with an increase in the COD concentration, which hinders the capacity of the adsorbent.

Furthermore, the slight increase and decrease in the adsorption capacity for each set of COD concentrations and varying carbon doses can be potentially attributed to the presence of other interfering moieties which compete for the surface-active sites on the carbon matrix. This can also be explained when more mass is added to the COD solution sample; more sorption sites are available, making the amount of adsorbate constant. It may also be well established that there is a decrease in the contact surface area between the AC and the adsorbate with increasing dose. This is evident in the adsorption processes when dealing with real-life raw samples apart from model pollutants. In addition, the maximum adsorption capacity for treated and untreated leachate was found to be 55 mg/g and 40 mg/g.

9.3.3 Degradation Kinetics

Degradation kinetics was observed for three tested methods: microbial consortia, fenton, and photo-fenton followed the first-order equation. Regression plots for all the three tested methods were linear, showing a significant correlation (Figure 9.6). Regression coefficients (R^2), sum square of error (SSE), and root mean square error (RSME) were also observed to predict the degree of similarity. The microbiological treatment method showed the least SSE (<0.5) and RSME (<0.5) and hence predicted the maximum similarity with the predicted model (Table 9.3). The observed pattern showed comparatively lesser similarity with the other two methods, but the results were significant. The regression coefficient (R^2) was significant at *p<0.01*.

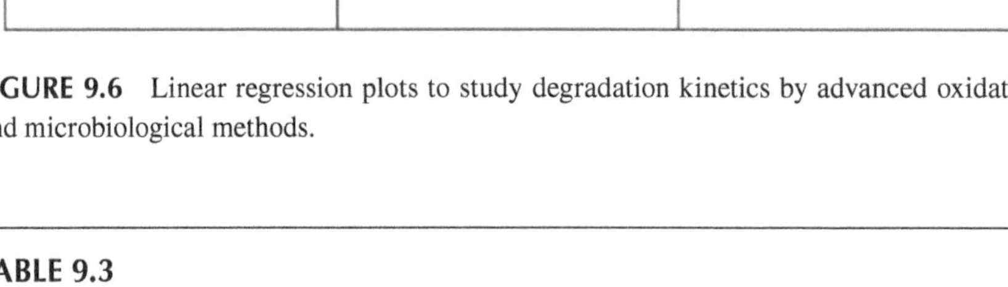

FIGURE 9.6 Linear regression plots to study degradation kinetics by advanced oxidation and microbiological methods.

TABLE 9.3
Regression coefficients of degradation

	k_{COD}	R^2	SSE	RSME
Fenton	0.417	0.989*	974.591	13.961
Photo-fenton	0.699	0.998*	2738.545	26.166
Microbial consortia	0.156	0.995*	0.025	0.079

Source: *Significant at *p<0.01*

9.3.4 Comparative Efficiency of Different Technologies

Sanitary landfilling is a step-forward waste management technique to avoid the direct discharge of toxic leachate into the environment. But process efficiency depends upon the treatment technology being applied at the landfill sites; the problem is solved if the treatment effectively brings the pollutant load down to the standard discharge limit. Hence it is essential to select a suitable treatment method to maintain environmental sustainability both in process efficiency and cost-effectiveness.

As discussed in the previous section, landfill leachate is a highly toxic dark liquid containing a high concentration of organic and inorganic contaminants and different heavy metals that need immediate attention to maintain environmental sustainability. Hence other primary and secondary treatment technologies are being tested and applied to remove contaminants from the sites. In the present study, we have assessed the potential of four different physicochemical and microbiological methods for treating sanitary landfill leachate. Here we have used four methods, namely (a) aeration, (b) advanced oxidation (AOPs: fenton and photo-fenton), (c) adsorption (activated charcoal, AC), and (d) bioaugmentation (using indigenous microbial consortia) to analyze their comparative treatment potential under *in-situ* leachate matrix. Each treatment method has its advantages and disadvantages based on environmental constrictions as well as end-user requirements, as tabulated in Table 9.4.

All the tested methods effectively reduced the pollutant load within a reasonable time. The study results demonstrate that activated charcoal adsorption is the most efficient tested treatment method considering percentage reduction and treatment time, followed by bioaugmentation and advanced oxidation methods. But other certain factors are quite important when choosing the appropriate treatment method.

9.3 CONCLUSION

This chapter presents a comparative assessment of the globally practiced leachate treatment methods. Also, the treatment efficiency of a few commonly used processes for treatment of real-time leachate collected from a sanitary landfill has been compared on a lab scale. The following aspects were pivotal in selecting the right technology or combination of treatment technologies for the treatment of any leachate.

a. **General factors**: Geographical regime, the composition of dumped waste, and landfill infrastructure are all the major environmental factors

TABLE 9.4
Comparative treatment potential of applied technologies

	Applied treatment method				
	Aeration	Advanced AOPs		Adsorption	Bio-augmentation
Parameters		Fenton	Photo-fenton		
COD reduction (%)	79	83	95	98	96
Treatment time (hr)	36	<1	<1	<1	504

that must be considered while designing. For example, if the waste consists of high amount of organics, then biological methods will be suitable; in the case of inorganics abundance, physico-chemical methods can be used.

b. **Environmental sustainability**: The choice of method must be environmentally friendly. It must have minimum after-effects like secondary product formation, sludge production, etc., additional cleaning requirement. AOPs generate a lot of sludge, which restricts their field application on a large scale.
c. **Cost**: Cost plays a significant role in designing any treatment strategy; regulatory authorities prefer to have a cost-effective but efficient method. AOPs and adsorption are high cost-intensive processes. They involve the use of costly chemicals and have some recurring costs (regeneration of AC), followed by aeration in which the cost incurred on the high-power aerators is included. But bioaugmentation is the low-cost technique among the discussed ones. Still, many times the indigenous microbes are pathogenic, and hence additional disinfection is required to maintain environmental sustainability and human health.
d. **End use**: Ultimate use of the treated water also impacts treatment choice. If the treated water is for drinking purposes, it must satisfy the stringent water quality standard, which requires a tertiary level of treatment (such as activated charcoal adsorption). If it is needed to treat the water up to surface standards, there is no need for such an exhaustive and expensive treatment; AOPs or bioaugmentation can work. Also, it is essential to specifically target the pollutants and accordingly choose the treatment method.
e. **Long-term impact**: Nutrient imbalance, accumulation of recalcitrant toxic chemicals, and effect on marine biota are some of the long-term implications of the physico-chemical methods; hence a proper after-care plan must be designed to avoid such uncertainties.
f. The area of application should also be considered while choosing the optimum treatment method.

All the globally practiced treatment methods are efficient but have restrictions that depend on certain driving factors. In the present study, aeration followed by bioaugmentation was quite a suitable treatment method for the observed leachate matrix in terms of process efficiency and cost implication. Also, strategical combinations of treatment technologies are more effective in terms of removal efficiency as well as required time.

ACKNOWLEDGEMENTS

The authors gratefully acknowledge the support of the Department of Biotechnology, Government of India, under grant no. BT/PR2730S/BCE/8/1432/2018. The guidance and support of the director, CSIR-NEERI and HOD, CSIR-NEERI, Kolkata Zonal Center, are highly appreciated. We also thank our associates, Ms. Shreyasi and Ms. Prateeti for their analytical support. The KRC (Knowledge Resource Center of

CSIR, National Environmental Engineering Research Institute) number for the manuscript is CSIR-NEERI/KRC/2022/. MARCH/KZC/1.

REFERENCES

Alves, A.T., Lasmar, D.J., Miranda, I.P.A., Chaar, J.S., Reis, J.S., 2021. The Potential of Activated Carbon in the Treatment of Water for Human Consumption, a Study of the State of the Art and Its Techniques Used for Its Development. 143–153. https://doi.org/10.4236/abb.2021.126010

APHA, 2017. Standard Methods for the Examination of Water and Wastewater, 23rd edition. American Public Health Association, Washington, DC.

Arunbabu, V., Indu, K.S., Ramasamy, E.V., 2017. Leachate Pollution Index as an Effective Tool in Determining the Phytotoxicity of Municipal Solid Waste Leachate. *Waste Management* 68. https://doi.org/10.1016/j.wasman.2017.07.012

Bernard, E., Jimoh, A., Odigure, J.O., 2013. Heavy Metals Removal from Industrial Wastewater by Activated Carbon Prepared from Coconut Shell. *Research Journal of Chemical Sciences* 3(8): 3–9.

Bhomick, P.C., Supong, A., Baruah, M., Pongener, C., Sinha, D., 2018. Pine Cone Biomass as an Efficient Precursor for the Synthesis of Activated Biocarbon for Adsorption of Anionic Dye from Aqueous Solution: Isotherm, Kinetic, Thermodynamic and Regeneration Studies. *Sustainable Chemistry and Pharmacy*. https://doi.org/10.1016/j.scp.2018.09.001

Chuah, T.G., Jumasiah, A., Azni, I., Katayon, S., Thomas Choong, S.Y., 2005. Rice Husk as a Potentially Low-Cost Biosorbent for Heavy Metal and Dye Removal: An Overview. *Desalination*. https://doi.org/10.1016/j.desal.2004.10.014

Jain, R., Majumdar, D., Mondal, R., 2021. Temporal and Seasonal Variation in Leachate Pollution Index (LPI) in Sanitary Landfill Sites: A Case Study of Baidyabati Landfill, West Bengal, India. In: Shit, P.K., Adhikary, P.P., Sengupta, D. (eds) Spatial Modeling and Assessment of Environmental Contaminants. Environmental Challenges and Solutions. Springer, Cham. https://doi.org/10.1007/978-3-030-63422-3_3

Li, H., Sun, Z., Zhang, L., Tian, Y., Cui, G., Yan, S., 2016. A Cost-Effective Porous Carbon Derived from Pomelo Peel for the Removal of Methyl Orange from Aqueous Solution. *Colloids and Surfaces A: Physicochemical and Engineering Aspects*. https://doi.org/10.1016/j.colsurfa.2015.10.041

Liu, J., Zhao, P., Xu, Y., Jia, X., 2019. Mg-Al Mixed Oxide Adsorbent Synthesized Using FCT Template for Fluoride Removal from Drinking Water. Edited by Ian S Butler. *Bioinorganic Chemistry and Applications* 5840205. https://doi.org/10.1155/2019/5840205

Mahdavi, A., Ghoreyshi, A.A., Rahimpour, A., Younesi, H., and Pirzadeh, K., 2018. COD Removal from Landfill Leachate Using a High-Performance and Low-Cost Activated Carbon Synthesized from Walnut Shell. *Chemical Engineering Communications* 205: 1193–1206. https://doi.org/10.1080/00986445.2018.1441831

MoEF, 2000. Municipal Solid Waste Management and Handling Rules, Ministry of Environment and Forests, Government of India.

Moradi, M., Farshid, G., 2014. Application of Response Surface Method for Coagulation Process in Leachate Treatment as Pretreatment for Fenton Process: Biodegradability Improvement. *Journal of Water Process Engineering* 4: 67–73.

Pajoudoro, D.N., Daniel, L., Baruch, A.A., Joseph, Z.M., Abdallah, A.E.B., Alfred, A.F.T., Desire, B.M., 2020. Antioxidant Properties of Lapachol and Its Derivatives and Their Ability to Chelate Iron (II) Cation: DFT and QTAIM Studies. *Bioinorganic Chemistry and Applications* 2020: 2103239. https://doi.org/10.1155/2020/2103239

Peng, C., Yan, X.B., Wang, R.T., Lang, J.W., Ou, Y.J., Xue, Q.J., 2013. Promising Activated Carbons Derived from Waste Tea-Leaves and Their Application in High Performance Supercapacitors Electrodes. *Electrochimica Acta*. https://doi.org/10.1016/j.electacta.2012.09.082

Poddar, K., et al., 2019. Construction of Potential Bacterial Consortia for Efficient Hydrocarbon Degradation. *International Biodeterioration and Biodegradation* 144: 104770.

Rajoo, K.S., Daljit, S.K., Ismail, A., Abdu, A., 2020. Evaluating the Leachate Contamination Impact of Landfills and Open Dumpsites from Developing Countries Using the Proposed Leachate Pollution Index for Developing Countries (LPIDC). *Environmental Nanotechnology, Monitoring & Management* 14: 100372. https://doi.org/10.1016/j.enmm.2020.100372

Rich, C., Jan, G., Nikolaos, V., 2008. The Potential for Aeration of MSW Landfills to Accelerate Completion. *Waste Management* 28(6): 1039–1048. https://doi.org/10.1016/j.wasman.2007.03.022

Sema, A.I., Bhattacharyya, J., 2021. A Brief Commentary on Activated Carbon Material as an Efficient Low-Cost Material for Water Decontamination. *International Journal of Chemical and Environmental Sciences* 2(2): 37–55. https://doi.org/10.15864/ijcaes.2204

Supong, A., Bhomick, P.C., Sinha, U.B., Sinha, D., 2019. A Combined Experimental and Theoretical Investigation of the Adsorption of 4-Nitrophenol on Activated Biocarbon Using DFT Method. *Korean Journal of Chemical Engineering*. https://doi.org/10.1007/s11814-019-0382-z

Wang, K., Li, L., Tan, F., Wu, D., 2018. Treatment of Landfill Leachate Using Activated Sludge Technology: A Review. *Archaea* 2018: Article ID 1039453, 10 pages. https://doi.org/10.1155/2018/1039453

Wang, Q., Zhou, C., Kuang, Y., Jiang, Z., Yang, M., 2019. Removal of Hexavalent Chromium in Aquatic Solutions by Pomelo Peel. *Water Science and Engineering*. https://doi.org/10.1016/j.wse.2019.12.011

Wang, Q., Wu, X., Jiang, L., Fang, C., Wang, H., Chen, L., 2020. Effective Degradation of Di-n-Butyl Phthalate by Reusable, Magnetic Fe_3O_4 Nanoparticle-Immobilized Pseudomonas Sp. W1 and Its Application in Simulation. *Chemosphere* 250: 126339. https://doi.org/10.1016/j.chemosphere.2020.126339

Zainal, N.H., Aziz, A.A., Idris, J., Jalani, N.F., Mamat, R., Ibrahim, M.F., 2018. Reduction of POME Final Discharge Residual Using Activated Bioadsorbent from Oil Palm Kernel Shell. *Journal of Cleaner Production*. https://doi.org/10.1016/j.jclepro.2018.02.110

10 East Kolkata Wetlands
Unique Case of Treatment of Municipal Wastewater of Kolkata City

Susmita Mukherjee

10.1 INTRODUCTION

Generation of wastewater is an outcome of developmental activities; hence, the handling and management of wastewater is a major concern. According to the report by United Nations (2015) only 20% of global wastewater is treated, leaving low-income countries badly hit by contaminated water supply and diseases. This encourages governments to consider treated wastewater as a valuable resource, especially in the context of scarcity of fresh water resources. The wastewater reuse and the implementation of the same have been practiced since the 1990s. But the Hyderabad Declaration on Wastewater Use in Agriculture (IWMI 2002) and, furthermore, Article 12, European Wastewater Directive (EU 1991), gave impetus to policymakers and planners to focus on wastewater reuse in agriculture and industrial purposes.

In a report by the Centre for Science and Environment (2010), it has been shown that the treatment capacity of sewage treatment plants (STP), which is the conventional system of wastewater treatment, is only 19% of total sewage generation. This clearly shows that the wastewater treatment facility is much less than the requirement.

In India, the total wastewater generated is 81% of the total water supply. On top of that, on average, 70% of the population residing in major cities are getting wastewater treatment facilities (CPCB 2007), putting the country under the pressure of a huge volume of untreated wastewater. In the city of Kolkata, the problem of wastewater treatment is partially mitigated by a stretch of wetlands on the eastern part of the city. The East Kolkata Wetlands (EKW) is centered on latitude 22^0 34' N and longitude 88^0 24' E, this wetlands can be subdivided into the following four types of land use patterns (Kundu and Chakraborty 2017):

- Substantial water body area (46%)
- Agricultural area (39%)
- Farming area (5%)
- Settlement area (10%)

DOI: 10.1201/9781003297901-10

Here the wastewater treatment primarily happens by the stretch of substantial water body area, in terms of fish cultivation. These ponds, where wastewater gets treated, are termed as *bheri* (fish pond) in the local language. This whole stretch of wetlands (125 sq. km.) has been designated as a Ramsar site (Bunting *et al.* 2011), giving this area international significance. Ecologically engineered wetlands provide different ecosystem services along with wastewater treatment that are highly valued. Treatment cost is low, and treatment is fueled directly by solar radiation. EKW offers a lot of ecosystem services and livelihood support to the people residing in and around (Kundu & Chakraborty, 2017). The fish ponds, known as waste stabilization ponds, follow a completely biological process which is inexpensive and eco-friendly.

In a study on the sewage cleaning efficiency of STPs in Delhi, it was found that efficiency depends on factors such as pond aeration, percentage flow treated, to name a few (Jamwal *et al.* 2009). A sample survey by Central Pollution Control Board (2007) classified the performance of the sewage treatment plants in India with 10% being good and 54% falling in poor to very poor categories. It is obvious that the treatment facility in the STPs is inefficient and unable to meet the requirement. In a work on municipal wastewater recycling in Chennai, India (Hingorani 2011), it was found that even if STPs exist, the collection network is insufficient, and hence most of the wastewater remains untreated and causes trouble to the ecosystem of the surface water bodies. The treatment facility by STPs is insufficient compared to the requirement, and the presence of an inexpensive and efficient system of wastewater treatment has become the demand of the day.

10.1.1 Evolution of EKW as a Wastewater-Fed Aquaculture System

The development of EKW as a wastewater-fed aquaculture system is connected with the city's sewage cleaning requirement. The whole process is developed by the local fish farmers. The eastern part of the city of Kolkata was the sewage basin during the colonial period, which is actually marked as the beginning of development of sewage disposal and management at the eastern part of the city and the consequent development of East Kolkata Wetlands. The present wetlands of East Kolkata comprises of 261 water bodies and an agricultural area from the administrative districts of North and South 24 Parganas. Since August 2002, this area has gotten international significance and is recognized as a Ramsar site (EKWMA website). In EKW, domestic wastewater is utilized in fish farming. The nutrients in the wastewater get converted into fish biomass. Almost 947 million litres of wastewater per day is being fed into around 261 fish ponds, generating 4.5 tons per hectare per year of fish (Little 2002).

10.1.2 Sewage Cleaning Process at EKW and STP

It has been found that the waste stabilization pond method followed at EKW in sewage cleaning is a single pond system. The system of resource recovery followed here is developed by the fish farmers long back. The city wastewater, through two main canals, the dry weather flow (DWF) and storm weather flow (SWF) canal, and the associated canal system enters into the main ponds located throughout the wetlands area. These ponds can be considered as maturation ponds. The wastewater in this wetland area, containing complex organic matters, gets degraded into simpler forms by a cascade of microorganisms. These nutrients either settle at the bottom of the pond or remain dissolved in the pond water. These nutrients, through the food chain, enter into fishes' bodies, thus waste-to-resource conversion. The ponds are fed with wastewater on a regular basis, and this depends on pond location, volume of waste, availability of wastewater and the season (Edwards 2000). The fish farmers maintain a schedule for feeding the ponds with wastewater. There is also a schedule for cleaning the ponds. It takes around 15–20 days for the treatment of wastewater (first-hand information collected by the author).

The STPs are operated by a combination of physical, mechanical and biological processes. The wastewater in any STP is collected in the collection tank, and from there, through two/three screenings, it goes to a grit chamber, and from there it goes to a primary sedimentation tank, where 30–40% biological oxygen demand (BOD) gets removed, then it goes to an aeration tank, where biological treatment occurs. Finally, the treated water goes to a secondary sedimentation tank and then is discharged as STP effluent (first-hand information collected by the author).

10.2 MATERIALS AND METHOD

For the purpose of understanding the efficiency of EKW in comparison to the conventional STP-driven waste cleaning system, one STP under Kolkata Municipal Corporation was chosen for sample collection, and for the same purpose, both the inlet and outlet points of EKW was chosen. National River Conservation Directorate, Ministry of Environment and Forests, has emphasized the (i) removal of organic matter, (ii) removal of pathogenic bacteria and (iii) removal of nutrients, in that order of priority, for determining the quality of the treated effluent (NRCD 2005). Besides, the city wastewater contains mostly domestic waste, and the organic loading is high. Hence the water quality parameters considered were biological oxygen demand (BOD), chemical oxygen demand (COD), total available nitrogen, dissolved phosphate and coliform bacterial count. Each of these parameters was estimated for both EKW and STP and was compared with the standard value set by Central Pollution Control Board (CPCB) to understand whether the wastewater treatment process is effective.

The objective of the work was to understand the efficiency of wastewater treatment by EKW in comparison to STP and to comment on the importance of EKW as part of a sewage management system of Kolkata. To understand the treatment efficiency, the water sample was collected from the inlet and outlet points in both the cases. For EKW, the inlet point was Topsia Pumping Station and *Ghosher khal* was the outlet point; for the STP, the inlet was the collection tank, and the outlet was the effluent discharge point from the STP.

The analyses of water sample were initiated on the day of the collection in the laboratory. The samples were equilibrated to the normal temperature, then pH of the water sample was measured by digital pH meter (model no: Systronics-802) after standardization. Biochemical oxygen demand (BOD), chemical oxygen demand

(COD), total available nitrogen and phosphorus in water were estimated following the standard methods as outlined in APHA [9]. The coliform bacterial count was estimated by the most probable number (MPN) method. The treatment efficiency of all the parameters both for STP and EKW was calculated in terms of percentage by the application of standard method of efficiency calculation.

10.3 RESULT

All the chosen parameters of the collected wastewater sample from EKW and STP were assayed in the laboratory following the standard method. The five parameters of the wastewater samples were assayed both at the entry and exit points of EKW and STP. The treated effluent both from EKW and STP were finally discharged into the surface water bodies; hence, these effluents must maintain the discharge standard for all the parameters set by Central Pollution Control Board (CPCB 2015, 2017). For this purpose, the value of the estimated parameters at the exit point of both STP and EKW was compared with the standard value set by CPCB (Figure 10.1).

Experimental results plotted in Figure 10.1 showed the comparison of BOD, COD and total available nitrogen content between the STP and EKW outlet with the effluent standard values set by Central Pollution Control Board. This clearly showed that for all the parameters, EKW effluent was much better, and in comparison to the standard value, it is much lower in case of total available nitrogen content and COD; however, for BOD both are the same.

The nitrogen and phosphorus ratio is an important parameter to understand the health of any aquatic system. Figure 10.2 shows that the N:P ratio was much less compared to the standard value at the EKW outlet.

A close observation of both Figure 10.1 and Figure 10.2 shows that the quality of the treated effluent coming out from EKW is better than that of the STP, and it

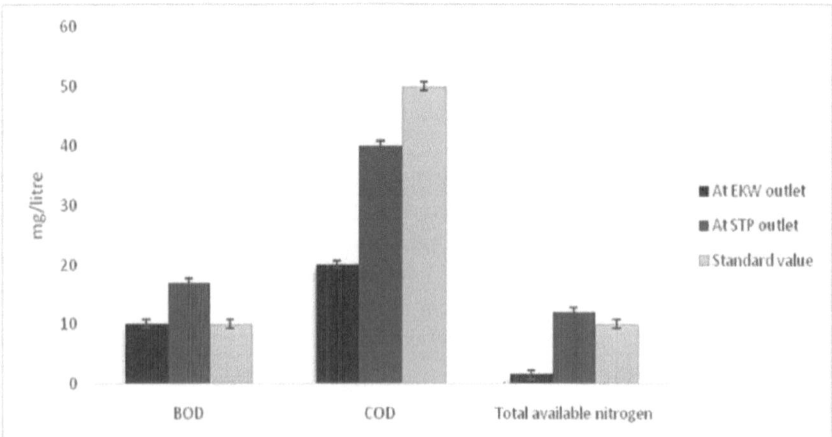

FIGURE 10.1 Comparison of water quality parameters of EKW and STP with the standard value.

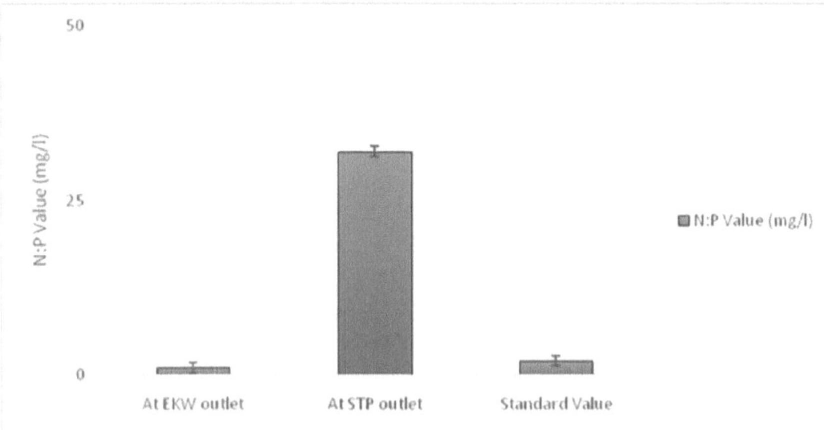

FIGURE 10.2 Comparison of nitrogen: phosphorus (N:P) reduction with that of the standard value.

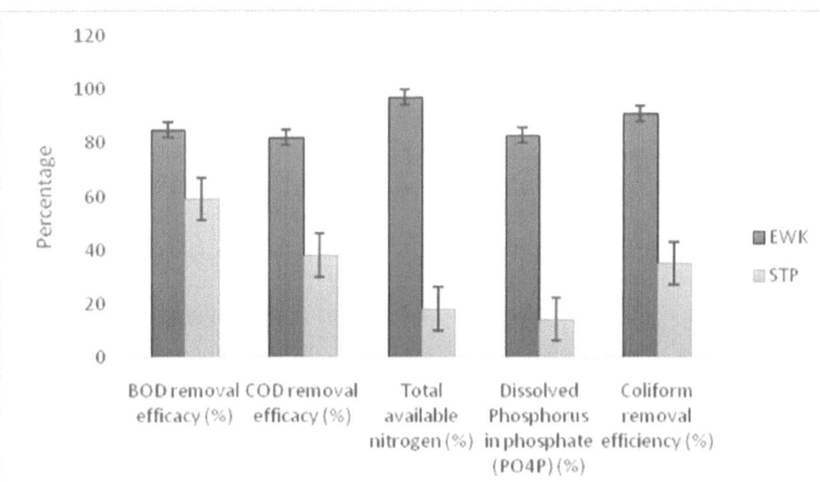

FIGURE 10.3 Comparison of wastewater cleaning efficiency on few selected parameters.

complies with the standard values as set by CPCB, but at the same time, it is also important to check which one of the two systems is more efficient. For efficiency calculation, the experimental results for all the parameters at both the inlet and outlet points were put in the general formula for efficiency calculation, and the values were plotted in Figure 10.3.

The efficiency of cleaning of each parameter, as plotted in Figure 10.3 shows that in all the parameters, EKW is cleaning the wastewater more efficiently compared to the conventional sewage treatment plant. In EKW, efficiency of BOD removal is 1.5 times more, whereas COD removal is two times more. Nitrogen and phosphate removal efficiency in EKW is more than five times compared to the STP value.

10.4 DISCUSSION

Constructed wetlands like EKW are engineered systems involving a natural process taking the benefit of varied wetland vegetations and microbial consortium in treating the wastewater. Constructed wetlands for wastewater treatment may be classified into different categories, but the basic treatment principle remains same.

Constructed wetlands are significant in removing the organic matter from dairy wastes, and in such a wetland system, BOD5 concentration was reduced significantly (Tuncsiper *et al.* 2015). Ghosh (2005) iterated the fact that the BOD reduction efficiency of EKW is above 80%, making it a great success story. The present work shows that the BOD at exit point of EKW is 10mg/ml, which is within the standard set by CPCB, and that of STP is 17mg/ml. Also, the BOD reduction efficiency of EKW is 1.5 times more compared to the STP system.

Distribution of inorganic nitrogen and phosphorus shows that nitrogen is the critical limiting factor to algal growth and eutrophication in an aquatic system (Li *et al.* 2014); hence, it is important for the health of an aquatic body to maintain proper N-P ratio. In the present work, the N-P ratio at the EKW exit point was estimated at 0.27, and at the STP exit, it was 32. The nitrogen and phosphorus removal efficiency at EKW is five times more compared to that of the STP.

Nature is slow, requiring a long detention period, which in turn implies a large land requirement.

At EKW, the time taken for the treatment of wastewater by the fish ponds is, on an average, nine to ten days (Sarkar *et al.* 2009). At the STPs, waste purification time is, on an average, two to three days (as reported by those working there). It has been found that coliform and other pathogens are present in high concentration in raw sewage, and the conventional treatment method (in STPs) does not include the removal of pathogens (Koivunen *et al.* 2003). The experimental results showed that EKW removed coliform bacteria three times more than STP.

At EKW the sewage when arrives the fish pond, under sunlight by algae-bacteria symbiosis the organic waste gets biodegraded into fish food. This exclusivity of EKW definitely is the reason for better treatment of wastewater (Dey & Banerjee, 2016).

10.5 CONCLUSION

The purpose of this work was to understand the efficacy of wastewater cleaning by EKW in comparison to the conventional wastewater treatment system to get convinced on the requirement of EK. The results and analysis of the data showed that wastewater cleaning at EKW is far better than the conventional STP. On top of this, the wastewater treatment facility in the city by the STPs is much less compared to the generation volume. Hence, presence of a complementary system is an absolute requirement. In such situation, governments must take appropriate measures to sustain the wetlands. There must be some incentive scheme for the fish farmers to sustain this system, which will benefit humankind at large.

REFERENCES

Bunting SW, Edwards P, Kundu N. 2011. Environment Management Manual: East Kolkata Wetlands. Manak Publications, Kolkata.

Central Pollution Control Board. 2007. Evaluation of Operation and Maintenance of Sewage Treatment Plants in India (Control of Urban Pollution Series CUPS/68/2007). Central Pollution Control Board, New Delhi.

Central Pollution Control Board. 2015. Effluent Discharge Standards for Sewage Treatment Plants. Central Pollution Control Board, New Delhi.

Central Pollution Control Board. 2017. Effluent Discharge Standards for Sewage Treatment Plants. Central Pollution Control Board, New Delhi.

Centre for Science and Environment. 2010. Do-it Yourself, Recycle and Reuse Wastewater. Central Pollution Control Board, New Delhi.

Dey D, Banerjee S. 2016. Wastewater management and the east-bound growth of the city of Kolkata: a compatibility analysis, *Environ Dev Sustain*, Springer, *19*, 1911–1932.

Edwards P. 2000. Wastewater-Fed Aquaculture: State-of-the-Art. In: Waste Recycling and Resource Management in the Developing World, Ecological Engineering Approach: 37–49, Jana B. B., Banerjee R. D., Guterstam B., Heeb J. (eds.), University of Kalyani, Kalyani, India and International Ecological Engineering Society, Wadenswil, Switzerland.

European Union. 1991. Council Directive Concerning Urban Wastewater Treatment. 91/271 European Economic Community, Brussels.

Ghosh D. 2005. Ecology and Traditional Wetland Practice: Lessons from Wastewater Utilization in the East Calcutta Wetlands. Worldview, Calcutta.

Hingorani P. 2011. The Economics of Municipal Sewage Water Recycling and Reuse in India. Chapter 21 in India Infrastructure Report. Oxford University Press, New Delhi, India

International Water Management Institute (IWMI). 2002. The Hyderabad Declaration on Wastewater Use in Agriculture. International Water and Sanitation Center, Natural Resource Institute and International Water Management Institute, Hyderabad.

Jamwal P, Mittal AK, Mouchel JM. 2009. Efficiency Evaluation of Sewage Treatment Plants with Different Technologies in Delhi, India. *Environmental Monitoring Assessment* 153: 293–305.

Koivunen J, Siitonen A, Helvi HT. 2003. Elimination of Enteric Bacteria in Biological-Chemical Wastewater Treatment and Tertiary Filtration Units. *Water Research* 37: 690–698.

Kundu N, Chakraborty A. 2017. Dependence on Ecosystem Goods and Services: A Case Study on East Kolkata Wetlands, West Bengal, India. *Wetland Science* 381–405.

Li RZ, Liu KF, Qian J, Yang JW, Zhang PP. 2014. Nitrogen and Phosphate Pollution Characteristics and Eutrophication Evaluation for Typical Urban Landscape Waters in Hefei City. *Huan Jing Ke Xue* 35(5): 1718–1726.

Little DC. 2002. Marketing of Fish in Peri-Urban Kolkata. University of Stirling, Stirling.

National River Conservation Directorate (NRCD). 2005. Annual Report. Ministry of Environment & Forest, New Delhi .

Sarkar S, Ghosh PB, Mukherjee K, Sil AK, Saha T. 2009. Sewage Treatment in a Single Pond System in East Kolkata Wetlands, India. *Water Science & Technology* 60(9): 2309–2317.

Tuncsiper B, Drizo A, Twohig E. 2015. Constructed Wetlands as a Potential Management Practice for Cold Climate Dairy Effluent Treatment-VT, USA. *Catena* 175: 184–192.

United Nations. 2015. UN Report Calls for Wastewater Focus in Post-2015 Agenda. As 80% of Worlds Watewater Discharged Untreated. UN Programme on Environment, Nairobi.

11 Role of Plant Polyphenolic Compounds in Alleviation of Metal-Induced Stress in Plants

Pratik Talukder, Sourish Bhattacharjee, Indrani, Elija Ghosh, Debankita Dutta

11.1 INTRODUCTION

Metabolism is a popular term that defines all chemical transformations involved in maintaining the living state of cells and the organism. Plant metabolism constitutes the physical and chemical events of photosynthesis, respiration and the synthesis of organic compounds (Kisa et al., 2016). The end products and intermediates formed as a result of metabolic processes are called metabolites. There are two types of metabolites:

Primary metabolites: These metabolites are compounds that are directly responsible for normal growth, development and reproduction in plants. In higher plants, they are stored in seeds and vegetative parts and are required for physiological development because of their vital role in basic cell metabolism (Shahzad et al., 2018).

Secondary metabolites: They produce certain components that help in the normal growth and development of plants indirectly under a given set of conditions. These metabolites are derived biosynthetically from primary metabolites. Plant secondary metabolites are necessary in various conditions such as for defence purposes, giving signals, communication and regulating primary metabolite pathways. They bring about adaptations towards environmental stress and serve as defence against pathogen attack. The phenylpropanoid pathway is one of the most efficient metabolic routes among secondary metabolites (Villiers et al., 2011).

Plants, being a sessile organism, are bound to encounter a plethora of environmental stresses throughout life. Plants are subjected to two types of environmental stresses: biotic and abiotic. The biotic stress includes pathogen attack such as attack by fungi, bacteria and herbivores. The abiotic stress includes extreme temperature fluctuations,

 DOI: 10.1201/9781003297901-11

water deficit, high light intensity, radiation, pesticides, salinity, drought and exposure to heavy metals. The effect of these stresses can lead to poor plant growth, lesser crop yield, permanent damage or even death if the stress exceeds the plant tolerance level. Plants, being sessile, need to opt for several intrinsic mechanisms at the time of stress (Handa et al., 2019).

Accumulation of heavy metals brings about toxic effect on plants such as stunted growth, root, leaf browning, chlorosis, alteration in photosynthesis and water balance and also senescence which can ultimately cause plant death (Poonam et al., 2015). The highly toxic heavy metals responsible for this stress include arsenic (As), lead (Pb), cadmium (Cd) and mercury (Hg) (Zafari et al., 2016). Metal stress brings about oxidative stress in plants. It can occur by triggering the production of harmful reactive oxygen species (ROSs) which ultimately leads to toxicity and poor or retarded plant growth (Pandey et al., 2009; Villiers et al., 2011). When there is any metal stress in plants, there is a mechanism in the organism called antioxidant activity, leading to the formation of secondary metabolites. The free radicals are also destroyed this way. This mechanism is developed in plants in order to prevent ROS accumulation. Enhanced biosynthesis of phenolic in plants under metal stress helps in protecting plants from oxidative stress (Kohli et al., 2018; Kaur et al., 2017a, 2017b). Flavonoids, belonging to the family of polyphenols, can be found in seeds, flowers, fruits and vegetables and are of high value (Williams et al., 2004). Flavonoids are responsible for regulating the antioxidant balance of plants and lessening the chances of mutation. They have the ability to reduce the levels of harmful hydroxyl radicals in plant cells by enhancing the metal chelation process (Ferreira and Slade, 2002). So an inference can be drawn that metal excess causes the level of flavonoids to rise in plants (Kaur et al., 2018; Handa et al., 2019) to cope with the enhanced production of ROS due to the metal toxicity. Certain specific flavonoids, for example anthocyanins and flavanols, are essentially related to the plant's defence mechanism and are enhanced during metal toxicity (Handa et al., 2019; Trejo-Tapia et al., 2001). This review focuses on the role of polyphenolic compounds as a response to heavy metal–induced stress.

11.2 MAJOR PATHWAYS RELATED TO POLYPHENOL PRODUCTION IN PLANTS

There are mainly two major pathways by which phenolic compounds are synthesized:

1. The shikimic acid pathway/phenylpropanoid pathway, where the products are phenylpropanoids (as the name suggests)
2. The acetic acid pathway, which produces simple phenols

(Fraser and Chapple, 2011)

Among these two pathways, the shikimic acid/phenylpropanoid pathway is the most studied pathway of secondary metabolism in plants (Fraser and Chapple, 2011). Usually, in plants growing under challenging environments, the accumulation of enhanced phenolic compounds is observed, which helps the plant to cope with the altered environmental condition (Sharma et al., 2016). The phenylpropanoid pathway

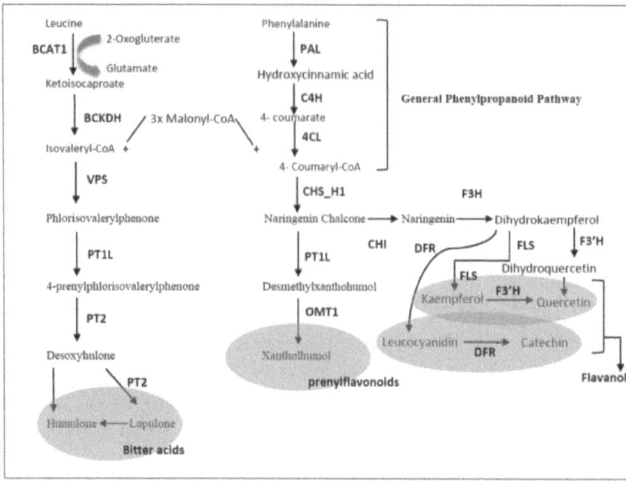

FIGURE 11.1 Shematic representation of the phenylpropanoid pathway.

is a major source of metabolites in plants, being required for the biosynthesis of lignin, and also produces many other important compounds, like flavonoids, coumarins and lignans (Sharma et al., 2019). Phenylpropanoids are a diverse group of compounds derived from the carbon skeleton of phenylalanine that are involved in plant defence, structural support and survival (Fraser and Chapple, 2011). The presence of large multiples of phenol structural units in the polyphenolic compounds gives them unique physical, chemical and biological properties. The phenylpropanoid pathway starts within the cytosol of the plant cells, with the precursor phenylalanine (Sharma et al., 2016). Another major pathway for polyphenolic production is Acetic Acid Pathway (Ferreira and Slade, 2006).

11.3 TYPES OF MAJOR POLYPHENOL IN PLANTS

Polyphenols are the water-soluble molecules bearing large number of di and tri hydroxyphenyl units. They must be oligomeric in nature or have multiple repeat of their phenolic motif in their monomeric forms (Kohli et al., 2017). Few major types of polyphenols are listed in Table 11.1.

11.3.1 Major Polyphenols in Stress Control

Environmental stress is the most important factor for the growth of a plant, and there are two types of environmental stress—biotic and abiotic stress. Both stresses cause heavy damage in plant growth and in their productivity.

11.3.2 Biotic Stress

This type of stress is caused by virus, weeds, insects, fungus or by other organisms. Plants have a very well-equipped response system to combat biotic stress. Specific chemical components as β-aminobutyric acid (βABA) or benzothiazole (BTH), some

TABLE 11.1
Few classes of plant polyphenols

Types	Example	Derived form
The proanthocyanidins (condensed tannins)	Procyanidins, anthrocyanidins, fisetinidins, delphinidins	Oligomerization of flavan-3-ol units as catechin, epi Gallocatechin, etc.
The gallo- and ellagitannins (hydrolysable tannin)	3, 4, 5-trihydroxybenzoicacid	Metabolism of shikimate-derived gallic acid, that goes through esterification and phenolic oxidative coupling reaction to (nearly 1000) monochromatic and oligomeric polyphenol galloyl derivative of sugar.
The phlorotannins	1, 3, 5-trihydroxybenzene	Oligomerizing dehydrogenative coupling of phloroglucinol.

plant hormones and salicylic acid (SA) play a significant role in the induced resistance of biotic stress (Rejeb et al., 2014).

11.3.3 Abiotic Stress

This stress is caused by drought, salinity, heat, cold, metal toxicity and heavy metals. As we know that heavy metals are of no use to plants, plants need to evolve strategies to withstand the negative effect of heavy metals. They have developed some response of secondary metabolite production, and phenolics play a pivotal role against abiotic stress, including heavy metal stress (Monni et al., 2000).

11.3.4 Metal Stress and Its Negative Impact on Plants

As we know, metals are resilient. Therefore, when the concentration level exceeds optimal levels, then they indirectly and directly harm the plants (Mira et al., 2002). The major ways by which damage is caused are as follows:

- By the high metal concentration, direct toxic effects are caused.
- Cytoplasmic enzymes are inhibited.
- Due to oxidative stress, the intracellular structures are damaged.

11.4 ROLE OF MACRO AND MICROELEMENTS

11.4.1 Microelements

In plants, microelements are the elements required in small quantities. If the level of microelements increases, the plants are being harmed in many ways. The negative impacts of higher accumulation of microelements are discussed in the following table (Table 11.3).

The impacts of excess of microelements at the cellular and molecular level are also mentioned in Table 11.4.

TABLE 11.2
Role of polyphenols in mitigating heavy metal–induced stress in plants

Heavy metals	Plant response
Cu	Increases the content of total phenolics like caffeic acid, kaempferol, etc. and enhances the transcription of the phenol productive enzyme genes like PAL, F3H, C4H, DFR, μFGT, etc.
Cd	Increases the content of flavonoids, phenols, polyphenols, anthocyanins and enhances the expression of PAL and CHS gene.
Zn	Levels of total phenolics are increased by activating the major phenol metabolic enzyme like shikimate dehydrogenase, polyphenol oxidase, etc.
Cr	Increases the content of flavonoids, phenols, polyphenols, anthocyanins and enhances the expression of PAL and CHS gene.
Pb	Increases the content of flavonoids, phenols, polyphenols, anthocyanins and enhance the expression of PAL and CHS gene and also the production of ferulic acid, daidzein, caffeic acid, myricetin, luteolin and diosmin.

TABLE 11.3
Negative impacts of microelements on plants

Name of microelements	About that element	Symptoms	Examples
1. Iron	When the plants are taking too much iron, leading to toxicity.	1. Bronzing 2. Stippling 3. Leaves become discoloured	Tomato, basil
2. Zinc	When plants are growing in acidic soil, it will face zinc toxicity. There is also excess magnesium in the soil.	1. Leaves become darker than normal colour	Eggplant
3. Manganese	Impaired fatty acid production is reported with high level of manganese in plants	1. Adversely affects the cuticular wax deposits	Crops
4. Boron	Naturally acid-loving plants are badly affected by a high level of boron	1. Perennials are often more susceptible to damage	Azaleas, camellias

TABLE 11.4
Impacts of excess of microelements at the cellular and molecular level

Name of microelements	Impact
• Iron	• Disrupts cellular homeostasis • Inhibits cell division
• Zinc	• Reduced biomass production • Retardation in certain metabolic activities: photosynthesis, transpiration and nitrogen metabolism.
• Manganese	• Decline in photosynthetic rate • Inhibits biosynthesis of auxin-reducing meristematic cell division
• Boron	• Inhibits expansion of cells • Affects vegetative and reproductive growth of plants

TABLE 11.5
Some essential macroelements and their effects

Name of Macroelements	Effects
• Calcium	1. Hollow stems and branches 2. Plant growth slowed down 3. Lower leaves twist and curled
• Phosphorus	1. Slow growth 2. Bluish leaves 3. Plants become weak
• Magnesium	1. Limited fruit production 2. Low storage roots
• Potassium	1. Nitrogen deficiency 2. Chlorosis 3. Veins with red tint

TABLE 11.6
Impacts of deficiency of macroelements at molecular level

Name of Macroelements	Impact
• Calcium	• Prevents seed germination • Reduces growth rate of plants
• Phosphorus	• Delays plant maturity, resulting in a reduction in yield • Alters the metabolism and translocation of carbohydrates
• Magnesium	• Retardation in growth • Chlorosis; impairment in carbon metabolism
• Potassium	• Rate of photosynthesis, transpiration and stomatal conductance decreases • Affects plant growth and metabolism

11.4.2 Macroelements

The essential minerals are referred to as macroelements, which are some basic elements that help the plants to maintain balance (Djingova and Kuleff, 2000). The stress of metal occurs when the level of macroelements decreases or increases above the optimal level. Some essential macroelements and their effects are mentioned subsequently (Table 11.5). The impacts of deficiency of macroelements at the molecular level are also mentioned in Table 11.6.

11.5 ROLE OF HEAVY METALS IN INDUCING STRESS IN PLANTS

Plants require certain heavy metals in minute amounts; higher concentrations of these metals are harmful (Jadia and Fulekar, 1999). There are several effects of heavy metals on plants, as they damage cell structure and have a negative impact to growth (Handa et al., 2019). Also it has been witnessed that the essential nutrients that are required for plant growth are damaged by heavy metal toxicity. This leads to major oxidative stress. Overall, plant metabolism is hampered (Assche and Clijsters, 1990).

Some specific heavy metals which are responsible for inducing heavy toxicity in plants are lead (Pb), mercury (Hg), cadmium (Cd), etc. (Wintz et al., 2002).

11.5.1 Lead (Pb)

Pb is one of the leading toxic heavy elements in nature. It brings about inhibition of seed germination and also retards the growth of seedlings. Some visible effects are blackening of roots and chlorosis. It leads to phytotoxicity of the cell membrane and alters permeability of the cell membrane. All this leads to normal physiological activities, even molecular events such as transpiration of the plant being hampered. Accumulation of Pb brings about the following changes at the molecular level: It causes the inhibition of photosynthesis. It hampers several enzyme activities (for example: Calvin cycle enzymes). Fresh biomass production is negatively affected. Due to excessive ROS production, there are chances of DNA damage. Also, ATP production and lipid peroxidation are inhibited (Kaur et al., 2019)

11.5.2 Mercury (Hg)

Hg brings about toxic effects on the growth and development of plants, its nodulation and nitrogen fixation of legumes. This accumulation often leads to distortion in cellular structure. Cell membrane permeability is altered. It ultimately leads to poor plant metabolism. It affects the dark and light reactions of photosynthesis. Also brings about substitution of the central atom of chlorophyll; that is, magnesium and itself take its place, leading to a breakdown in the photosynthesis process.

11.5.3 Cadmium (Cd)

Chlorosis, leaf rolls and stunted growth are the most common visible symptoms of cadmium toxicity. It can hinder the uptake of minerals and transport of water and essential nutrients. This also affects certain physiological processes such as stomatal opening, transpiration and photosynthesis (Pandey and Singh, 2012). Its presence brings about negative effects by causing structural, physiological and chemical changes in plants. It is also responsible for inhibiting the rate of chlorophyll production. It reduces fresh and dry masses of plants. It alters antioxidant production in plants. Also, it hinders several enzymatic activities such as the synthesis and activity of the enzymes involved in the Calvin cycle and disturbs carbohydrate metabolism (Kaur et al., 2019).

11.5.4 Metalloids

Metalloids are chemical elements whose physical and chemical properties fall in between the metal and the non-metal categories. There are many metalloids present in plants, out of which some show the maximum toxic effect like arsenic (As), antimony (Sb), etc.

11.5.4.1 Arsenic (As)

Arsenic accumulates in plants to such levels that lead to cellular toxicity. It inhibits plant growth, root extension and also its proliferation. Also, it affects photosynthesis and reproduction in plants. When accumulated in excess, its impact at the molecular level includes very limited biomass production, leading to a reduction in yield. Also, it causes overproduction of ROS which ultimately leads to carbohydrate, protein and DNA damage (Chandrakar et al., 2016).

11.5.4.2 Antimony (Sb)

Sb is a toxic trace element and is generally considered to be the pollutant of high interest. It is easily absorbable by plants, which might induce toxic effects in plants such as retardation in plant growth. It can also bring about different extents of damage such as compromising the anatomy and ultrastructure of plants and adversely affecting its physiology and biochemistry. When accumulated in excess, the molecular level impact shows a reduction in biomass production. It brings about a repression in photosynthesis and also a reduced synthesis of certain metabolites. These are some negative impacts of metal stress on plants. Metals are very important for plants, but they have their own permissible limit. Microelements, macroelements, heavy metals and metalloids are essential parts of plants. But when all these are present in huge amounts, the plants are negatively affected in many ways (Luo et al., 2021).

11.6 CONCLUSION

Plant polyphenolic compounds play an intricate role in reducing oxidative stress brought about by heavy metals and metalloids in plants. Overproduction of ROS is one of the most common and obvious issue in plants as a result of stress induction. Excessive ROS synthesis hampers several plant physiological processes. It also inhibits plant growth and development. Since plants are sedentary and unable to escape from stress situations, these polyphenolic compounds come in aid for plants. This develops resistance in plants against stresses, which proves the efficacy of these phenolic compounds. This is done either by scavenging ROSs (generated due to exposure of plants to various stresses) in plants or by regulating the antioxidant balance of plants and thus helping in plant defence. Hence, it can be said that when there is an increase in metal toxicity level in plants, there is an increased synthesis of plant secondary metabolites. Thus, this review reports the vital role that the plant phenolic compounds can play in alleviating metal-induced stress in plants.

REFERENCES

Assche F, Clijsters H. Effects of metals on enzyme activity in plants. 1990. *Plant, Cell and Environment* 13(3): 195–206.

Chandrakar V, Naithan SC, Keshavkant S. Arsenic-induced metabolic disturbances and their mitigation mechanisms in crop plants: A review. 2016. *Biologia* 71(4): 367–377.

Djingova R, Kuleff I. Instrumental Techniques for Trace Analysis. In: Vernet JP, Ed., Trace Elements: Their Distribution and Effects in the Environment, Elsevier Science, Netherlands, 2000, pp. 137–185.

Ferreira D, Slade D. Oligomeric proanthocyanidins: naturally occurring O-heterocycles. 2002. *Natural Product Reports* 5: 517–541.

Ferreira D, Slade D, Marais JPJ. In Flavanoids—Chemistry, Biochemistry and Applications (Eds. ØM Andersen, KR Markham), CRC/Taylor & Francis, Boca Raton, FL, 2006, pp. 553–616.

Fraser CM, Chapple C. The phenylpropanoid pathway in Arabidopsis. 2011. *Arabidopsis Book* 9: e0152. doi: 10.1199/tab.0152

Handa N, Kohli SK, Sharma A, Thukral AK, Bhardwaj R, AbdAllah EF, Alqarawi AA, Ahmad P. Selenium modulates dynamics of antioxidative defence expression, photosynthetic attributes and secondary metabolites to mitigate chromium toxicity in *Brassica juncea* L. plants. 2019. *Environmental and Experimental Botany* 161: 180–192.

Jadia CD, Fulekar MH. Phytoremediation of heavy metals: Recent techniques. 1999. *African Journal of Biotechnology* 8(6): 921–928.

Kaur P, Bali S, Sharma A, Vig AP, Bhardwaj RE. Role of earthworms in phytoremediation of cadmium (Cd) by modulating the antioxidative potential of *Brassica juncea* L. 2018. *Applied Soil Ecology* 124: 306–316.

Kaur P, Handa N, Verma V, Bakshi P, Kalia R, Sareen S, Nagpal A, Vig A, Mir BA, Bhardwaj R. Cross Talk Among Reactive Oxygen, Nitrogen and Sulfur During Abiotic Stress in Plants. In: Hasanuzzaman M, Fotopoulos V, Nahar K, Fujita M, Eds., Reactive Oxygen, Nitrogen and Sulfur Species in Plants: Production, Metabolism, Signaling and Defense Mechanisms, John Wiley & Sons, USA, 2019, pp. 857–871.

Kisa D, Elmasta M, Öztürk L, Kayır Ö. Responses of the phenolic compounds of Zea mays under heavy metal stress. 2016. *Applied Biological Chemistry* 59: 813–820.

Kohli SK, Handa N, Sharma A, Gautam V, Arora S, Bhardwaj R, Wijaya L, Alyemeni MN, Ahmad P. Interaction of 24-epibrassinolide and salicylic acid regulates pigment contents, antioxidative defense responses, and gene expression in *Brassica juncea* L. seedlings under Pb stress. 2018. *Environmental Science and Pollution Research* 25: 15159–15173.

Kohli SK, Handa N, Sharma A, Kumar V, Kaur P, Bhardwaj R. Synergistic effect of 24-epibrassinolideand salicylic acid on photosynthetic efficiency and gene expression in *Brassica juncea* L. under Pb stress. 2017. *Turkish Journal of Biology* 41: 943–953.

Luo W, He L, Li F, Li J. Exogenous salicylic acid alleviates the antimony (Sb) toxicity in rice (*Oryza sativa* L.) seedlings. 2021. *Journal of Plant Growth Regulation* 40: 1327–1340.

Mira L, Fernandez MT, Santos M, Rocha R, Florencio MH, Jennings KR. Interactions of flavonoids with iron and copper ions: A mechanism for their antioxidant activity. 2002. *Free Radical Research* 36: 1199–1208.

Monni S, Salemma M, Millar N. The tolerance of *Empetrum nigrum* to copper and nickel. 2000. *Environmental Pollution* 109: 221–229.

Pandey N, Singh GK. Studies on antioxidative enzymes induced by cadmium in pea plants (Pisum sativum). 2012. *Journal of Environmental Biology* 33(2): 201.

Pandey V, Dixit V, Shyam R. Chromium effect on ROS generation and detoxification in pea (Pisum sativum) leaf chloroplasts. 2009. *Protoplasma* 236: 85–95.

Poonam RK, Bhardwaj R, Sirhindi G. Castasterone regulated polyphenolic metabolism and photosyntheticsystem in *Brassica juncea* plants under copper stress. 2015. *Journal of Pharmacognosy and Phytochemistry* 4: 282–289.

Rejeb IB, Pastor V, Mani BM. Plant responses to simultaneous biotic and abiotic stress. 2014. *Molecular Plants* 3(4): 458–475.

Shahzad B, Tanveer M, Che Z, Rehman A, Cheema SA, Sharma A, Song H, Ur Rehman S, Zhaorong D. Role of 24-epibrassinolide (EBL) in mediating heavy metal and pesticide induced oxidative stress in plants: A review. 2018. *Ecotoxicology and Environmental Safety* 147: 935–944.

Sharma A, Kumar V, Thukral AK, Bhardwaj R. Epibrassinolide-imidaclopridinteraction enhancesnon-enzymatic antioxidants in Brassica juncea L. 2016. *Indian Journal of Plant Physiology* 21: 70–75.

Sharma A, Shahzad B, Rehman A, Bhardwaj R, Landi M, Zheng B. Response of phenylpropanoid pathway and the role of polyphenols in plants under abiotic stress. 2019. *Molecules* 24(13): 2452.

Trejo-Tapia G, Jimenez-Aparicio A, Rodriguez-Monroy M, De Jesus-Sanchez A, Gutierrez-Lopez G. Influence of cobalt and other microelements on the production of betalains and the growth of suspension cultures of Beta vulgaris. 2001. *Plant Cell, Tissue and Organ Culture* 67: 19–23.

Villiers F, Ducruix C, Hugouvieux V, Jarno N, Ezan E, Garin J, Junot C, Bourguignon J. Investigating the plant response to cadmium exposure by proteomic and metabolomic approaches. 2011. *Proteomics* 11: 1650–1663.

Williams RJ, Spencer JP, Rice-Evans C. Flavonoids: Antioxidants or signalling molecules. 2004. *Free Radical Biology and Medicine* 36: 838–849.

Wintz H, Fox T, Vulpe C. Responses of plants to iron, zinc and copper deficiencies. 2002. *Biochemical Society Transactions* 30(4): 766–768.

Zafari S, Sharifi M, Ahmadian Chashmi N, Mur LA. Modulation of Pb-induced stress in prosopis shoots through an interconnected network of signaling molecules, phenolic compounds and amino acids. 2016. *Plant Physiology and Biochemistry* 99: 11–20.

12 Aquatic Plant-Mediated Bioremediation Strategies

Falguni Barman, Shreya Nath, Swarnali Dey, Md. Tauhid Alam Ansari, Rita Kundu, Subhabrata Paul

12.1 INTRODUCTION

Water is one of the basic needs of all living beings. Industrialization, global population rise, urbanization, climatic change, utilization of natural resources has led to severe water scarcity worldwide. About one-fifth of the world's population is living in an area with physical water scarcity, and one-fourth of the world's population is going to face economic water scarcity (Chakraborti et al., 2019). The aquatic ecosystem becomes invaded by various types of waste, leading to water pollution, posing serious threats towards marine life and causing destruction of natural water support. Extensive usage of freshwater for industrial and agricultural purposes adds up to the already-increased water demand.

The aquatic floras are the primary producers of the food chain. So if the aquatic environment is contaminated by any toxic pollutant, such contaminants might be adsorbed or internalized alongside essential elements required by flora or fauna, resulting in bioaccumulation and biomagnification of the pollutant. Thus, remediation strategies become a necessary step for reclaiming the polluted water for further use. The various chemical and physical remediation processes available are either costly, time-constraining or have some types of drawbacks that prevent them from fulfilling the purpose of remediation.

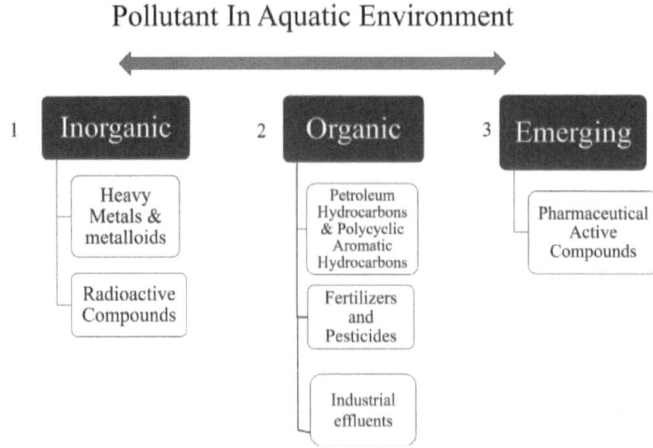

FIGURE 12.1 Major pollutants in the aquatic system.

 DOI: 10.1201/9781003297901-12

Phytoremediation is one such way of repurposing the polluted sources of water. Phytoremediation is inexpensive, requires less maintenance, eco-friendly, making it a better alternative to traditional remediation practices. Phytoremediation has been applied for the removal of heavy metals (Abdallah, 2012; Harguinteguy et al., 2015), radioactive wastes (Azmoonfar et al., 2019; Kawai et al., 2014) organic pollutes like PAH (Fathi et al., 2013), pesticides (Cáceres et al., 2008), azo compounds (Jinqi & Houtain, 1992), etc. from aquatic environments. In this chapter, we discuss the aquatic plant-based remediation strategies that help in amelioration of various toxic compounds from water bodies. The chapter mainly focuses on the types of pollutants, their sources and the plants that can help in their removal, particularly emphasizing on the mechanism of amelioration.

12.2 INORGANIC POLLUTANTS

12.2.1 Heavy Metals and Metalloids

Heavy metals (HMs) can be defined as metals with significantly high atomic weight and high density, at least five times more than that of water, whereas metalloids are a small group of elements possessing characteristics of both metals and non-metals (Tchounwou et al., 2012). HM/metalloid contamination or pollution can give rise to deadly diseases like Minamata, Itai-Itai, arsenic acid poisoning, etc. (Matsuo, 2003). The main source of such elements in aquatic ecosystems is anthropogenic activities, although geochemical events and electronic industries also contribute to some of them. Few HMs/metalloids and their sources in aquatic ecosystems are mentioned in Table 12.1.

HMs or metalloids become toxic to an organism when they fail to metabolize it, leading to accumulation in the body. Owing to the incessant release of HMs/metalloids

TABLE 12.1
Sources of different heavy metals in aquatic bodies

Sl. no.	Heavy metal	Source
1	Cadmium	Wastewater from industries for electroplating, batteries, TV sets, photography, metal finishing, insecticides, wastewater with cadmium containing phosphate fertilizer, detergents and refined petroleum products also water from metal-ore refining
2	Copper	Water source from metallurgy, mining and other industries
3	Zinc	Water source from metallurgy, mining
4	Nickel	Water source from the battery, aircraft part, cosmetic, coin and the stainless steel industry
5	Arsenic	Water source from mining and smelting, pesticide industry and also natural weathering of rocks
6	Mercury	Water source from the mining and battery industry
7	Chromium	Water source from plastic manufacturing, electroplating as well as weathering of rocks and volcanic eruption

in the environment, there lies a greater chance of entry of such elements into the food chain, culminating in bioaccumulation. As humans are situated at the highest peak of the food chain, it becomes most precarious for them. In the human body, arsenic poisoning can lead to respiratory and skin cancer, whereas cadmium is known to cause osteomalacia, proteinuria, etc. (Mahurpawar, 2015). As aquatic bodies become the repository of such toxic elements, it is necessary to employ remediation strategies to combat toxicity and hazardous effects of such contaminants. But most of the mitigation strategies come with several impediments (Carolin et al., 2017) (Table 12.2). Phytoremediation is the usage of plants to remove HMs/metalloids from aquatic bodies either through uptake and accumulation by roots or direct translocation and storage in aerial parts of the plants (Sharma et al., 2015). Phytoremediation is not only cheap but also an *in situ* recuperation of the environment in a sustainable way.

Various aquatic plants have been identified as potential candidates for HM/metalloid remediation from aquatic systems (Table 12.3). *Azolla pinnata*, used to extract Hg, Cd, Zi, Fe, resulted in a 30% decrease of such metals in wastewater (Rai, 2008). *Alternanthera bettzickiana* can accumulate Cd and Pb in the shoot and can be used as a potential candidate for phytoremediation (Tauqeer et al., 2016).

TABLE 12.2
Different strategies available for the removal of heavy metals/metalloids from aquatic bodies

Sl. no.	Name of the process	Description	Disadvantage
1	Coagulation/ flocculation	Increasing the density of particles of a suspension to settle them down and then separate them. In order to achieve bigger particles, sometimes agglomerating flocculants are being added.	Formation of sludge
2	Membrane filtration	Separates the particles with the help of a membrane based on size, pH, the concentration of the solution and applied pressure	Expensive and complex process
3	Adsorption	Waste can be transferred to the active site of the adsorbent	Desorption
4	Electrochemical treatment	Combination of several techniques which incorporates electrics like electrocoagulation, electrodeposition, electroflotation, electrodeionization	High investment, the requirement of high electric supply
5	Electrodialysis	Transfer of metal ions across membranes using electrics	Clogging
6	Ion exchange	Substitution of one metal ion for another	Limited removal
7	Photocatalysis	Both heavy metals and organic pollutants can be eliminated	Long process
8	Biological treatment	Algal biomass, fungal biomass, bacteria are used as bio-adsorbent to remove heavy metals	Underdeveloped
9	Oxidation	Removal of heavy metals without the use of electricity	Rusting in system

TABLE 12.3
Phytoremediation strategies for the removal of hms/metalloids

Sl. no.	Metal remediated	Nature of metal	Name of plant	Plant group	Brief mechanism of remediation	Reference
1.	Chromium (Cr)	HM	*Lemna minor* (Araceae) Natural colonization of algae	Angiosperm Algae	Uptake and accumulation of heavy metals from the wastewater of textile industry in plant system, with 94% removal rate; efficiency can be increased after 24 hours light and a pre-treatment step.	(Sekomo et al., 2012)
			Lemna gibba (Araceae) *Ceratophyllum demersum* (Ceratophyllaceae)	Angiosperm	Accumulation of heavy metals (at four concentrations: 2, 4, 10 and 15 mg/L) in the plant system without any detectable toxicity; range of removal percentages for chromium are 86.2–94.8% and 13.0–84.3% respectively.	(Abdallah, 2012)
			Landoltia Punctata (Araceae), *Elodea Canadensis* (*Hydrocharitaceae*), *Marsilea quadrifolia* (Marsileaceae)	Angiosperm Pteridophyte	Uptake and accumulation of heavy metals from the selenium-rich mining wastewater; removal efficiencies for chromium are 54.41%, 66.17 % and 15.92% respectively per gm of dry weight.	(Miranda et al., 2014)
2.	Lead (Pb)	HM	*Myriophyllum aquaticum* (Haloragaceae) *Egeria densa* (Hydrocharitaceae)	Angiosperms	The accumulation capacity of *M aquaticum* for lead was higher than that of *Egeria densa* at a concentration of 15 mg/L.	(Harguinteguy et al., 2015)
			Lemna minor (Lemnaceae) *Elodea Canadensis* (Hydrocharitaceae) *Leptodictyum riparium* (Amblystegiaceae)	Angiosperms Moss	All three plant species can accumulate and withstand the toxic effects of lead (plants grown in the presence of 10^{-3}, 10^{-4}, 10^{-5}, 10^{-6}, 10–^{7}M Pb. Among them, *L. riparium* was the most resistant and most efficient in accumulating lead. Lead treatment lead to swollen chloroplast and changes in the arrangement of thylakoids.	(Basile et al., 2012)
			Scirpus grossus (Cyperaceae)	Angiosperm	Plants were grown in 600L water spiked with three concentrations of lead—10, 30 and 50 mg/L. After 28 days, the lead concentration in water bodies was reduced to 100%, 99.9% and 99.7% respectively.	(Tangahu et al., 2013)

(Continued)

TABLE 12.3 (*Continued*)
Phytoremediation strategies for the removal of hms/metalloids

Sl. no.	Metal remediated	Nature of metal	Name of plant	Plant group	Brief mechanism of remediation	Reference
3	Pb(II)	HM	*Pistia stratiotes* (Araceae)	Angiosperms	Only 0.125 gm of biomass was efficient to accumulate 96% of lead from 25L water body within one hour through biosorption.	(Volf et al., 2015)
4	Iron (Fe)	HM	*Limnocharis flava*, (Alismataceae) *Thalia geniculata* (Marantaceae) *Typha latifolia* (Typhaceae)	Angiosperm	The highest accumulation was found in *Thalia geniculata* (570.8 mg/kg) after being grown in contaminated wastewater for six months.	(Anning et al., 2013)
5	Mercury (Hg)	HM	*Lemna minor* (Araceae)	Angiosperm	Removal of mercury occurred by uptake and accumulation; the percentage of removal was 30% after growing the plant in contaminated water (0.1mg/L) for 21 days.	(Parra et al., 2012)
			Elodea canadensis, (*Hydrocharitaceae*) *Myriophyllum spicatum*, (*Haloragaceae*) *Ceratophyllum demersum* (*Ceratophyllaceae*) *Eichhornia crassipes* (*Pontederiaceae*)	Angiosperms	The submerged plants showed 13 times more accumulation of mercury than the natant plant; the highest bioaccumulation was found in *Ceratophyllum demersum.*	(Pelcová et al., 2022)
			Eichhornia crassipes (*Pontederiaceae*) *Pistia stratiotes* (Araceae) *Scirpus tabernaemontani* (Cyperaceae) *Colocasia esculenta* (Araceae)	Angiosperms	All of them can reduce mercury in water (water spiked with 0.5 and 2 mg/L mercury) after growing for 30 days by root uptake and accumulation. However, *Eichhornia crassipes* and *Pistia stratiotes* were more effective than the other two.	(Skinner et al., 2007)

6	Molybdenum (Mo)	Transition Metal	*Eichhornia crassipes* (*Pontederiaceae*)	Angiosperms	The plant could efficiently uptake the heavy metal and was able to translocate it to the shoots and leaves. Bioconcentration factor for Mo was 24,360. Though the distance of tailing increased, the amount of Mo accumulation in the plants was not affected by the concentration of heavy metal in the plant.	(Romanova et al., 2016)
			Eleocharis acicularis (Cyperaceae)	Angiosperms	The plants could accumulate 18 mg Mo per kg of dry weight when they were allowed to grow in Mo contaminated water for 3 months. However, the accumulation rate decreased with time.	(Yamazaki et al., 2017)
7	Cobalt (Co)	HM	*Landoltia Punctata* (Araceae), *Elodea Canadensis* (*Hydrocharitaceae*), *Marsilea quadrifolia* (Marsileaceae)	Angiosperms Pteridophyte	Uptake and accumulation of heavy metals from the selenium-rich mining wastewater; removal efficiencies for cobalt were 86.72%, 20.37 % and 39.78% respectively per gm of dry weight.	(Miranda et al., 2014)
			Pistia stratiotes (Araceae)	Angiosperms	Plants could effectively uptake cobalt and deposit it on the root surface. However, it was not a hyperaccumulator for cobalt.	(Lu et al., 2011)
8	Selenium (Se)	HM	*Landoltia punctata* (Araceae), *Elodea canadensis* (*Hydrocharitaceae*), *Marsilea quadrifolia* (*Marsileaceae*)	Angiosperms Pteridophyte	Uptake and accumulation of heavy metals from the selenium-rich mining wastewater (SMW); removal efficiencies for total selenium were 12.52%, 18.74 % and 23.11% respectively per gm of dry weight from 100% SMW.	(Miranda et al., 2014)

(Continued)

TABLE 12.3 (*Continued*)
Phytoremediation strategies for the removal of hms/metalloids

Sl. no.	Metal remediated	Nature of metal	Name of plant	Plant group	Brief mechanism of remediation	Reference
	Arsenic (As)	HM	*Micranthemumum brosum* (Linderniaceae)	Angiosperm	After growing the plants in water containing 02, 0.45 and 1.0μg/mL arsenic for seven days, the removal efficiencies were found to be 80.48, 89.56, 79.3 respectively. In all three concentrations, the plants can efficiently translocate the heavy metal from root to shoot and further to the leaf, making it a hyperaccumulator for arsenic.	(Islam et al., 2013)
			Vallisneria natans (Hydrocharitaceae) *Potamageton crispus* (Potamogetonaceae) *Myriophyllum spicatum* (Haloragaceae) **Ceratophyllum demersum (Ceratophyllaceae) Hydrilla verticillata** (Hydrocharitaceae)	Angiosperms	*Vallisneria natans* could accumulate arsenic after exposure to 5mg/L As (V) to the highest degree (~1700mg/kg of dry weight). Inside the plant As (V) was reduced to As (III). The plant health and biomass remained unaltered even in the presence of arsenic due to the activity of non-enzymatic antioxidants which further helped in arsenic detoxification in the plant system.	(G. Chen et al., 2015)
10	Nickel (Ni)	HM	*Salvinia minima* (Salviniaceae)	Angiosperms	Plants were able to grow in water having a low concentration of heavy metal (0.020 mg/L), observed via an increase in leaf area and growth area. However, at a higher concentration (0.40 mg/L), plant can efficiently uptake and accumulate the heavy metal (BCF 900 after 60 days), though its growth was ceased.	(Iha & Bianchini, 2015)
			Phragmite scummunis (Poaceae) *Typha angustifolia* (Typhaceae)	Angiosperms	*Phragmites cummunis* was a shoot accumulator, whereas *Typhaangusti folia* was found to be a root accumulator of the heavy metal. However, between them, *Phragmites cummunis* was able to accumulate nickel more than *Typhaangusti folia.*	(Chandra & Yadav, 2011)

11	Copper (Cu)	HM	*Trapa natans* (Lythraceae)	Angiosperm	The plant can accumulate cadmium in the root, shoot and fruit (15.626, 12.3 and 9.46 µg copper per gm of dry weights of the tissue) after growing in contaminated water (0.226mg/L).	(Sweta et al., 2015)
	Cadmium (Cd)	HM	*Trapa natans* (Lythraceae)	Angiosperm	The plant can accumulate cadmium in the root, shoot and fruit (2.656, 2.081 and 1.189 µg cadmium per gm of dry weights of the tissue) after growing in contaminated water (0.015mg/L).	
			Elodea Canadensis (*Hydrocharitaceae*) *Salvinia natans* (Salviniaceae) *Lemna minor* (Araceae)	Angiosperm	In the case of both monometallic treatment (only Cd at 65µM/L) and multi-metallic treatment (Cd+Cu+Zn: 4.10 mg/L Cu, 4.30 mg/L Zn and 7.30 mg/L Cd), *Lemna minor* showed the highest removal efficiencies for heavy metals among three plants. The phytoremediation property of the plant could be attributed to the increased level of ascorbate and phytochelatin for detoxification.	(Török et al., 2015)
			Lemna minor (Araceae) *Spirodela Polyrhiza* (Araceae)	Angiosperm	After growing both plants in different concentrations of Cd (0.5, 1.0, 1.5, 2.0, 2.5 and 3.0 mg/L) for 22 days, it was observed that *Spirodela polyrhiza* (52–75%) can remove Cd more or less equally to *Lemna minor* (42–78%). But *Lemna minor* could tolerate Cd level up to 2mg/L, which in the case of *Spirodela polyrhiza* was 3mg/L.	(Chaudhuri et al., 2014)
	Zinc (Zn)	HM	*Lemna minor* (Lemnaceae) *Elodea canadensis* (Hydrocharitaceae) *Leptodictyum riparium* (Amblystegiaceae)	Angiosperm Moss	All plant species accumulate zinc. *L. riparium* was most resistant and efficient in accumulating zinc. Zinc treatment damaged chloroplast ultra-structure and cytoplasmic vesicles.	(Basile et al., 2012)
14	Manganese (Mn)	HM	*Glyceria maxima* (Poaceae) *Phragmites australis* (Poaceae) *Typha latifolia* (Tyhaceae) *Phalaris arundinacea* (Poaceae)	Angiosperm	Accumulates Mn. *Phragmites australis* and *Typha latifolia* mainly accumulate it in leaves. *Glyceria maxima* showed highest degree of sequestration.	(Fletcher et al., 2020)

12.2.2 Radioactive Wastes

Radioactive wastes (RW) are the wastes that are produced due to the presence of radioactive-emitting radionuclide in any environment. RW are persistent, making it a major concern for living organisms (International Atomic Energy Agency, 2009). RWs are classified based on the duration of radiation they emit. The main sources of RW in the aquatic environment are nuclear power, plants, which are generally situated near coastal land due to the large requirement of water. Other sources include mining of uranium, thorium, medical equipment like X-rays, MRI, etc. The use of a radioactive weapon or nuclear submarine can also serve as a source of RW. This inclusion of radionuclide is very detrimental for the aquatic ecosystem as well as for human beings. RW polluted water affects global fish migration, human health and ecological balance (Chow, 2011).

Phytoremediation plays a crucial role in RW removal by transforming them into less toxic forms or depositing them in different parts of the plant without inducing any toxicity. Many microorganisms live in a mutualistic relationship with the plant, and this allows the microorganism to further detoxify the radioactive waste in the water bodies. Advantages of phytoremediation include sustainability, effectiveness against low-level waste and non-destructive nature. Besides, this technique is very cost-effective and requires a low amount of resources (De Filippis, 2015). Various plants and their role in RW removal from aquatic bodies have been established and are summarized in Table 12.4.

Low level waste	Intermediate level waste	High level waste
• Produce short life radiation by self decaying property.	• Contains more radioactive material than low level waste.	• Produce very high radioactivity along with heat due to more decaying of radionuclide
• Example- rags, papers, clothing	• Example- sludge, metal, nuclear fuel	• Example- Iodine (I-131), Plutonium, Uranium, Cesium (Cs-137)

FIGURE 12.2 Types of radioactive wastes based on disposal time.

Source: International Atomic Energy Agency, 2009

TABLE 12.4
Phytoremediation strategies for radioactive waste removal from aquatic bodies

Sl. no.	Name of the RW	Name of the plant	Type of plant	Brief mechanism	Reference
1	^{60}Co and I-131	*Dunaliella salina* (Dunaliellaceae) *Chlorella vulgaris* (Chlorellaceae)	Microalgae	These species of algae remediate water-soluble radionuclides by absorbing and bioaccumulating them within the cell.	(Azmoonfar et al., 2019)
2	^{137}Cs	*Desmodesmus armatus* (Scenedesmaceae)			(Kim et al., 2019)
3	Sr-90 and Cs-137	*Eichhornia crassipes* (Pontederiaceae)	Angiosperm	Adsorb and accumulate RW present in contaminated water bodies within the cells of root and shoot and reduce them to less toxic form.	(Saleh, 2012)
4	^{137}Cs, ^{134}Cs and ^{238}Pu	*Ulva pertusa* (Ulvaceae) *Saccharina religiosa* (Laminariaceae) *Sargassum horneri* (*Sargassaceae*)	Seaweeds or macroalgae	Since seaweeds depend on seawater for the uptake of nutrients, due to contamination of seawater with Cs^+ (soluble), they uptake radionuclide by K^+ transporting channel due to structural similarity.	(Kawai et al., 2014)
5	Uranium	*Azolla imbricate* (Salviniaceae)	Pteridophyte	Azolla roots help in rhizofilteration to absorb and precipitate the radionuclides of uranium from wastewater within the root system.	(Sood et al., 2012)
		Hydrilla verticillata (Hydrocharitacae)	Angiosperm	Radioactive uranium mimics the nutrients like Ni, Fe, Ca and Mg for entry into the plant cell by divalent ion transporter and, due to nutrient deficiency, accumulation and reduction of uranium, take place within the cells.	(Srivastava & Bhainsa, 2016)

Since phytoremediation is an effective technique for RW removal, genetic engineering advancements are being applied to enhance the capability of plants for remediation. Desired genes can be inserted into the plant of interest through recombinant DNA technology and synthesize enzymes required for uptake, translocation, sequestration, detoxification, the transformation of radionuclide, due to which the plant posses higher accumulating capacity (Mello-Farias et al., 2011). This transgenic

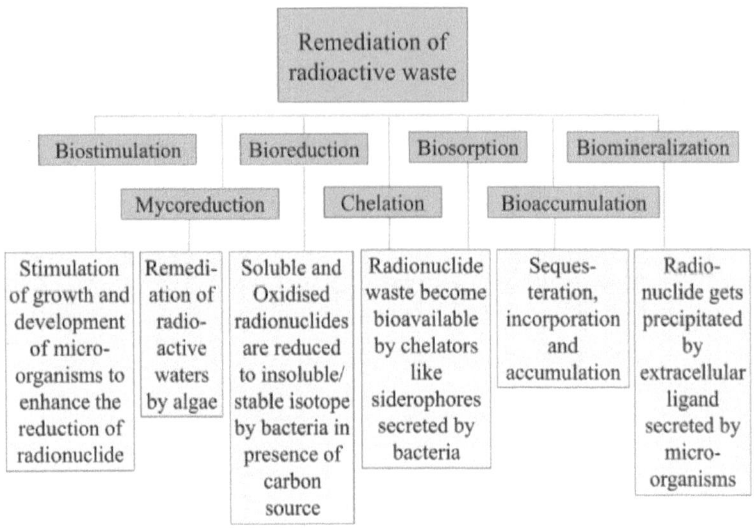

FIGURE 12.3 Different modes of bioremediation of radioactive wastes.

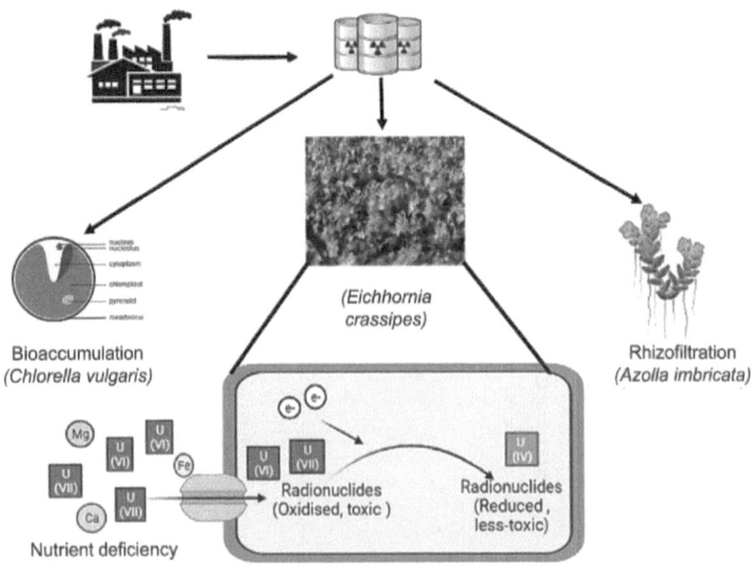

FIGURE 12.4 Different modes of phytoremediation of radioactive wastes.

approach of inserting ATP sulfurylase, glutathione synthetase or *γ-ECS* in *Brassica juncea* leads to the accumulation of selenium more than its wild type (Doty, 2008). Another aspect is the selection of hyperaccumulator plants that can accumulate and tolerate more concentrations of radionuclide than conventional phytoremediation. *P. australis* hyperaccumulates uranium and thorium within tissues and detoxifies them through root sequestration (Singh et al., 2022).

12.3 ORGANIC POLLUTANTS

The emergence of organic pollutants in the environment affects the stability of both terrestrial and aquatic ecosystems, which further leads to human health hazards. Organic pollutants include petroleum hydrocarbons, pesticides and tributyltin (TBT) that are deposited in the aquatic systems as industrial effluents, household waste and agricultural run-offs. Organic waste derived from industrial activities and agricultural run-offs are highly toxic and termed as persistent organic pollutants (POP). As mentioned by Stockholm Convention 12, chemicals that are listed under POPs include pesticides (aldrin, chlordane, DDT, dieldrin, endrin, heptachlor, hexachlorobenzene, mirex, toxaphene), industrial chemicals (hexachlorobenzene, polychlorinated biphenyls, that is, PCBs) and several by-products (hexachlorobenzene; polychlorinated dibenzo-*p*-dioxins and polychlorinated dibenzofurans, that is, PCDD/PCDF). Characteristically, these POPs are highly lipophilic, thus accumulating in animal fatty tissues through the food chain, leading to potential human health risks including cancer, genetic defects, weakening of the endocrine and immune system, and also threaten aquatic life forms. Under aerobic conditions, organic pollutants can degrade most rapidly. Figure 12.5 reveals the main principle of aerobic degradation of hydrocarbons (Das & Chandran, 2011). Phytoremediation is a well-established method that can be used for effective removal of organic pollutants from water bodies.

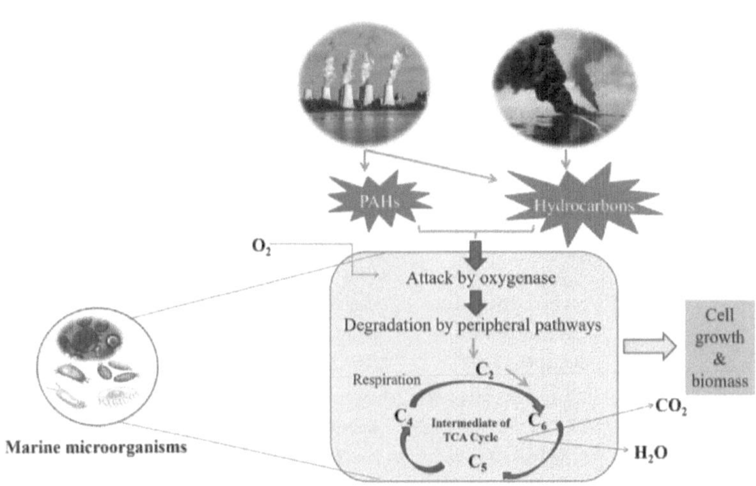

FIGURE 12.5 Main principle of aerobic degradation of hydrocarbons and PAHs by microorganisms.

12.3.1 Petroleum Hydrocarbons and Polycyclic Aromatic Hydrocarbons (PAHs)

The major energy source in industry and daily life are petroleum-based products. Leakage and accidental spills are common incidents during the exploration, production, refining, transport and storage of petroleum and petroleum products, serving as main sources of release of such hydrocarbons into the environment, resulting in soil and water pollution (Holliger et al., 1997). Oil spills have prolonged adverse effects on marine life. The key component of petroleum sewage is polycyclic aromatic hydrocarbons (PAHs) which are joined by two or more aromatic rings arranged in an angular, linear or clustered manner. It is found mainly in crude oil and generated through combustion of fuel, incineration of waste materials or petroleum exploration. PAHs are the key pollutant in seawater due to their water solubility, toxicity, bioaccumulation potential and recalcitrant nature. Industrial PAH emission takes place through the cement industry, rubber tire and coal-tar pitch fabrication, manufacture of iron, steel, aluminum and the waste materials of dye, asphalt, fungicide and insecticide industry; whereas coal coking, burning of garbage and wood, gas burners, kerosene stoves, etc. are the common source of domestic PAHs. Burning of plant parts, crop residue, and waste material are the source of agricultural PAHs. Although more than 400 PAHs have been reported, based on toxicity and abundance, the U.S. Environmental Protection Agency (EPA) has enlisted 16 of them as primary pollutants (Table 12.5).

TABLE 12.5
Common PAHs as considered by U.S. Environmental protection agency (EPA)

Sl. no.	Name	Common sources	Toxic effects on humans, animals and plants	References
1	Naphthalene	Common air pollutant from metal industry, gasoline, biomass burning, oil combustion, tobacco smoking.	Toxicity was first reported on *Tribolium castaneum* which triggered genes associated with oxidative stress, metabolism and reproduction. Causes respiratory despair, haemolysis, renal failure, methemoglobineia and brain dysfunction.	(Pannu & Singla, 2020)
2	Acenaphthene	Found in crude oil, coal tar, car exhaust, cigarette smoke and some foods.	Long-term exposure in animals can shrink the weight of the adrenal glands and increase liver size.	(Mojiri et al., 2019)
3	Acenaphthylene	Indigenous component of coal tar, used in the synthesis of dyes, plastics, insecticides and fungicides.	Not considered as a carcinogen but irritates the skin and mucous membrane.	(Cairns & Nebeker, 1982)

TABLE 12.5 (*Continued*)
Common PAHs as considered by U.S. Environmental protection agency (EPA)

Sl. no.	Name	Common sources	Toxic effects on humans, animals and plants	References
4	Anthracene	Found in coal tar, cigarette smoke, engines exhaust, smoked foods, drinking water and edible aquatic organisms.	Reduces growth and photosynthesis of phytoplanktons and toxic for other aquatic organisms.	(Kottuparambil & Park, 2019)
5	Phenanthrene	Coal tar, diesel fuel exhaust, and contaminated marine water.	Hampers different metabolic processes in the life cycle of aquatic organisms like growth, development and reproduction of free-living nematode (*Diplolaimelloidesdelyi*).	(Pontes et al., 2021)
6	Fluorene	Mostly found in the vapour phase of various PAH emissions.	Irritates and burns the eyes and skin. In plants, it induces various oxidative stresses.	(Lisar et al., 2015)
7	Fluoranthene	Fluoranthene is extensively disseminated in marine and soil environments. Mostly comes from organic raw materials like coal and petroleum.	Induces various oxidative stresses.	(Sun et al., 2020)
8	Benzo[a] anthracene (tetraphene)	Coal tar and pitch, diesel exhaust, coal combustion emissions, tobacco smoke, broiled foods using charcoal, wood soot smoke, creosote, asphalt and mineral oils.	Human carcinogenic and animal studies have been proved it shows genotoxicity, micro-nuclear breakage, lung, skin and liver tumour.	(Partila & Mohammed, 2019)
9	Chrysene	Highly persistent PAH in seawater.	Chrysene converted into 2- and 6-hydroxychrysene (OHCHR), which initiate and provoke the estrogen receptor (ER) in zebrafish.	(Diamante et al., 2017)
10	Pyrene	Found in aquatic bodies.	Neurotoxicity, membrane peroxidation of aquatic organisms.	(Oliveira et al., 2012)

(Continued)

TABLE 12.5 ***(Continued)***
Common PAHs as considered by U.S. Environmental protection agency (EPA)

Sl. no.	Name	Common sources	Toxic effects on humans, animals and plants	References
11	Benzo[a]pyrene	Benzo(a)pyrene is generally found in high concentrations in petroleum products, tobacco smoke, charbroiled foods, and polluted water.	Increases cell apoptosis and shrink migration as a result of dysfunction of trophoblast and, ultimately, pregnancy loss.	(Ye et al., 2020)
12	Benzo[b] fluoranthene	Derived from gasoline exhaust, tobacco and cigarette smoke, coal tar also comes from amino and fatty acid pyrolysis.	Carcinogen, potent immunotoxin. Induces glomerular podocytes injury in mouse.	(Guo et al., 2020)
13	Benzo[k] fluoranthene	Combustion of organic matter at high temperature; most common natural source is a forest fire and different anthropogenic sources.	One, 10, 25 and 50 μg/L of B (k) F for 96h showed a significant increment of SOD and MDA level that ultimately led to genotoxicity of marine gastropod.	(Bhagat et al., 2017)
14	Dibenz[a,h] anthracene	Gasoline exhaust, tobacco and cigarette smoke, coal tar.	Interacts with receptorspresent in the ES-D3cells.	(Kamelia et al., 2020)
15	Benzo[ghi] perylene	Crude oil, coal and also found in incomplete combustion.	Its exposure to the bronchial cell line causes oxidative stress, morphological alterations in the cytoplasm and activation of the Ah receptor (AhR).	(Zaragoza-Ojeda et al., 2016)
16	Indeno[1,2,3-cd] pyrene	Fossil fuel and incomplete combustion	Enhances allergic lung inflammation in mice.	(Wong et al., 2018)

Emission of PAHs in the environment and the subsequent bioaccumulation in humans has led to drastic effects in lungs, kidneys, liver, and skin; reduces immunity and increases the chance of carcinogenicity and mutagenicity. Thermal and photolytic remediation strategies have been used for PAHs removal but are limited due to partial removal. Therefore, bioremediation through aquatic microorganisms stands as an eco-friendly and communally adequate substitute to the conservative methods.

Microalgae (primary producers) can degrade hydrocarbons through carbon fixation. The notable hydrocarbon degrading microalgae are *Chlorella vulgaris*, *Scenedesmus acutus*, *Selenastrum capricornutum*, *Dunaliella* sp., *Nitzschia* sp., *Euglena gracilis*, *Chlamydomonas* sp. and *Scenedesmus obliquus*. Under photoautotrophic conditions, these algal species can oxidize PAHs to form hydroxylated intermediates (Mueller et al., 1996). Especially cyanobacteria have been reported to oxidize the two most common PAHs like naphthalene and phenanthrene (Yadav et al., 2020). Since past few years, the utility of microalgae has been considered for biodegradation and bioremediation of petroleum-related contaminants and wastewater treatment respectively. Simultaneously, the algal biomass is used as the raw materials of biofuel production (Huang et al., 2010). Micro algal chlorophyll catalyzes the photo-transformation of mutagenic and carcinogenic benzo[a]pyrene (BaP) to less toxic compounds (Luo et al., 2015). In aquatic environment, algae and heterotrophic bacteria can coexist and synergistically affect one another's physiological and metabolic processes. Bacterium (*Azosprillum brasilense*)-derived indole-3-acetic acid and cobalamin (vit. B_{12}) were used as a growth-promoting factor by algae *(Chlamydomonas reinhardtii)* in algae bacteria consortia culture (Peng et al., 2020). As a result, algae and bacteria consortia gain the priority for bioremediation of PAHs. Due to poor water solubility and less availability of activated O_2, the degradation of PAHs is quite difficult because activated O_2 is required for the initial breakdown of PAHs. This problem is solved by using algae bacteria consortia, as algae generate O_2 which can be used by bacteria to degrade the PAH (Abdallah, 2012). In return, algae can use CO_2 released by bacterial respiration. The major sources of oil are crude oil released from tankers, offshore platforms, drilling rigs, heavier fuels used by large ships such as bunker fuel and refined petroleum products and their by-products (H. Chen et al., 2015). To cope with toxicity resulting from oil spills, several clean-up techniques have been adopted recently. One of the major areas of research is utilization of several algal species for wastewater treatment. Previously, researchers had shown that algal species such as Botryococcus, Chlamydomonas, Chlorella and Phormidium were capable of *phycoremediation*. They can oxidize and degrade toxic hydrocarbons into less harmful components (El-Sheekh et al., 2013). The mixotrophic algae are excellent agents for bioremediation and carbon sequestration (Subashchandrabose et al., 2011). By photosynthesis, algae can generate O_2 and fix CO_2, which ultimately increases the BOD level in the polluted marine water and removes excess nutrients (Fathi et al., 2013).

12.3.2 Pesticides and Fertilizers

The main reason behind pesticide and fertilizer contamination in the aquatic environment is the agricultural run-off (Topp et al., 2008). Besides this, several agricultural activities like silage effluents and processing wastes from plantation crops and veterinary pharmaceuticals (especially antibiotics) used in animal farming (such as cattle farming) are also important contributors to aquatic pollution. Focusing on pesticides, the annually estimated pesticide active ingredients are 3 million metric tons. These are extensively applied to croplands, and the rate of consumption has increased from 15 to 20 fold in only 30 years (Popp et al., 2013). In this way, the entrance of toxic substances leads to subsequent potential risks for soil and water ecosystems, and eventually for human health.

For the treatment of wastewater or effluents from agrochemical industries, the application of a low-cost algal-based system has been recommended (Avila et al., 2021). Microalgae can assimilate different types of pesticide on the basis of lipid content, strain and the chemical structure of the pesticide through the process of biosorption and bioaccumulation. Moreover, microalgae can utilize pesticides and cyanide to fulfil their carbon and nitrogen requirement (Sutherland & Ralph, 2019). The combination of *Chlorella vulgaris* and biosurfactants is an acceptable approach for nutrient removal from wastewaters (Madadi et al., 2016). Diazinon is a highly poisonous insecticide and can be digested by *Chlorella vulgaris* and transformed into the less toxic 2-isopropyl-6-methyl-4-pyrimidinol (IMP) metabolite (Kurade et al., 2016). It is also reported that microalgae *Scenedesmus* sp. and *Chlorococcum* sp. can convert endosulfan (a cyclodiene pesticide) to endosulfan sulfate (Sethunathan et al., 2004). Blue green algae were also found to be capable of degradation of the toxic pesticide known as fenamiphos (Cáceres et al., 2008).

12.3.3 Other Organic Compounds

A frequently released industrial chemical is dimethyl phthalate (DMP), which is widely used in construction, automobile, chemical engineering, etc. and has been accumulated into the environment at a growing rate throughout the globe. Due to its large quantity and widespread distribution, DMP is known as a ubiquitous environmental pollutant that has a toxic effect on aquatic organisms. It was found that microalgae are key players for the biodegradation of phthalate ester. Jinqi and Houtain (1992) reported that algal members like *Chlorella pyrenoidosa*, *Chlorella vulgaris* and *Oscillateria tenuis* were capable of degrading more than 30 azo compounds into simpler aromatic amine by breaking the azo linkage. Phenolic compound is another type of organic compound which can be degraded by cyanobacteria. Shashirekha et al. (1997) reported that polyphenol oxidase and laccase enzymes were first isolated from cyanobacterium *Phormidium valderianum*. Klekner and Kosaric (1992) reported that *Spirulina maxima* could degrade a variety of phenolic compounds. The freshwater blue green algae *Anabaena cyalindrica* and *Phormidum foveolarum* were the first algal species reported as capable of bio-remediating phenolic compounds without production of metabolites (Ellis, 1977).

12.4 EMERGING CONTAMINANTS (ECs)

Emerging contaminants (ECs) are pharmaceuticals, personal care products (PCPs), nanomaterials and perfluorinated compounds that have a noxious effect on human health and the environment (Kurade et al., 2021). These ECs pose risks including endocrine disruption, chronic ecotoxicity and antibiotic resistance (Arnold et al., 2013). To reduce these ECs, large-molecular-weight ECs are converted into smaller ones by microbial degradation. Among several plants that have been reported till date, *Ipomoea aquatic* has been effective in the removal of antibiotics of Sulfamethoxazole and Tetracyclin group (H.-R. Chen et al., 2017), *Lactuca sativa* has been effective in removal of various drugs like bisphenols, caffeine compounds, propranolol, tonalide, etc. (Hurtado et al., 2016) and *Echinodorus sp.* has been helpful in the removal of antifungal and antibacterial agents like triclosan (Pi et al., 2017). With time, new contaminants are being identified that open the room for future investigations to find suitable candidates for phytoremediation.

12.5 CONCLUSION

Phytoremediation, alternatively known as green remediation or agro-remediation, is the removal of toxic pollutants from contaminated soil or water by non-edible plants. Phytoremediation is now preferred over conventional remediation strategies, owing to its cost-effectiveness, sustainability, easy implementation and maintenance. However, it faces some impediments like dependency of plant growth on environmental factors like climate, rainfall, temperature or other abiotic and biotic factors, prolonged time requirement for absolute remediation, variable result and lack of information about the ultimate fate of the contaminant (Kamusoko & Jingura, 2017). Phytoremediation is a propitious technique and an innovative tool for detoxification of water bodies. With advancement and amalgamation of science and technology, the natural phytoremediation strategy can be strengthened by identification of more species and implication of transgenic approaches in order to upscale and expedite the remediation process. This shall be beneficial for rendering the aquatic ecosystem free of pollutants in an environment-friendly yet effective technique.

ACKNOWLEDGEMENTS

FB acknowledges UGC for his fellowship. SN (DBT-SRF) and SD (CSIR SPM) are grateful to DBT and CSIR respectively for providing them fellowship. The authors are grateful to Presidency University and University of Calcutta for the institutional support.

REFERENCES

Abdallah, M. A. M. (2012). Phytoremediation of heavy metals from aqueous solutions by two aquatic macrophytes, Ceratophyllum demersum and Lemna gibba L. *Environmental Technology*, *33*(14), 1609–1614.

Anning, A. K., Korsah, P. E., & Addo-Fordjour, P. (2013). Phytoremediation of wastewater with Limnocharis flava, Thalia geniculata and Typha latifolia in constructed wetlands. *International Journal of Phytoremediation*, *15*(5), 452–464.

Arnold, K. E., Boxall, A. B. A., Brown, A. R., Cuthbert, R. J., Gaw, S., Hutchinson, T. H., Jobling, S., Madden, J. C., Metcalfe, C. D., & Naidoo, V. (2013). *Assessing the exposure risk and impacts of pharmaceuticals in the environment on individuals and ecosystems*. The Royal Society.

Avila, R., Peris, A., Eljarrat, E., Vicent, T., & Blánquez, P. (2021). Biodegradation of hydrophobic pesticides by microalgae: Transformation products and impact on algae biochemical methane potential. *Science of the Total Environment*, *754*, 142114.

Azmoonfar, R., Najafi, M., Rezaeyan, A., Khodamoradi, E., & Rasouli, N. (2019). Adsorption of radioactive materials by green microalgae dunaliella salina from aqueous solution. *Iranian Journal of Medical Physics*, *16*(6), 392–396.

Basile, A., Sorbo, S., Conte, B., Cobianchi, R. C., Trinchella, F., Capasso, C., & Carginale, V. (2012). Toxicity, accumulation, and removal of heavy metals by three aquatic macrophytes. *International Journal of Phytoremediation*, *14*(4), 374–387.

Bhagat, J., Ingole, B. S., & Shyama, S. K. (2017). Effects of benzo(k)fluoranthene, a polycyclic aromatic hydrocarbon on DNA damage, lipid peroxidation and oxidative stress in marine gastropod Morula granulata. *Chemistry and Ecology*, *33*(9), 869–882. https://doi.org/10.1080/02757540.2017.1384470

Cáceres, T. P., Megharaj, M., & Naidu, R. (2008). Biodegradation of the pesticide fenamiphos by ten different species of green algae and cyanobacteria. *Current Microbiology*, *57*, 643–646.

Cairns, M. A., & Nebeker, A. V. (1982). Toxicity of acenaphthene and isophorone to early life stages of fathead minnows. *Archives of Environmental Contamination and Toxicology*, *11*(6), 703–707. https://doi.org/10.1007/BF01059158

Carolin, C., Kumar, P., Saravanan, A., Janet Joshiba, G., & Naushad, M. (2017). *Efficient techniques for the removal of toxic heavy metals from aquatic environment: A review*. Elsevier. Retrieved February 8, 2023, from www.sciencedirect.com/science/article/pii/S2213343717302208

Chakraborti, R. K., Kaur, J., & Kaur, H. (2019). Water shortage challenges and a way forward in India. *Journal—American Water Works Association*, *111*(5), 42–49. https://doi.org/10.1002/awwa.1289

Chandra, R., & Yadav, S. (2011). Phytoremediation of Cd, Cr, Cu, Mn, Fe, Ni, Pb and Zn from aqueous solution using Phragmites cummunis, Typha angustifolia and Cyperus esculentus. *International Journal of Phytoremediation*, *13*(6), 580–591. https://doi.org/10.1080/15226514.2010.495258

Chaudhuri, D., Majumder, A., Misra, A. K., & Bandyopadhyay, K. (2014). Cadmium removal by Lemna minor and Spirodela polyrhiza. *International Journal of Phytoremediation*, *16*(7–12), 1119–1132. https://doi.org/10.1080/15226514.2013.821446

Chen, G., Liu, X., Brookes, P. C., & Xu, J. (2015). Opportunities for phytoremediation and bioindication of arsenic contaminated water using a submerged aquatic plant: Vallisneria natans (lour.) hara. *International Journal of Phytoremediation*, *17*(3), 249–255. https://doi.org/10.1080/15226514.2014.883496

Chen, H., An, W., You, Y., Lei, F., Zhao, Y., & Li, J. (2015). Numerical study of underwater fate of oil spilled from deepwater blowout. *Ocean Engineering*, *110*, 227–243.

Chen, H.-R., Rairat, T., Loh, S.-H., Wu, Y.-C., Vickroy, T. W., & Chou, C.-C. (2017). Assessment of veterinary drugs in plants using pharmacokinetic approaches: The absorption, distribution and elimination of tetracycline and sulfamethoxazole in ephemeral vegetables. *PLoS One*, *12*(8), e0183087.

Chow, S. (2011). *Radioactive contamination and the food chain*. Food Safety Focus (58th Issue, May 2011) – Incident in Focus.

Das, N., & Chandran, P. (2011). Microbial degradation of petroleum hydrocarbon contaminants: An overview. *Biotechnology Research International*, *2011*.

De Filippis, L. F. (2015). Role of phytoremediation in radioactive waste treatment. *Soil Remediation and Plants: Prospects and Challenges*, 207–225.

Diamante, G., do Amaral E Silva Müller, G., Menjivar-Cervantes, N., Xu, E. G., Volz, D. C., Dias Bainy, A. C., & Schlenk, D. (2017). Developmental toxicity of hydroxylated chrysene metabolites in zebrafish embryos. *Aquatic Toxicology (Amsterdam, Netherlands)*, *189*, 77–86. https://doi.org/10.1016/j.aquatox.2017.05.013

Doty, S. L. (2008). Enhancing phytoremediation through the use of transgenics and endophytes. *New Phytologist*, *179*(2), 318–333.

Ellis, B. E. (1977). Degradation of phenolic compounds by fresh-water algae. *Plant Science Letters*, *8*(3), 213–216.

El-Sheekh, M. M., Hamouda, R. A., & Nizam, A. A. (2013). Biodegradation of crude oil by Scenedesmus obliquus and Chlorella vulgaris growing under heterotrophic conditions. *International Biodeterioration & Biodegradation*, *82*, 67–72.

Fathi, A., Mahgoub Azooz, M., Fathi, A. A., Azooz, M. M., & Al-Fredan, M. A. (2013). Phycoremediation and the potential of sustainable algal biofuel production using wastewater. *Researchgate.Net*, *10*(2), 189–194. https://doi.org/10.3844/ajassp.2013.189.194

Fletcher, J., Willby, N., Oliver, D. M., & Quilliam, R. S. (2020). *Phytoremediation using aquatic plants* (Issue April). Springer International Publishing. https://doi.org/10.1007/978-3-030-00099-8_7

Guo, J., Zhang, L., Li, C., Qin, Y., Liu, W., Li, J., & Huang, G. (2020). Benzo[b]fluoranthene impairs mouse oocyte maturation via inducing the apoptosis. *Frontiers in Pharmacology, 11*. www.frontiersin.org/articles/10.3389/fphar.2020.01226

Harguinteguy, C. A., Pignata, M. L., & Fernández-Cirelli, A. (2015). Nickel, lead and zinc accumulation and performance in relation to their use in phytoremediation of macrophytes myriophyllum aquaticum and egeria densa. *Ecological Engineering, 82*, 512–516. https://doi.org/10.1016/j.ecoleng.2015.05.039

Holliger, C., Gaspard, S., Glod, G., Heijman, C., Schumacher, W., Schwarzenbach, R. P., & Vazquez, F. (1997). Contaminated environments in the subsurface and bioremediation: Organic contaminants. *FEMS Microbiology Reviews, 20*(3–4), 517–523.

Huang, G., Chen, F., Wei, D., Zhang, X., & Chen, G. (2010). Biodiesel production by microalgal biotechnology. *Applied Energy, 87*(1), 38–46.

Hurtado, C., Domínguez, C., Pérez-Babace, L., Cañameras, N., Comas, J., & Bayona, J. M. (2016). Estimate of uptake and translocation of emerging organic contaminants from irrigation water concentration in lettuce grown under controlled conditions. *Journal of Hazardous Materials, 305*, 139–148.

Iha, D. S., & Bianchini, I. J. (2015). Phytoremediation of Cd, Ni, Pb and Zn by Salvinia minima. *International Journal of Phytoremediation, 17*(10), 929–935. https://doi.org/10.1080/15226514.2014.1003793

International Atomic Energy Agency. (2009). *Classification of Radioactive Waste* (Issue GSG-1), IAEA Safety Standards Series No. GSG-1, IAEA, Vienna.

Islam, M. S., Ueno, Y., Sikder, M. T., & Kurasaki, M. (2013). Phytofiltration of arsenic and cadmium from the water environment using micranthemum umbrosum (J.F. Gmel) S.F. blake as a hyperaccumulator. *International Journal of Phytoremediation, 15*(10), 1010–1021. https://doi.org/10.1080/15226514.2012.751356

Jinqi, L., & Houtain, L. (1992). Degradation of azo dyes by algae. *Environmental Pollution, 75*, 273–278.

Kamelia, L., de Haan, L., Spenkelink, B., Bruyneel, B., Ketelslegers, H. B., Boogaard, P. J., & Rietjens, I. M. C. M. (2020). The role of metabolism in the developmental toxicity of polycyclic aromatic hydrocarbon-containing extracts of petroleum substances. *Journal of Applied Toxicology: JAT, 40*(3), 330–341. https://doi.org/10.1002/jat.3906

Kamusoko, R., & Jingura, R. M. (2017). Utility of Jatropha for phytoremediation of heavy metals and emerging contaminants of water resources: A review. *CLEAN–Soil, Air, Water, 45*(11), 1700444.

Kawai, H., Kitamura, A., Mimura, M., Mimura, T., Tahara, T., Aida, D., Sato, K., & Sasaki, H. (2014). Radioactive cesium accumulation in seaweeds by the Fukushima 1 Nuclear Power Plant accident—two years' monitoring at Iwaki and its vicinity. *Journal of Plant Research, 127*, 23–42.

Kim, I., Yang, H. M., Park, C. W., Yoon, I.-H., Seo, B.-K., Kim, E.-K., & Ryu, B.-G. (2019). Removal of radioactive cesium from an aqueous solution *via* bioaccumulation by microalgae and magnetic separation. *Scientific Reports, 9*, 10149. https://doi.org/10.1038/s41598-019-46586-x

Klekner, V., & Kosaric, N. (1992). Degradation of phenols by algae. *Environmental Technology, 13*(5), 493–501.

Kottuparambil, S., & Park, J. (2019). Anthracene phytotoxicity in the freshwater flagellate alga Euglena agilis Carter. *Scientific Reports, 9*(1), 15323. https://doi.org/10.1038/s41598-019-51451-y

Kurade, M. B., Ha, Y.-H., Xiong, J.-Q., Govindwar, S. P., Jang, M., & Jeon, B.-H. (2021). Phytoremediation as a green biotechnology tool for emerging environmental pollution: A step forward towards sustainable rehabilitation of the environment. *Chemical Engineering Journal, 415*, 129040.

Kurade, M. B., Kim, J. R., Govindwar, S. P., & Jeon, B.-H. (2016). Insights into microalgae mediated biodegradation of diazinon by Chlorella vulgaris: Microalgal tolerance to xenobiotic pollutants and metabolism. *Algal Research*, *20*, 126–134.

Lisar, S. Y., Deljoo, S., & Moral, M. (2015). The physiological effect of fluorene on Triticum aestivum, Medicago sativa, and Helianthus annus. *Cogent Food & Agriculture*, *1*, 1020189. https://doi.org/10.1080/23311932.2015.1020189

Lu, Q., He, Z. L., Graetz, D. A., Stoffella, P. J., & Yang, X. (2011). Uptake and distribution of metals by water lettuce (Pistia stratiotes L.). *Environmental Science and Pollution Research*, *18*(6), 978–986. https://doi.org/10.1007/s11356-011-0453-0

Luo, L., Lai, X., Chen, B., Lin, L., Fang, L., Tam, N. F. Y., & Luan, T. (2015). Chlorophyll catalyse the photo-transformation of carcinogenic benzo [a] pyrene in water. *Scientific Reports*, *5*(1), 1–11.

Madadi, R., Pourbabaee, A. A., Tabatabaei, M., Zahed, M. A., & Naghavi, M. R. (2016). Treatment of petrochemical wastewater by the green algae Chlorella vulgaris. *International Journal of Environmental Research*, *10*(4), 555–560.

Mahurpawar, M. (2015). Effects of heavy metals on human health. *International Journal of Research*, *530*, 1–7.

Matsuo, T. (2003). Japanese experiences of environmental management. *Water Science and Technology*, *47*(1), 7–14.

Mello-Farias, P. C. D., Soares Chaves, A. L., & Leoneti, C. (2011). Transgenic plants for enhanced phytoremediation—physiological studies. *Genetic Transformation*. https://doi.org/10.5772/24355

Miranda, A., Muradov, N., Gujar, A., Stevenson, T., Nugegoda, D., Ball, A., & Mouradov, A. (2014). Application of aquatic plants for the treatment of selenium-rich mining wastewater and production of renewable fuels and petrochemicals. *Journal of Sustainable Bioenergy Systems*, *4*(1), 97–112.

Mojiri, A., Zhou, J. L., Ohashi, A., Ozaki, N., & Kindaichi, T. (2019). Comprehensive review of polycyclic aromatic hydrocarbons in water sources, their effects and treatments. *The Science of the Total Environment*, *696*, 133971. https://doi.org/10.1016/j.scitotenv.2019.133971

Mueller, J. G., Cerniglia, C. E., & Pritchard, P. H. (1996). Bioremediation of environments contaminated by polycyclic aromatic hydrocarbons. *Biotechnology Research Series*, *6*, 125–194.

Oliveira, M., Gravato, C., & Guilhermino, L. (2012). Acute toxic effects of pyrene on Pomatoschistus microps (Teleostei, Gobiidae): Mortality, biomarkers and swimming performance. *Ecological Indicators*, *19*, 206–214. https://doi.org/10.1016/j.ecolind.2011.08.006

Pannu, A. K., & Singla, V. (2020). Naphthalene toxicity in clinical practice. *Current Drug Metabolism*, *21*(1), 63–66. https://doi.org/10.2174/1389200220666191122110036

Parra, L.-M. M., Torres, G., Arenas, A. D., Sánchez, E., & Rodríguez, K. (2012). *Phytoremediation of low levels of heavy metals using duckweed (Lemna minor) BT—abiotic stress responses in plants: Metabolism, productivity and sustainability* (P. Ahmad & M. N. V Prasad, Eds., pp. 451–463). Springer. https://doi.org/10.1007/978-1-4614-0634-1_24

Partila, A. M., & Mohammed, M. R. (2019). Evaluation of the benzo[a]anthracene biodegradation by animal bioassays. *Bulletin of the National Research Centre*, *43*(1), 72. https://doi.org/10.1186/s42269-019-0115-9

Pelcová, P., Kopp, R., Ridošková, A., Grmela, J., & Štěrbová, D. (2022). Evaluation of mercury bioavailability and phytoaccumulation by means of a DGT technique and of submerged aquatic plants in an aquatic ecosystem situated in the vicinity of a cinnabar mine. *Chemosphere*, *288*, 132545. https://doi.org/10.1016/j.chemosphere.2021.132545

Peng, H., de-Bashan, L. E., Bashan, Y., & Higgins, B. T. (2020). Indole-3-acetic acid from Azosprillum brasilense promotes growth in green algae at the expense of energy storage products. *Algal Research*, *47*, 101845.

Pi, N., Ng, J. Z., & Kelly, B. C. (2017). Bioaccumulation of pharmaceutically active compounds and endocrine disrupting chemicals in aquatic macrophytes: Results of hydroponic experiments with Echinodorus horemanii and Eichhornia crassipes. *Science of the Total Environment*, *601*, 812–820.

Pontes, L. P., Vafeiadou, A. M., de França, F. J. L., Cavalcante, R. A., de Araújo França, D. A., Brito, C. M., Alves, R. N., de Carvalho, P. S. M., & dos Santos, G. A. P. (2021). Toxic effects of phenanthrene intensify with an increase of temperature for the populations of a free-living nematode. *Ecological Indicators*, *120*. https://doi.org/10.1016/j.ecolind.2020.106868

Popp, J., Pető, K., & Nagy, J. (2013). Pesticide productivity and food security. A review. *Agronomy for Sustainable Development*, *33*, 243–255.

Rai, P. K. (2008). Phytoremediation of Hg and Cd from industrial effluents using an aquatic free floating macrophyte Azolla pinnata. *International Journal of Phytoremediation*, *10*(5), 430–439.

Romanova, T. E., Shuvaeva, O. V., & Belchenko, L. A. (2016). Phytoextraction of trace elements by water hyacinth in contaminated area of gold mine tailing. *International Journal of Phytoremediation*, *18*(2), 190–194. https://doi.org/10.1080/15226514.2015.1073674

Saleh, H. M. (2012). Water hyacinth for phytoremediation of radioactive waste simulate contaminated with cesium and cobalt radionuclides. *Nuclear Engineering and Design*, *242*, 425–432.

Sekomo, C. B., Rousseau, D. P. L., Saleh, S. A., & Lens, P. N. L. (2012). Heavy metal removal in duckweed and algae ponds as a polishing step for textile wastewater treatment. *Ecological Engineering*, *44*, 102–110. https://doi.org/10.1016/j.ecoleng.2012.03.003

Sethunathan, N., Megharaj, M., Chen, Z. L., Williams, B. D., Lewis, G., & Naidu, R. (2004). Algal degradation of a known endocrine disrupting insecticide, α-endosulfan, and its metabolite, endosulfan sulfate, in liquid medium and soil. *Journal of Agricultural and Food Chemistry*, *52*(10), 3030–3035.

Sharma, S., Singh, B., & Manchanda, V. K. (2015). Phytoremediation: Role of terrestrial plants and aquatic macrophytes in the remediation of radionuclides and heavy metal contaminated soil and water. *Environmental Science and Pollution Research*, *22*, 946–962.

Shashirekha, S., Uma, L., & Subramanian, G. (1997). Phenol degradation by the marine cyanobacterium Phormidium valderianum BDU 30501. *Journal of Industrial Microbiology and Biotechnology*, *19*(2), 130–133.

Singh, B. S. M., Dhal, N. K., Kumar, M., Mohapatra, D., Seshadri, H., Rout, N. C., & Nayak, M. (2022). Phytoremediation of 137Cs: Factors and consequences in the environment. *Radiation and Environmental Biophysics*, *61*(3), 341–359.

Skinner, K., Wright, N., & Porter-Goff, E. (2007). Mercury uptake and accumulation by four species of aquatic plants. *Environmental Pollution*, *145*(1), 234–237. https://doi.org/10.1016/j.envpol.2006.03.017

Sood, A., Uniyal, P. L., Prasanna, R., & Ahluwalia, A. S. (2012). Phytoremediation potential of aquatic macrophyte, Azolla. *Ambio*, *41*(2), 122–137. https://doi.org/10.1007/s13280-011-0159-z

Srivastava, S., & Bhainsa, K. C. (2016). Evaluation of uranium removal by Hydrilla verticillata (L.f.) Royle from low level nuclear waste under laboratory conditions. *Journal of Environmental Management*, *167*, 124–129. https://doi.org/10.1016/j.jenvman.2015.11.018

Subashchandrabose, S. R., Ramakrishnan, B., Megharaj, M., Venkateswarlu, K., & Naidu, R. (2011). Consortia of cyanobacteria/microalgae and bacteria: Biotechnological potential. *Biotechnology Advances*, *29*(6), 896–907.

Sun, K., Song, Y., Liu, Z., Jing, M., Wan, J., Tang, J., & Liu, R. (2020). Toxicity assessment of Fluoranthene, Benz(a)anthracene and its mixed pollution in soil: Studies at the molecular and animal levels. *Ecotoxicology and Environmental Safety*, *202*, 110864. https://doi.org/10.1016/j.ecoenv.2020.110864

Sutherland, D. L., & Ralph, P. J. (2019). Microalgal bioremediation of emerging contaminants—Opportunities and challenges. *Water Research*, *164*, 114921. https://doi.org/10.1016/j.watres.2019.114921

Sweta, Bauddh, K., Singh, R., & Singh, R. P. (2015). Short communication. *Ecological Engineering*, *83*(C), 39–42. https://doi.org/10.1016/j.ecoleng.2015.06.003

Tangahu, B. V., Abdullah, S. R. S., Basri, H., Idris, M., Anuar, N., & Mukhlisin, M. (2013). Phytoremediation of wastewater containing lead (Pb) in pilot reed bed using Scirpus grossus. *International Journal of Phytoremediation*, *15*(7), 663–676.

Tauqeer, H., Ali, S., Rizwan, M., Ali, Q., Saeed, R., Iftikhar, U., Ahmad, R., Farid, M., & Abbasi, G. H. (2016). *Phytoremediation of heavy metals by Alternanthera bettzickiana: Growth and physiological response*. Elsevier. Retrieved February 8, 2023, from www.sciencedirect.com/science/article/pii/S0147651315302190

Tchounwou, P. B., Yedjou, C. G., Patlolla, A. K., Sutton, D. J. (2012). Heavy metal toxicity and the environment. In: Luch, A. (ed.), *Molecular, clinical and environmental toxicology. Experientia supplementum*, vol. 101. Springer. https://doi.org/10.1007/978-3-7643-8340-4_6

Topp, E., Monteiro, S. C., Beck, A., Coelho, B. B., Boxall, A. B. A., Duenk, P. W., Kleywegt, S., Lapen, D. R., Payne, M., & Sabourin, L. (2008). Runoff of pharmaceuticals and personal care products following application of biosolids to an agricultural field. *Science of the Total Environment*, *396*(1), 52–59.

Török, A., Gulyás, Z., Szalai, G., Kocsy, G., & Majdik, C. (2015). Phytoremediation capacity of aquatic plants is associated with the degree of phytochelatin polymerization. *Journal of Hazardous Materials*, *299*, 371–378. https://doi.org/10.1016/j.jhazmat.2015.06.042

Volf, I., Rakoto, N. G., & Bulgariu, L. (2015). Valorization of pistia stratiotes biomass as biosorbent for lead(II) ions removal from aqueous media. *Separation Science and Technology*, *50*(10), 1577–1586. https://doi.org/10.1080/01496395.2014.978018

Wong, T.-H., Lee, C.-L., Su, H.-H., Lee, C.-L., Wu, C.-C., Wang, C.-C., Sheu, C.-C., Lai, R.-S., Leung, S.-Y., Lin, C.-C., Wei, Y.-F., Wang, C.-J., Lin, Y.-C., Chen, H.-L., Huang, M.-S., Yen, J.-H., Huang, S.-K., & Suen, J.-L. (2018). A prominent air pollutant, Indeno[1,2,3-cd]pyrene, enhances allergic lung inflammation via aryl hydrocarbon receptor. *Scientific Reports*, *8*(1), 5198. https://doi.org/10.1038/s41598-018-23542-9

Yadav, A. P. S., Dwivedi, V., Kumar, S., Kushwaha, A., Goswami, L., & Reddy, B. S. (2020). Cyanobacterial extracellular polymeric substances for heavy metal removal: A mini review. *Journal of Composites Science*, *5*(1), 1.

Yamazaki, S., Okazaki, K., Kurahashi, T., & Sakakibara, M. (2017). Phytoremediation of arsenic- and molybdenum-contaminated alkaline wastewater by Eleocharis acicularis in winter in Japan. *IOP Conference Series: Earth and Environmental Science*, *71*, 012018. https://doi.org/10.1088/1755-1315/71/1/012018

Ye, Y., Jiang, S., Zhang, C., Cheng, Y., Zhong, H., Du, T., Xu, W., Azziz, R., Zhang, H., & Zhao, X. (2020). Environmental pollutant benzo[a]pyrene induces recurrent pregnancy loss through promoting apoptosis and suppressing migration of extravillous trophoblast. *BioMed Research International*, *2020*, 8983494. https://doi.org/10.1155/2020/8983494

Zaragoza-Ojeda, M., Eguía-Aguilar, P., Perezpeña-Díazconti, M., & Arenas-Huertero, F. (2016). Benzo[ghi]perylene activates the AHR pathway to exert biological effects on the NL-20 human bronchial cell line. *Toxicology Letters*, *256*, 64–76. https://doi.org/10.1016/j.toxlet.2016.05.023

13 Microbial Advancement towards Strategic Bioremediation Techniques
An Overview

Archisman Bhunia, Dibyajit Lahiri, Moupriya Nag

13.1 INTRODUCTION

The world of the twenty-first century has been under the cover of pollution since the beginning of the (r)evolution, with the modernization in the urban and industrial sectors, supported by introduction of innovative technologies. But this modernization also raises global concern about the disposal of end products/by-products into the environment. It has been observed that heavy metals in complexation with diverse toxic substances are deposited in soil, water, and air thereby entering the biotic system via food web (Abdullah et al., 2019; Manisalidis et al., 2020; Zamora-Ledezma et al., 2021). The result of such entry of contaminants, specifically heavy metal residues of Fe, Cr, Zn, Hg, into the food web via direct or indirect route of exposure has brought a devastating impact upon the living beings, affecting human health. This includes neurotoxicity to mutation and damage of nucleotides, protein misfolding and hampering of metabolic pathways (Briffa et al., 2020).

The source of heavy metals as pollutants in nature have been some natural occurrences like weathering and erosion of rock morphology, forest fires, volcanic activities, and/or seamounts. Alongside nature, anthropogenic activities have also contributed to pollution worldwide. Industrial chimney vents, automobiles, storage/disposal site leakage, and industrial and household effluents have been among the list of anthropogenic routes to nature (Paul et al., 2021). In order to delimit the harmful effects of pollution, search for novel remediation techniques is on a rise (Khan et al., 2004). In this aspect Zamora-Ledezma et al. (2021) proposed the implementation of conventional techniques first and then non-conventional techniques. A wide range of techniques like the chemical techniques, ion-exchange technique, membrane technology, coagulation, and electrochemical technology have been categorized under the conventional techniques, depending upon their operational specificity. But their limitations in efficiency resulted in bringing forth the use of techniques like adsorption, nanotechnology, and microbial fuel cell into the scenario (Zamora-Ledezma

DOI: 10.1201/9781003297901-13

et al., 2021). In spite of bringing modern strategic practices into play, limitations have always been a speed breaker in the path of smooth flow chain towards novel remediation. This opened the domain of exploring the unexplored path of pollution eradication. Phytoremediation and mycoremediation have been brought into the limelight as new alternatives that gained recognition for their economical, eco-friendly, and easy in-practice operation (Ali et al., 2013; Awa and Hadibarata, 2020; Jin et al., 2018).

As an efficient alternative strategy, recently, a surge in microbe-assisted bioremediation methods has been observed, the reason being the nature of strategy adopted by the microbes in remediation. Usually a three-dimensional consortium of planktonic cells of same and/or different species forms the 'biofilm', functioning as a single unit of eradicating agent (Myckatyn et al., 2016). Intense research has reported the biofilm consortium to reprogram itself from multifunctioning to unifunctioning after the biological components like eDNA, proteins, and lipids cluster under one structure embedded by EPS (Rabin et al., 2015; Asri et al., 2019). This dramatic phenomenon, along with the surrounding environmental parameters like varying pH, temperature, pressure gradient, concentration of pollutants, available nutrients, and aeration for development and proliferation of microbial colony, has been among the key regulating parameters deciding the degree of efficiency of the strategy (López et al., 2010).

This dramatic phenomenon of biofilm formation and remediation potential of microbes in remediating pollutants from the environment opened a new domain of implementing techniques like biopile, bioventing, land farming, bioslurping, and windrows in practice (Paul et al., 2021). It is further supported by *in silico* studies involving metabolomics, proteomics, and genomics data overcoming the present limitations. The present chapter brings forth the current scenario of implemented techniques in remediation and the microbial advancement as an alternative for a promising outcome henceforth.

13.2 CONVENTIONAL TECHNIQUES OF REMEDIATION

Some of the techniques that are categorized under conventional and non-conventional strategies are as follows:

a. Ion exchange: It is based on the principle of exchange interaction of ionic form of pollutants from soil and the aquatic environment. Categorized under conventional strategies, zeolite, organic resins, inorganic metals, and hybrid fabrications have been reported to be used for the purpose (Sharma et al., 2018; Zamora-Ledezma et al., 2021). Though possessing efficient kinetics in remediation, ion exchange often has drawbacks due to pH and temperature sensitivity alongside limitation in operational concentration (Sharma et al., 2018).
b. Membrane technology: A traditional method of separation, membrane technology works upon the principle of separation upon size of the substrate passing through the membrane pores (Abdullah et al., 2019).

A further modification of the technology includes Donnan exclusion, that is, charge-charge repulsion, and specific adsorption of the contaminants upon the membrane. Depending upon the target specific charges, organic and/or inorganic fabrications like silica, cellulose acetate, and ceramics are often used in membrane preparation (Ezugbe and Rathilal, 2020). The mechanism of membrane separation is often regulated by the rate of effluent flow across the cross section of the membrane. Based upon the pressure gradient, membrane separation has been later classified under the categories of microfiltration (MF), nanofiltration (NF), ultrafiltration (UF), reverse osmosis (RO), direct osmosis (DO), forward osmosis (FO), and electrodialysis (ED) (Zamora-Ledezma et al., 2021). Among these, nanofiltration (NF) has been identified to show the highest degree of remediation, followed by reverse osmosis (RO), forward osmosis (FO), and others (Abdullah et al., 2019), though the separation also depends on various microenvironment parameters (Ezugbe and Rathilal, 2020).

c. Chemical technology: Being operational under a vast pH range, the metal pollutants are reported to undergo hydrolysis following precipitation as secondary deposits in few scenarios. Zamora-Ledezma et al. (2021), in their work, classified this precipitation into two categories following the reagents used: Sulfide precipitation—the metal sulfide precipitated along the flow chain are separated under gravitational settlement or filtration; Hydroxide precipitation—the hydroxide precipitated are removed from the efflux by filtration or sedimentation in their downstream. Literatures have reported the use of various chelating agents like Diaion resin, H_2O_2, HNO_3, EDTA, tartrate, and oxalate to perform scavenging at varying temperature, pH and at different ionic concentration and contact time (Sharma et al., 2018).
d. Vitrification: Another lesser practice technique in use since the 1980s for contaminant remediation is vitrification. The technique operates on the principle of heating of the contaminated soil at a temperature of 1700–2000°C, followed by rapid freezing, resulting in formation of glasslike substances. They contribute in immobilization and encapsulate the metal contaminants due to their restricted pore size, separating the specific pollutants from the soil biomass (Liu et al., 2018). Further modern-day research to improvise the method of vitrification has witnessed the use of nano-metallic additives and activated carbon mixed with the sample to boost the effectivity of immobilization to ~95% alongside reducing the temperature to ~1100–1200°C (Mallampati et al., 2015).
e. Coagulation/flocculation: A physicochemical method of remediation, coagulation involves the use of various coagulants, causing clusters of pollutants that precipitate and are separated in the course of time. Examples include polyaluminum chloride (PACl), aluminum sulfate ($Al_2(SO4)_3$), polyacrylamide (PAM), etc. used for clustering the metal pollutants.

13.3 NON-CONVENTIONAL TECHNIQUES IN BIOREMEDIATION

a. Adsorption: The technology involves efficient removal capacity due to large exposure of surface area, potential target-specific and stable adsorbents in use, as well as low secondary deposits, which have significantly kept the technique well preferred (Zamora-Ledezma et al., 2021). It uses a range of structured and non-structured materials such as activated carbon, zeolite, clay minerals, industrial solid by-products, carbon-nanotubes, and graphene-based composites for the purpose of promising metal pollutant eradication. An effective remediation of different metal ions at specific range of pH, temperature, and concentration has been observed. The efficiency of scavenging metal ions using activated carbon is in the sequence of $Cr^{6+} > Cd^{2+} > Co^{2+} > Cr^{3+} > Ni^{2+} > Cu^{2+} > Pb^{2+}$, and that for using clinoptilolite is $Pb^{2+} > Cd^{2+} > Cu^{2+} > Co^{2+} > Cr^{6+} > Zn^{2+} > Ni^{2+} > Hg^{2+}$ (Burakov et al., 2018). Biological wastes comprising lignin, cellulose, and industrial wastes like ash, sludge, and red mud have also served the purpose of remediating contaminants from the ecosystem (Zamora-Ledezma et al., 2021).
b. Microbial fuel cells: Stepping towards the future, electrokinetic remediation has been further upgraded with microbial collaboration to perform two-front benefits of generating bioelectricity alongside remediation (Zamora-Ledezma et al., 2021). It follows the principle of microbial scavenging, generating protons and electrons stored in the microbial cells as adenosine triphosphate (ATP). Further pipeline of utilization is decided by the microenvironment (aerobic and/or anaerobic) of occurrence via redox reaction (oxidation and reduction). For example, anaerobic condition results in release of electrons through chemical mediators, while in the presence of oxygen (aerobic), the protons and electrons combine with oxygen. The anaerobic environment is maintained in the anode chamber loaded with microbes to conduct biological oxidation, and is parted from the cathode chamber by a proton exchange membrane, that is, cation exchange membrane, (CEM) restricting the passage of oxygen. The aerated cathode chamber further carries the reaction of reduction involving electrons and protons, forming water and electricity generation along the external wires during electron migration (Bagchi and Behera, 2019).
c. Nanotechnology: Nanoparticles are widely for their large surface area and surface-to-volume ratio, electronic, magnetic, and optical properties (Zamora-Ledezma et al., 2021). Some of the known nanoscale components used are alumina nanofibers, non-metal oxides, magnetic nanoparticles, nano-sheets, nano-spheres, and nano-rods of ZnO (Mallikarjunaiah et al., 2020).

13.4 PLAYERS OF BIOREMEDIATION

In the quest for exploring alternative strategies of remediation that can bypass and come forth as significantly economical, eco-friendly, simple, procedural, and has easily available components, intense research has brought forth the concept of phytoremediation and microbial remediation into the scenario.

13.4.1 Phytoremediation

Though heavy metal pollutants have been reported to affect plant species upon accumulation in access, researchers have identified certain plant breeds that possess genetic makeup that is tolerant to metal pollutants. Being a green technology, phytoremediation became well known for its efficient, economical, eco-friendly, and environmental in situ application (Ali et al., 2013). Phytoremediation has been grouped depending on few mechanisms like phytodegradation, phytoextraction/phytoaccumulation, phytovolatilization, phytofiltration, and phytostabilization that are performed by the specific plants to carry out the method of remediation. Their description has been summarized in Table 13.1.

Some of the noteworthy plants include rice (*Oryza sativa* L. var. *Khandagiri*), wheat (*Triticum aestivum* L. cv. UP 262), green gram (*Vigna radiate* L. *Wilczek*. var K-851), Indian mustard (*Brassica juncea*), *Trifolium* sp., alpine pennycress (*Thlaspi caerulescens*), *Alyssum bertolonii* (Ali et al., 2013; Mohanty and Patra, 2012; Chen et al., 2021).

Though economical and eco-friendly, phytoextraction, phytovolatilization, and phytostabilization mechanisms of phytoremediation face certain limitations such as optimization of media and plant species, ready availability of chelating agents and presence of sufficient concentration of metals in soil (Awa and Hadibarata, 2020).

TABLE 13.1
Summary of the mechanisms of phytoremediation

Mechanism	Description
Phytodegradation	Specific plant enzymes are responsible for the degradation of xenobiotic components within the plant tissue.
Rhizodegradation	Degradation of xenobiotics in association with class of rhizobacteria in the rhizosphere topography.
Phytoextraction	High rate of intake resulting in accumulation of metal pollutants in harvestable biomasses of the plants, that is, shoot.
Phytostabilization	Restricting the easy mobility of the metal pollutants in the soil, mainly by roots performing sorption.
Phytovolatilization	Targeting organic pollutants and metals like Se and Hg, pollutant uptakes are converted to volatile forms, releasing them into the atmosphere.
Phytofiltration	Sequestration of pollutants by absorption/adsorption by roots and seedlings from wastewater bodies.
Phytodesalination	Halophytes mostly perform desalination of excess metal salts from the saline-contaminated soil.

Source: Ali et al., 2013

13.4.2 Microbial Remediation

After the conventional and non-conventional remediation and phytoremediation strategies for scavenging the pollutants, focusing metal pollutants from the environment, a group of researchers dived deep into exploring the microbes for pollutant elimination. Microbes have been found to have significant efficiency in remediation. Examples include *Aureobacterium esteroaromaticum*, *Escherichia* sp., *Pseudomonas putida*, *Klebsiella oxytoca*, *Bacillus licheniformis*, *Enterobacter* sp., and *Acinetobacter calcoaceticus* that have been documented towards metal pollutant degradation and removal (Kumar and Bharadvaja, 2020). These niches are well recognized in remediation, using mechanisms like bioaccumulation, biotransformation, and biosorption (Jin et al., 2018). A table has been presented that summarizes some of the microbial contributions towards the heavy metal pollutants (Table 13.2).

TABLE 13.2
Summarizing microbes for heavy metal remediation

Group of microbes	Microbial agents	Heavy metals	Efficiency of sorption (%)	Reference
Bacteria	*Sporosarcina saromensis (M52)*	Cr	82.5	Zhao et al. (2016)
	Bacillus subtilis		99.6	Kim et al. (2015)
	Bacillus cereeus		81	Nayak et al. (2018)
	Pseudomonas aeruginosa (PCP2)		72	Kumar et al. (2011)
	Bacillus sp. *B*		47	
	Bacillus circulans MN1		96	Chaturvedi (2011)
	Methylobacterium organophilum	Pb	62.28	Bharagava and Mishra (2018)
	Bacillus firmus		98.3	Salehizadeh and Shojaosadati (2003)
	Cellulosimicrobium sp. *(KX710177)*		84.62–99.3	Bharagava and Mishra (2018)
	Desulfovibrio desulfuricans		90.3	Kumar et al. (2011)
	Acinetobacter sp. *B9*		68.94	Bhattacharya and Gupta (2013)
	Klebsiellap neumoniae	Hg	29.83	Jafari et al. (2015)
	Vibrio parahaemolyticus (PG02)		80	
	Vibriofluvialis	Co	8	
	Desulfovibrio desulfuricans (KCTC5768)	CuNi	78.2–97.4	Congeevaram et al. (2007)
Fungi	*Candidapara psilosis*	Hg	80	Muneer et al. (2013)
	Aspergillus versicolor	Ni	30.05	Taştan et al. (2010)
Algae	*Spirogyra* sp.	Cr	98.3	Mane and Bhosle (2012)
	Nostoc sp.	Cd	95.5	Kumaran et al. (2011)
	Spirogyra sp.	Cu	89	Mane and Bhosle (2012)

Though significantly effective in remediation of pollutants by its mechanisms at a varying degree, the main strategy of the microbes in the scenario is formation of the consortium of planktons to perform under one single and effective colony.

13.5 MICROBIAL STRATEGY COUNTERING TOXIC POLLUTANTS

The consortium of planktons of microbes mechanically performing as a single functional unit has been coined as 'biofilm'. But, since its appearance into the limelight of science, biofilms have been the foremost topic of controversies among researchers for its formation and dramatic social behavior (Costerton, 1999). This strategy has benefitted microbes in bringing about a novel scenario of environmental remediation and helped in exploring beyond the boundaries of scientific understanding.

13.5.1 Factors Regulating the Biofilm Strategy

Formation of biofilms are prominently reported in the microenvironment of desired nutrients/ions and at specific concentrations. A slight variation in the concentration may result to a drastic inhibition of the biofilm formation. Being a consortium of planktons/eukaryotes, the pH for exopolysaccharide (EPS) matrix synthesis varies upon the species. Alongside pH, temperature too decides the efficiency of biofilm formation and degree of enzymatic activities of the consortia. Though temperature-sensitive, biofilm formation has been observed in many bacterial species at an optimum temperature of 40°C, but is not limited to that temperature and has also been reported to occur in cold water environments.

With maturity and altered genetic expression within the consortia, alteration has also been observed in nutrient gradient and electron acceptance, resulting in quorum sensing mechanisms up-regulating genes encoding alcohol dehydrogenase, triphosphate, phosphoglycerate mutase within the biofilm (López et al., 2010). Surfactin and other secondary metabolites, siderophores, divalent cations, and antibiotics have also influenced the quorum-sensing mechanisms in regulating the biofilm synthesis and function. Works of Renner and Weibel (2011) also focused upon the external interaction-regulating factors like turbulence, electrostatic interactions, hydrophobic and steric force, nature and topography of the surface, surface-charge, and antimicrobial property that decides the interaction, adhesion, and formation of the microbial consortia upon the surface support.

13.6 MICROBIAL COUNTER MECHANISM

The microbes primarily adapt four counter mechanisms to eradicate the pollutants from the lap of the environment: biodegradation, bioaccumulation, biosorption, and biomineralization.

13.6.1 Biodegradation

The term 'biodegradation', in simpler form, signifies the chain of mechanisms involved in degradation and/or reduction of complex pollutants into their simple and comparatively reduced toxic forms by means of degrading microbes. The biological pathways adopted in reduction results in 'mineralization' of the pollutants but are generally recognized as biodegradation. The mechanism of biodegradation takes place in two stages: growth and co-metabolism, where the pollutants primarily act as sources of carbon for microbial metabolism. Researchers have also reported bacteria, fungi, protozoa, and algae to release carbon dioxide as an end product alongside water and methane under aerobic and anaerobic environments (Borai et al., 2016; Al-Dahhan et al., 2018).

Use of microbes in biodegradation of plastic wastes is found to depend on factors like duration of exposure, concentration, and varying temperature of 50–75°C. *Streptomyces* S27, *Brassica rapa*, *E. gallinarum*, *hodococcus ruber* C208, *Phanerochaete chrysosporium*, and *Tenebrio molitor* have been among some of the microbial classes that are potent towards expression of biodegradation property (Varjani et al., 2020). Upon this prospect, enzymes like protease, cutinase, hydrolase mark special mention for polymer degradation. The mode of degrading follows in two phases by the microbes: the initial stage brings about colonization and attachment of planktons over the polymers forming biofilm, followed by hydrolysis of polymers by polymer-degrading proteins synthesized by the consortia (Ahmed et al., 2018).

13.6.2 Biosorption

A metabolically passive mechanism operating in correlation with bioaccumulation, known mostly to bring about selective sequestration of organic pollutants from the industrial effluents by means of sorption, is termed as biosorption. The mechanism of biosorption is based on techniques like surface complexity, chelation, ion exchange, and precipitation (Hansda et al., 2016; Fomina and Gadd, 2014) and comprises of two phases: the sorbents in the solid phase and target pollutants to be sorbed form the liquid phase (Fomina and Gadd, 2014). Vast species of microbes and flora like *Rhizopus arrhizus*, *Palmyra pamata*, *Sphagnum* sp., *Chlorella vulgaris*, and *Ascophyllum nodosum* are among the potential players of biosorption that have also contributed in target metal recovery from the industrial effluents.

Aside from time complexity, the changing solution, pH gradient, dosage, effluent concentration, fluid dynamics, and competitive interaction are also among the deciding factors in its execution (Fomina and Gadd, 2014). Hansda et al. (2016), in their work, highlighted the mechanism describing the rapid attachment in sorption to involve van der Waals force, surface binding protein like S-layer protein (SLP), adsorption, and microprecipitation alongside exposure to large surface-to-volume ratio. Microbes like *Chiarella vulgaris* and *Zoogloea ramigera* have been reported to perform recovery of copper by physical adsorption. Researchers also documented the change in ionic concentration and pH to affect the precipitation and chelating property across the biofilm cell membranes (Redha, 2020).

13.6.3 Bioaccumulation

Bioaccumulation can be defined as the net uptake, transformation, and precipitation of substances along the folds of membrane within the planktons and/or microbial consortium upon intake through various roots of exposure like water, air, and sediments.

This mechanism of accumulation of toxic pollutants by the microbes is a result of higher degree of absorption in contrast to permissible rate of elimination. Studies suggest the process to undergo in distinct stages: initial binding of the metabolically independent extracellular substances upon the biofilm surface, followed by slow metabolic response in binding and penetration through biofilm architecture (Hansda et al., 2016). Microbial selection is also a point of focus in this method of bioaccumulation, and researchers preferred selecting the ones that are well adapted to the environment from the site of contamination (Sarkar et al., 2017; Muhammad et al., 2020).

Apart from the primary regulating factors of pH, temperature, surrounding pressure and adjacent availability of ionic concentration, degree of bioaccumulation is also decided upon the size of the pollutant biomass, nutrient concentration in the medium, and upon the duration of contact with the pollutants during the operation (Hansda et al., 2016). Quorum sensing has been identified to play a vital role on altered gene expression, regulating the potential of biofilm formation and degree of bioaccumulation. The potentially toxic elements (PTEs) under potential gradient diffuse through the ATP-dependent transporters and further metabolize through the efflux system (Geng et al., 2019). The PET accumulation occurs due to biochemical immobilization, biosorption, intracellular accumulation upon the biofilm surface which are metabolized by the efflux pumps under an oxygen-controlled microenvironment. The metabolized residues henceforth transport and accumulate into the layers of biofilm architecture (Hansda et al., 2016). They are often potent in solubilizing PETs to facilitate the uptake rate (Geng et al., 2019).

13.6.4 Biomineralization

Biomineralization is a series of multidisciplinary mechanisms performed for the purpose of mineral recovery by a wide range of microbes. It comprises of heavy metal oxides, silicate, and carbonates (Benzerara, 2011). Biomineralization too, like other biologically mediated remediation technologies, is influenced by the changing temperature, pH, and pressure of the surroundings. Arias et al. (2017) focused upon the mechanism and have classified biomineralization into three categories depending upon their nature of remediation.

a. Biologically controlled biomineralization: The pollutants, post intercellular processing, accumulates within specific folds only under favorable conditions. The bacteria possess the potential to regulate the mineral phase and degree of biosynthesis.
b. Biologically induced biomineralization: The microbial consortia is potent enough to optimize the chemical microenvironment by virtue of biological activities, resulting in oversaturation and precipitation of specific minerals targeted.
c. Biologically influenced biomineralization: It occurs due to interaction between mineral-specific surface receptors and the minerals present in the microenvironment followed by metabolic activity. Considered to be a passive mechanism, biologically influenced biomineralization decides the physicochemical property of the biosynthesized deposits to some extent.

13.7 TECHNIQUES OF MICROBIAL BIOREMEDIATION

With the advancement in the domain of microbial remediation, researchers came up with implementing in situ and ex situ strategies into practice for a fruitful outcome. While the in situ bioremediation is well noted for its beneficial contribution in restoration of contaminated soil and water bodies without evacuation of the sample, ex situ bioremediation involves excavated polluted samples shifted to preferred destination for remedial treatment. This has added additional operational costs to the techniques and has also imposed certain challenges in terms of large-scale practice, while the in situ strategy is mainly targeted towards restoration and integrity of the natural resources (Paul et al., 2021).

13.7.1 In Situ Bioremediation

The technique centers around treating the pollutants at its residual site and does not require transportation and additional evacuation in the operational pipeline. For on-site operationality, in situ bioremediation is mainly focused for eradicating dyes, hydrocarbons, chlorinated solvents, and heavy metal contaminants. Though putting forth points of such economical advantage, the techniques are dependent on the environmental condition of the site and degree of potency to biologically degrade the contaminants (Frascari et al., 2015; Kim et al., 2014). Though economical and eco-friendly, in situ techniques lag in certainty of efficiency along with requirement of long time exposure with favorable nutrient and molecular oxygen supply, pH, temperature, and moisture content. Some of the in-practice in situ techniques are the following.

a. Bioattenuation: A comparatively economical technique in practice, bioattenuation results in immobilization and/or transformation of pollutants to its lesser toxic forms. Reports in literature have witnessed the effective removal of heavy metals like aluminum, nickel, arsenic, cobalt, cadmium, manganese, and copper biologically from the contaminated ground sites by bringing in use acinetobacteria and proteobacteria (Fauziah et al., 2017; Schmidt et al., 2005).
b. Bioaugmentation: A technique using a variety of mixed culture due to addition of microbial population into the pollutant site due to lack of specific consortium is termed bioaugmentation. It has been shown that calcium enriched landmass is formed using *Streptomyces* sp. and *Bacillus* sp. and heavy metal remediation occurs by filamentous consortium (Lebeau and Jézéquel, 2008). Though biotic and abiotic factors regulate the efficiency of the microbes in the mixed culture, consortium has often been observed to modify the microenvironment to make the mechanism of bioaugmentation thrive optimally for a long-term effect.
c. Bioventing: The optimum target of bioventing is conversion of toxic pollutants to its lesser harmful forms under aerobic condition with sensible degree of biodegradation upon additional supply of nutrients and moisture content.

d. Biostimulation: The efficiency of the biostimulation technique is solely decided by the bioavailability of nitrogen, carbon and phosphorous sources, oxygen supply, nature of pollutant, temperature, pH, and redox potential. Biodegradable components are also found to be added to enhance the mechanism of biostimulation, where the pollutants get degraded as secondary substances. Since temperature is a primary controller of the activity, a vessel system like slurry bioreactors are most preferred for the advantage of temperature regulation to optimize the remedial efficiency (Paul et al., 2021). Isolates from the contaminated sites have been found to effectively biostimulate heavy metals like Cd, Fe, and Cr using aerobic bacteria (Fulekar et al., 2012; Kanmani et al., 2012).
e. Bioslurping: A flow chain of techniques like soil vapor extraction, vacuum-enhanced recovery, and bioventing implemented simultaneously, facilitating supply of oxygen in an indirect fashion to render the remediation of pollutants in a soil and water system (Gidarakos and Aivalioti, 2007). The technique is primarily designed to focus upon recovery of light non-aqueous phase liquids (LNAPLs), avoiding extraction of large quantity of groundwater. The technique has been reported in literature to be used in the recovery of cadmium using *Arabidopsis helleri*, but faces limitation under low moisture content and permeability in the soil scenario.

13.7.2 Ex Situ Bioremediation

In contrast to in situ techniques, ex situ techniques are effective in a degree of remediation for being operated under a controlled environment, but it comes at an additional economic expense. The topography of pollutant site, transportation, favorable environmental setup, and procedural requirements according to the nature and degree of pollutants are among the primary criteria that decides the ex situ bioremediation techniques. Being executed under a controlled environment with good supplements and aeration, it holds the advantage of fast action in a shorter span of time over in situ techniques (Paul et al., 2021). Some of the ex situ techniques in practice are the following.

a. Land farming: A comparatively cheaper technique among others, land farming does not involve excavation; rather, tilling and involvement of leachates, avoiding groundwater contamination. Tilling facilitates efficient aeration movement within the soil topography, enhancing biological activity along with rise in count of heterotrophic bacteria. Though economical, land farming involves the requirement of a large open space, and the efficiency of the microbial colonies has also been noted to diminish under the environmental conditions.
b. Biopile: The technique involves piling of the contaminated soil sample following aeration throughout for an overall rise in the microbial activity of bioremediation. Some other forms like biocells, biomoulds, bioheaps are well-used scavenging petroleum contaminants from the soil samples. Researchers consider the rise in the use of biopiling being largely due

to its easy regulation of temperature, moisture content, pH, nutritional concentration alongside its high efficiency in terms of water transfer and homogenous distribution of aeration and nutrients (Whelan et al., 2015).

c. Windrow: Considered to be an extended mechanism to biopiling in order to boost the degree of bioremediation, windrows are often brought into practice periodically. This periodical turning results in homogenized distribution of water, air, pollutants, and nutrients across the sample as well as facilitates in active biological mechanism of mineralization, biotransformation, and assimilation to a higher extent. The development of anaerobic zones within synthesizes methane (CH_4) and greenhouse gases (Hobson et al., 2005). For being a periodic method of turning, windrow is not preferred against toxic volatile pollutants but is in practice for fractionation and removal of metal ions of Cu, Ni, Zn, and Cd (Paul et al., 2021).
d. Bioreactor: The mode of reaction is best carried out in vessel containers under controlled environment of optimized parameters in solid and liquid (slurry) phases. Depending upon the mode of operationality, bioreactors are classified into batch, fed-batch, multi-staged and continuous reactors. For optimum effectivity, natural environment for growth of cell biomass is also developed, regulating the pH, temperature, nutritional concentration, and aeration input, along with constant stirring, to maintain homogeneity within the reactor. These parameters are mutable and are altered to the requirement of the operation to enhance the rate of biological reaction as well as reduce the reaction time.

13.8 CONCLUSION AND FUTURE PERSPECTIVES

Irrespective of various limitations, intense research needs to be conducted about the conventional and non-conventional strategies, alongside biological strategies, for their age-old reliability over newly introduced techniques. Yet a vast domain of biological techniques remains unexplored, which, with future advancement, will get into the spotlight overcoming major limitations.

Recently, computational strategies are being used to decipher innovative remediation methods. Modern usage of theoretical tools such as artificial neural networking (ANN), gene editing tools, metabolomics, transcriptomics, and proteomics studies have taken the first initial step towards achieving this goal. Such bioinformatics, mathematical, and computational modellings have also cut short the duration of research to a greater extent, minimizing the probability of error and obtaining highly accurate outputs. In silico gene editing and multiomics approaches have also helped in development of recombinant variants having much higher degree of efficient remediation (Paul et al., 2021).

REFERENCES

Abdullah, N., Yusof, N., Lau, W.J., Jaafar, J., & Ismail, A.F. 2019. Recent trends of heavy metal removal from water/wastewater by membrane technologies. *Journal of Industrial and Engineering Chemistry*, 76, 17–38.

Ahmed, T., Shahid, M., Azeen, F., Rasul, I., Shah, A.A., & Muhammad, S. 2018. Biodegradation of plastics: Current scenario and future prospects of environmental safety. *Environmental Science and Pollution Research*, 25(1), 1–12.

Al-Dahhan, M., Al-Ani, F., & Al-Saned, A. 2018. Biodegradation of phenolic compounds in wastewater by microalgae: A review. *Environmental Engineering and Sustainable Development*, 162.

Ali, H., Khan, E., & Sajad, M.A. 2013. Phytoremediation of heavy metals-concepts and applications. *Chemosphere*, 91(7), 869–881.

Arias, D., Cisternas, L.A., & Rivas, M. 2017. Biomineralization mediated by ureolytic bacteria applied to water treatment: A review. *Crystals*, 7(11), 345.

Asri, M., Elabed, S., Koraichi, S.I., & Ghachtouli, N.E. 2019. Biofilm-based system for industrial wastewater treatment. In *Handbook of Environmental Materials Management*, (pp. 1767–1787). Springer International Publishing AG.

Awa, S.H., & Hadibarata, T. 2020. Removal of heavy metals in contaminated soil by phytoremediation mechanism: A review. *Water, Air, & Soil Pollution*, 231(47).

Bagchi, S., & Behera, M. 2019. Microbial fuel cells: A sustainable technology for pollutant removal and power generation. In Krishnaraj, N., & Sani, R.K. (Eds.), *Bioelectrochemical Interface Engineering* (pp. 91–116). Wiley Online Library.

Benzerara, K. 2011. Biomineralization. In Gargaud, M., et al. (Eds.), *Encyclopedia of astrobiology*. Springer.

Bharagava, R.N., & Mishra, S. 2018. Hexavalent chromium reduction potential of *Cellulosimicrobium* sp. isolated from common effluent treatment plant of tannery industries. *Ecotoxicology and Environmental Safety*, 147, 102–109.

Bhattacharya, A., & Gupta, A. 2013. Evaluation of *Acinetobacter* sp. B9 for Cr (VI) resistance and detoxification with potential application in bioremediation of heavy-metals-rich industrial wastewater. *Environmental Science and Pollution Research*, 20(9), 6628–6637.

Borai, A.M., Eltayeb, K.M., Mostafa, A.R., & El-Assar, S.A. 2016. Biodegradation of industrial oil-polluted wastewater in Egypt by bacterial consortium immobilized in different types of carriers. *Polish Journal of Environmental Studies*, 25(5), 1901–1909.

Briffa, J., Sinagra, E., & Blundell, R. 2020. Heavy metal pollution in the environment and their toxicological effects on humans. *Heliyon*, 6(9), E04691.

Burakov, A.E., Galunin, E.V., Burakova, I.V., Kucherova, A.E., Agarwal, S., & Gupta, V.K. 2018. Adsorption of heavy metals on conventional and nanostructured materials for wastewater treatment purposes: A review. *Ecotoxicology and Environmental Safety*, 148, 702–712.

Chaturvedi, M.K. 2011. Studies on chromate removal by chromium-resistant *Bacillus* sp. isolated from tannery effluent. *Journal of Environmental Protection*, 2(1), 76.

Chen, Y., Hu, W., Li, P., Liu, Y., Chen, X., & Zhang, Y. 2021. Phytoremediation of hexavalent chromium by mung bean through bio-accumulation and bio-stabilization in a short duration. *International Journal of Environmental Science and Technology*, 18, 3023–3034.

Congeevaram, S., Dhanarani, S., Park, J., Dexilin, M., & Thamaraiselvi, K. 2007. Biosorption of chromium and nickel by heavy metal resistant fungal and bacterial isolates. *Journal of Hazardous Materials*, 146(1–2), 270–277.

Costerton, J.W. 1999. Introduction to biofilm. *International Journal of Antimicrobial Agents*, 11, 217–221.

Ezugbe, E.O., & Rathilal, S. 2020. Membrane technologies in wastewater treatment: A review. *Membranes*, 10(5), 89.

Fauziah, S.H., Jayanthi, B., Emenike, C.U., & Agamuthu. 2017. Remediation of heavy metal contaminated soil using potential microbes isolated from a closed disposal site. *International Journal of Bioscience, Biochemistry and Bioinformatics*, 7(4), 230–237.

Fomina, M., & Gadd, G.M. 2014. Biosorption: Current perspectives on concept, definition and application. *Bioresource Technology*, 160, 3–14.

Frascari, D., Zanaroli, G., & Danko, A.S. 2015. 'In situ' aerobic cometabolism of chlorinated solvents: A review. *Journal of Hazardous Materials*, 283, 382–399.

Fulekar, M.H., Sharma, J., & Tendulkar, A. 2012. Bioremediation of heavy metals using biostimulation in laboratory bioreactor. *Environmental Monitoring and Assessments*, 184(12), 7299–7307.

Geng, N., Wu, Y., Zhang, M., Tsang, D.C.W., Rinklebe, J., & Ok, Y.S. 2019. Bioaccumulation of potentially toxic elements by submerged plants and biofilms: A critical review. *Environment International*, 131.

Gidarakos, E., & Aivalioti, M. 2007. Large scale and long term application of bioslurping: The case of a Greek petroleum refinery site. *Journal of Hazardous Materials*, 149, 574–581.

Hansda, A., Kumar, V., & Anshumali. 2016. A comparative review towards potential of microbial cells for heavy metal removal with emphasis on biosorption and bioaccumulation. *World Journal of Microbiology and Biotechnology*, 32(170).

Hobson, A.M., Frederickson, J., & Dise, N.B. 2005. CH_4 and N_2O from mechanically turned windrow and vermicomposting systems following in-vessel pre-treatment. *Waste Management*, 25, 345–352.

Jafari, S.A., Cheraghi, S., Mirbakhsh, M., Mirza, R., & Maryamabadi, A. 2015. Employing response surface methodology for optimization of mercury bioremediation by Vibrio parahaemolyticus PG02 in coastal sediments of Bushehr, Iran. *CLEAN–Soil, Air, Water*, 43(1), 118–126.

Jin, Y., Luan, Y., Ning, Y., & Wang, L. 2018. Effects and mechanism of microbial remediation of heavy metals in soil: A critical review. *Applied Sciences*, 8(8), 1336.

Kanmani, P., Aravind, J., & Preston, D. 2012. Remediation of chromium contaminants using bacteria. *International Journal of Environmental Science and Technology*, 9, 183–193.

Khan, F.I., Husain, T., & Hajazi, R. 2004. An overview and analysis of site remediation technologies. *Journal of Environmental Management*, 71(2), 95–122.

Kim, I.H., Choi, J.H., Joo, J.O., Kim, Y.K., Choi, J.W., & Oh, B.K. 2015. Development of a microbe-zeolite carrier for the effective elimination of heavy metals from seawater. *Journal of Microbiology, Biotechnology*, 25(9), 1542–1546.

Kim, S., Krajmalnik-Brown, R., Kim, J.O., & Chung, J. 2014. Remediation of petroleum hydrocarbon-contaminated sites by DNA diagnosis-based bioslurping technology. *Science of the Total Environment*, 497–498.

Kumar, L., & Bharadvaja, N. 2020. Microbial remediation of heavy Metals. In Shah, M.P. (Ed.), *Microbial Bioremediation & Biodegradation* (pp. 49–72). Springer.

Kumar, R., Bhatia, D., Singh, R., Rani, S., & Bishnoi, N.R. 2011. Sorption of heavy metals from electroplating effluent using immobilized biomass *Trichoderma viride* in a continuous packed-bed column. *International Biodeterioration & Biodegradation*, 65(8), 1133–1139.

Kumaran, N.S., Sundaramanicam, A., & Bragadeeswaran, S. 2011. Adsorption studies on heavy metals by isolated cyano bacterial strain (*Nostoc* sp.) from Uppanar estuarine water, southeast coast of India. *Journal of Applied Sciences Research*, 7(11), 1609–1615.

Lebeau, T., & Jézéquel, K. 2008. Performance of bioaugmentation-assisted phytoextraction applied to metal contaminated soils: A review. *Environmental Pollution*, 153(3), 497–522.

Liu, L., Li, W., Song, W., & Guo, M. 2018. Remediation techniques for heavy metal-contaminated soils: Principles and applicability. *Science of the Total Environment*, 633, 206–219.

López, D., Vlamakis, H., & Kolter, R. 2010. Biofilms. *Cold Spring Harbor Perspectives in Biology*, 7, a000398–a000401.

Mallampati, S.R., Mitoma, Y., Okuda, T., Simion, C., & Lee, B.K. 2015. Dynamic immobilization of simulated radionuclide 133Cs in soil by thermal treatment/vitrification with nanometallic Ca/CaO composites. *Journal of Environmental Radioactivity*, 139, 118–124.

Mallikarjunaiah, S., Pattabhiramaiah, M., & Metikurki, B. 2020. Application of nano-biotechnology in the bioremediation of heavy metals and wastewater management. In Thangadurai, D., Sangeetha, J., & Prasad, R. (Eds.), *Nanotechnology for Food, Agriculture, and Environment* (pp. 297–321). Springer.

Mane, P.C., & Bhosle, A.B. 2012. Bioremoval of some metals by living algae *Spirogyra sp.* and *Spirullina sp.* from aqueous solution. *International Journal of Environmental Research*, 6(2), 571–576.

Manisalidis, I., Stavropoulou, E., Stavropoulos, A., & Bezirtzoglou, E. 2020. Environmental and health impacts of air pollution: A review. *Frontiers in Public Health*, 8(14).

Mohanty, M., & Patra, H.K. 2012. Effect of chelate-assisted hexavalent chromium on physiological changes, biochemical alterations, and chromium bioavailability in crop plants-an in vitro phytoremediation approach. *Bioremediation Journal*, 16(3), 147–155.

Muhammad, N., Nafees, M., Khan, M.H., Ge, L., & Lisak, G. 2020. Effect of biochars on bio-accumulation and human health risks of potentially toxic elements in wheat (*Triticum aestivum* L.) cultivated on industrially contaminated soil. *Environmental Pollution*, 260.

Muneer, B., Iqbal, M.J., Shakoori, F.R., & Shakoori, A.R. 2013. Tolerance and biosorption of mercury by microbial consortia: Potential use in bioremediation of wastewater. *Pakistan Journal of Zoology*, 45(1), 247–254.

Myckatyn, T.M., Cohen, J., & Chole, R.A. 2016. Clarification of the definition of "biofilm". *Plastic and Reconstructive Surgery*, 137(1), 237e–238e.

Nayak, A.K., Panda, S.S., Basu, A., & Dhal, N.K. 2018. Enhancement of toxic Cr (VI), Fe, and other heavy metals phytoremediation by the synergistic combination of native *Bacillus cereus* strain and *Vetiveria zizanioides* L. *International Journal of Phytoremediation*, 20(7), 682–691.

Paul, O., Jasu, A., Lahiri, D., Nag, M., & Ray, R.R. 2021. In situ and ex situ bioremediation of heavy metals: The present scenario. *Journal of Environmental Engineering and landscape Management*, 29(4), 454–469.

Rabin, N., Zheng, Y., Opoku-Temeng, C., Du, Y., Bonsu, E., & Sintim, H. 2015. Biofilm formation mechanisms and targets for developing antibiofilm agents. *Future Medicinal Chemistry*, 7(4).

Redha, A.A. 2020. Removal of heavy metals from aqueous media by biosorption. *Arab Journal of Basic and Applied Sciences*, 27(1), 183–193.

Renner, L.D., & Weibel, D.B. 2011. Physicochcemical regulation of biofilm formation. *MRS Bulletin*, 36(5), 347–355.

Salehizadeh, H., & Shojaosadati, S.A. 2003. Removal of metal ions from aqueous solution by polysaccharide produced from *Bacillus firmus*. *Water Research*, 37(17), 4231–4235.

Sarkar, A., Bhagat, J., Sarker, M.S., Gaitonde, D.C.S., & Sarker, S. 2017. Evaluation of impact of bioaccumulation of PAH from the marine environment on DNA integrity and oxidative stress in marine rock oyster (*Saccostrea cucullata*) along the Arabian sea coast. *Ecotoxicology*, 26, 1105–1116.

Schmidt, A., Haferburg, G., Sineriz, M., Merten, D., Büchel, G., & Kothe, E. 2005. Heavy metal resistance mechanisms in actinobacteria for survival in AMD contaminated soils. *Geochemistry*, 65(S1), 131–144.

Sharma, S., Tiwari, S., Hasan, A., Saxena, V., & Pandey, L.M. 2018. Recent advances in conventional and contemporary methods for remediation of heavy metal-contaminated soils. *3 Biotech*, 8(216).

Taştan, B.E., Ertuğrul, S., & Dönmez, G. 2010. Effective bioremoval of reactive dye and heavy metals by *Aspergillus versicolor*. *Bioresource Technology*, 101(3), 870–876.

Varjani, S., Rakholiya, P., Ng, H.Y., You, S., & Teixeira, J.A. 2020. Microbial degradation of dyes: An overview. *Bioresource Technology*, 314, 123728.

Whelan, M.J., Coulon, F., Hince, G., Rayner, J., McWatters, R., & Snape, I. 2015. Fate and transport of petroleum hydrocarbons in engineered biopiles in polar regions. *Chemosphere*, 131, 232–240.

Zamora-Ledezma, C., Negrete-Bolagay, D., Figueroa, F., Zamora-Ledezma, E., Ni, M., & Guerrero, V.H. 2021. Heavy metal water pollution: A fresh look about hazards, novel and conventional remediation methods. *Environmental Technology & Innovation*, 22, 101504.

Zhao, R., Wagn, B., Cai, Q.T., Li, X.X., Liu, M., Hu, D., Guo, D.B., & Fan, C. 2016. Bioremediation of hexavalent chromium pollution by *Sporosarcina saromensis* M52 isolated from offshore sediments in Xiamen, China. *Biomedical and Environmental Sciences*, 29(2), 127–136.

14 Microbial Bioremediation
Importance, Features, Applications, and Future Perspectives

Soching Luikham and Jhimli Bhattacharya

14.1 INTRODUCTION

Human activities release massive quantities of organic and inorganic substances into the environment every year. These discharges can be intentional and appropriate under certain conditions (e.g., commercial pollutants), but they can also be unintentional (e.g., oil spills or chemicals). In both aquatic and terrestrial habitats, many of these chemicals are poisonous and persistent. The aggregation of such harmful chemicals in abundance leads to surface water, groundwater, and soil contamination. In several countries, the concerns linked with contaminated areas are becoming more prominent, and the contaminants are circulating in the environment (Figure 14.1). Water bodies that have been polluted are usually the product of earlier industrial operations. The threat is widespread, and the number of active contaminated locations is estimated to be large (Cairney 1993). Polluted environments are now increasingly recognized as a possible risk to human health. Oil spill is among the most significant sources of pollution in aquatic environments. Crude oil spills from containers, drilling rigs, spills of contaminants (such as gasoline and diesel) with their various by-products, and oil platforms can all cause oil spills. The consequences of oil pollution are significant. By transformation, mineralization, or alteration, the detoxification process targets the toxic compounds (Omokhagbor Adams et al. 2015).

The typical method just relocates the contamination and may pose considerable risks in hazardous material excavation, handling, and transportation. Rather than using these traditional practices, it is preferable to destroy the contaminants or convert them into harmless compounds. High-temperature burning and different types of chemical degradation were used in the past (e.g., UV oxidation, base catalysis, de-chlorination). It can be quite efficient for reducing the levels of various pollutants, but it has several disadvantages too.

Bioremediation is a process in which microorganisms play distinct and distinctive functions in the detoxification of contaminants in the environment. To break down the pollutants, bioremediation relies on metabolism of microbes with favourable environmental conditions and appropriate nutrition. As a result, it employs low-cost,

DOI: 10.1201/9781003297901-14

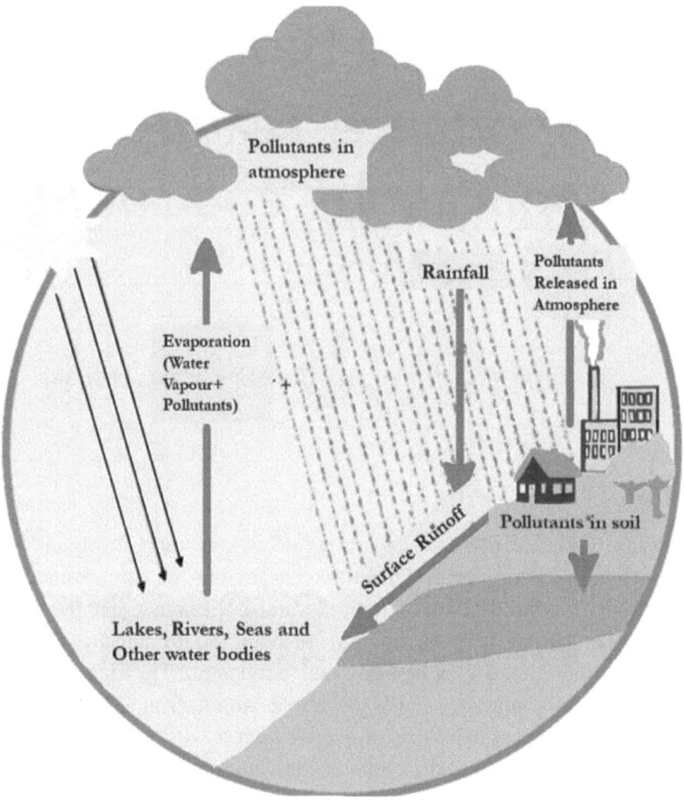

FIGURE 14.1 The circulation of toxic contaminants in the environment.

low-technology approaches that are widely accepted by the public and may often be conducted onsite. The majority of organic matter is mineralized CO_2 by microorganisms. The purpose of microbial bioremediation is to use biological systems to facilitate the cleanup or detoxification of man-made contaminants. Bioremediation appears to be an alternative to traditional clean-up procedures, according to recent studies. Bioremediation has been utilized successfully worldwide. As more knowledge and expertise are obtained, techniques improve. We aim to help by offering a clear, practical understanding of the processes involved in bioremediation, the benefits and drawbacks of the approach, and the concerns to consider when working with a bioremediation application (Omokhagbor Adams et al. 2015). Although bioremediation is a multidisciplinary endeavour, it is based on a full understanding and use of microbial ecology concepts.

Bioremediation is one such process of microbial cells producing energy through a redox reaction. Respiration and other biological activities required for cell survival and replication are included in these reactions. An energy source (electron donor), nutrients, and an electron acceptor are all required by most delivery systems. Different types of microbial electron acceptor classes, such as oxygen, manganese, nitrate, sulphate, iron, or carbon dioxide, and their corresponding redox potentials, can be involved in bioremediation. According to Venosa et al., the most crucial need for bioremediation is the existence of microorganisms with the proper metabolic capability (Venosa et al. 2002). We have explored different bioremediation methods in this chapter and found that microbiological methods (for remediation) have gained significant importance in recent years. The importance of such methods and the need/search for advanced and effective technologies will be more prominent in the near future, because the amount of environmental degradation we are facing these days is quite alarming, and its extent is increasing day by day.

14.2 METHODS

We sought to review the various reported data and methods on microbial bioremediation in the literature. The publications relevant to the title were considered mostly.

14.3 PRINCIPLES OF BIOREMEDIATION

Composting and wastewater treatment are two well-known examples of old environmental biotechnologies. Recent research in molecular biology and ecology suggests that biological processes can be made more efficient. Bioremediation is the method of biologically degrading organic wastes under controlled circumstances to a harmless state or values below relatively low concentrations (Kensa 2011). Bioremediation is one such process which degrades environmental contaminants into relatively less hazardous forms by utilizing microorganisms, typically bacteria. It degrades pollutants that may be harmful to humans and the surroundings using naturally occurring microorganisms, plants, or fungi. Microorganisms may be native to a polluted location or isolated from a different place and brought to polluted sites. Microorganisms should biochemically destroy contaminants and transform them into harmless compounds for bioremediation to be an efficient process. Bioremediation, like all other technologies, has inherent limitations. They degrade slowly or not at all, making it difficult to anticipate the rates of bioremediation clean-up; there are any standards for determining if a contaminant can be eliminated. Bioremediation procedures are often less expensive than traditional methods like incineration, and even some contaminants can be processed onsite, lowering exposure hazards for clean-up workers and potentially wider exposure due to transportation incidents. Bioremediation is much more acceptable to the public than other technologies because it is derived from natural origin. Usually, bioremediation systems are operated in aerobic circumstances, but operating a few under anaerobic conditions can allow microbial organisms to digest compounds that are otherwise resistant to degradation (Holliger and Zehender 1996). Bacteria and fungi are used to remove or degrade chemical pollutants due to their innate capacity to adapt and proliferate in adverse circumstances.

The origin of the contaminant, which might include chemical pesticides, dyes, chlorinated compounds, heavy metals, plastics, greenhouse gases, sewage, hydrocarbons, and nuclear waste determines the pollutant removal procedure (Azubuike, Chikere, and Okpokwasili 2016). Microbial bioremediation can be categorized based on the type of substrate that is being removed or degraded, namely (1) microbial bioremediation of inorganic pollutants and (2) microbial bioremediation of organic pollutants (Srivastava et al. 2014)

14.3.1 Microbial Bioremediation of Inorganic Pollutants

Toxic metals and their derivatives are major inorganic contaminants that come from a variety of industrial operations, including metallurgy, mining, chemical production, and power plants (Goyal, Jain, and Banerjee 2003). Microbiological metal ion removal from the environment is a novel bio method that is also the most cost-effective. It is critical to understand that microorganisms cannot destroy metal ions and can only convert them from higher to lower oxidation states to stabilize them (Shukla, Rai, and Dubey 2009). Ion exchange, physical adsorption, and complexation are the first steps in the interaction between microbial cells and metal ions, followed by physiochemical reactions like bio-sorption. During the microbial conversion of metal ions, particular enzymes for oxidation, methylation, reduction, precipitation, and de-alkylation are involved in biochemical reactions (Han and Gu 2004), for example, *Escherichia coli*, *Pseudomonas aeruginosa*, *Staphylococcus carnosus*, and *Staphylococcus xylosus* (Pazirandeh et al. 1995).

14.3.2 Microbial Bioremediation of Organic Pollutants

Organic substances such as polymers, biocides, flame retardants, solvents, crude oil, chlorinated organic, and explosives compounds are used widely and extensively. Microbes have demonstrated the ability to break down most organic chemicals in the environment, including PAHs, PCBs, and PBEs (Hiraishi 2008) (Figure 14.2). Organic chemical biodegradation is the catabolic action of bacteria that converts them into less or completely harmless wastes. The microorganisms utilize the

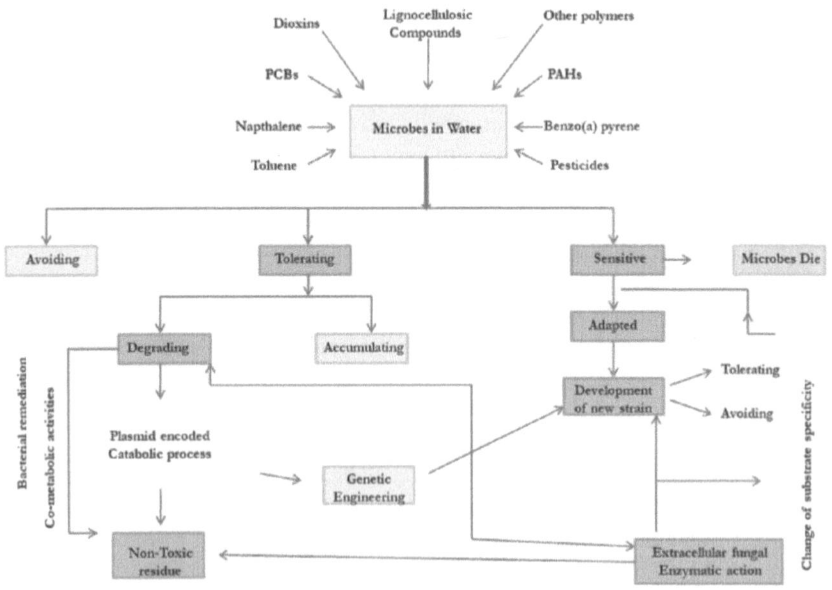

FIGURE 14.2 Strategy of microbes to degrade organic compounds.

organic substrate as the source of organic carbon and energy. Genetic engineering is one of the most advanced biotechnological branches, delivering a limitless number of genetic codes that can be used to build potentially degrading and substrate-specific microorganisms (Margesin, Zimmerbauer, and Schinner 2000).

14.4 REQUIREMENTS TO SUPPORT MICROBIAL BIOREMEDIATION

Requirements to support the biological function of bioremediation are as follows:

i) **The concentration of contaminants**: It is directly independent of microbial activity. Contaminants may have hazardous effects on the microorganisms present when concentrations are too high. Low contaminant concentrations may hinder the stimulation of bacterial degradation enzymes.

ii) **Bioavailability of contaminants**: For pollutants that are more strongly absorbed in solids, bioavailability for microbial responses is lower.

iii) **Characteristics of the site**: It also has a big influence on the efficacy of any bioremediation technique. For bioremediation applications, site environmental parameters such as pH, with an ideal range of 6–8, are vital to consider.

iv) **Oxygen content and redox potential**: The presence of electron acceptors such as manganese oxides, nitrate, sulphate, and iron oxides influences the oxidizing or reducing conditions, that is, redox potential.

v) **Nutrients present:** They are required for microbial cell division and proliferation. To induce bioremediation, appropriate amounts of trace nutrients can be given in a usable form or via an organic substrate that also functions as an electron donor.

vi) **The amount of moisture content**: Microbial growth necessitates a high level of water in the surrounding environment. Microorganisms require 12% to 25% moisture for optimal growth and proliferation (Mukherjee and Das 2005).

vii) **Temperature maintenance**: It directly impacts microbial metabolism and, as a result, microbial activity in the environment. The rate of biodegradation increases with higher temperature and decreases with lower temperature.

14.5 TYPES OF MICROBIAL POPULATIONS FOR REMEDIATION PROCESSES

Microorganisms have the potential to adapt and develop in sub-zero temperatures, intense heat, humid conditions, the presence of an excess of oxygen in water, and anaerobic environments with dangerous substances or different waste streams. The most important necessities are an energy source and carbon. Microorganisms and several other biologically active systems can be employed to decompose or remove environmental problems due to their adaptability. We can further divide it into the following:

i) **Aerobic bacteria:**
Bacteria that require oxygen to grow are known as aerobic bacteria. *Alcaligenes*, *Pseudomonas*, *Sphingomonas*, *Mycobacterium*, and *Rhodococcus* are some examples of aerobic bacteria, known for their degradation ability. Pesticides and hydrocarbons, including alkanes and poly-aromatic chemicals, can be degraded by these bacteria multiple times.

ii) **Methylotrophs bacteria (uses methane):**
Aerobic bacteria use methane as a source of energy and carbon. Methane mono-oxygenase, the first enzyme in the aerobic degradation pathway, has a vast range of substrates and can degrade a variety of molecules, which includes the compound that is chlorinated aliphatic trichloroethylene and 1,2-dichloroethane.

iii) **Anaerobic bacteria:**
In the non-existence of oxygen, anaerobic bacteria grow. Anaerobic microorganisms are less commonly used than aerobic bacteria. The use of anaerobic bacteria for the microbial bioremediation of PCBs (polychlorinated biphenyls) in water bodies, similarly de-chlorination of TCE (trichloroethylene) solvents, and chloroform, is gaining popularity.

iv) **Fungi (ligninolytic fungi):**
Fungi, like the white-rot fungus *Phanaero chaetechrysosporium*, can decompose a wide spectrum of persistent or harmful contaminants in the environment. Straw, sawdust, and corn cobs are common substrates.

v) **Nature of bacteria (gram-positive and gram-negative):**
Gram-positive bacteria have a wide metabolic range and survive in environments contaminated by aromatic hydrocarbons and hazardous metals (Narancic et al. 2012). Gram-positive bacteria include *Rhodococcus* sp. and *Staphylococcus* sp., among others. Gram-negative bacteria are more resistant to metal contamination in general (Nweke et al. 2007). Lăzăroaie (2010) also reported gram-negative bacteria are thought to be more resistant to a toxic mix of saturated hydrocarbons, mono-aromatic hydrocarbons, and poly-aromatic hydrocarbons (Lăzăroaie 2010).

14.6 VARIOUS FACTORS AFFECTING MICROBIAL BIOREMEDIATION

Various abiotic and biotic variables influence microbial cell activity and growth, affecting a variety of biological activities in a community of microbes. To improve microbiological action and identify the effective bioremediation process, a thorough understanding of microbial ecology is essential (Watanabe et al. 2000). There are two different types of factors that influence microbial processes: (1) environmental physicochemical features or abiotic factors and (2) biological factors or biotic factors.

14.6.1 Physicochemical or Abiotic Factors Affecting Bioremediation Processes

Redox potential, ionic strength, pH, solubility, temperature, etc. are all examples of physicochemical factors. Metal ions are reduced to insoluble forms by microorganisms from higher to lower oxidation states (pH-dependent) as soluble metal ions in bioremediation procedures. In the degradation of organic pollutants, solubility is crucial (Pieper and Reineke 2000). The solubility of metal ions is affected by pH values, which increase as the pH of the medium decreases, impacting microbial cell adsorption (Blázquez et al. 2009). *Pseudomonas* sp. and *Burkholderia acepacia*, which create organic acids, produce gluconic acid, while *Rhizobium* sp. and *Bacillus firmus* make ketogluconic acid (Robles-González, Fava, and Poggi-Varaldo 2008), which lowers the pH of the system and promotes metal ion solubility.

14.6.2 Biological or Biotic Factors Affecting Bioremediation Processes

Biological variables include a variety of microorganism characteristics that influence substrate decomposition. Chemotaxis is a beneficial bacterial activity for the breakdown of recalcitrant organic molecules (Pandey and Jain 2002). Maphosa et al. described it as a function of metagenomics or community genomics in which information from member species of consortia that facilitate substrate breakdown is provided by metagenomics sequencing (Maphosa et al. 2012). Furthermore, microbe optimal growing circumstances are unpredictably variable (Ingham et al. 2007). The toxic pressure exerted by pollutants on the microorganisms that cause enzymatic changes determines when the microbial breakdown of xenobiotics begins (Pandey and Jain 2002).

Some of the promising and successful bioremediation techniques are given subsequently (Figure 14.3).

i) **Intrinsic/in situ bioremediation:** This method has gained favor since the contaminant is treated at the pollution site, which reduces excavation costs. Intrinsic bioremediation takes place and is based on naturally occurring biological processes carried out by local microorganisms. Natural attenuation comprises both physical and chemical processes, and intrinsic bioremediation is one of them (Hinchee, Wilson, and Downey 1995). These methods should, in theory, be less expensive. Some in situ bioremediation procedures (bioventing, bio-stimulation, bio-sparging, bio-augmentation, and phytoremediation) could be improved, while others could be left alone (natural attenuation or intrinsic bioremediation). Dyes, chlorinated solvents, hydrocarbons, and heavy metals have all been successfully treated using in situ bioremediation approaches (Roy et al. 2015).

   a) **Bioventing**: The most common/popular intrinsic process is bioventing, which includes providing nutrients to contaminated soil via wells to promote the indigenous microorganisms. It is important and effective for hydrocarbons that are simple and can be utilized in situations where pollution is found far beneath the surface.

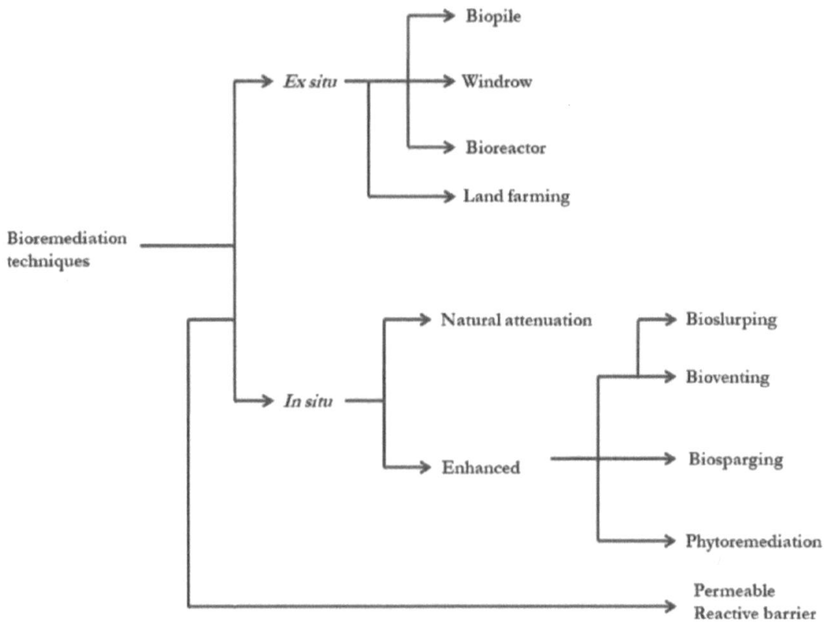

FIGURE 14.3 Various types of bioremediation techniques.

b) **Bio-sparging**: Bio-sparging is the process of inserting pressurized air under the water table to increase the groundwater level of oxygen and increase the biological degradation of pollutants by microorganisms.
c) **Bio-stimulation**: The nutrients (typically nitrogen, carbon, and phosphorus), oxygen, or other electron donors or acceptors are known as bio-stimulation. The purpose of these changes is to improve the number or activity of naturally occurring microbes that can be utilized in bioremediation.
d) **Bio-augmentation**: The inclusion of microbes which can bio-transform or biodegrade a specific pollutant is commonly known as bio-augmentation. The constant addition of microorganisms to a bioreactor for above-ground groundwater treatment can improve this process.

14.6.3 Extrinsic/Ex Situ Bioremediation

Ex situ bioremediation is an above-ground treatment in which soils are excavated and cleaned, or sediments are taken from the subsurface and decontaminated. Recently, ex situ bioremediation strategies have attempted to use genetically modified microbes. Contaminated groundwater is inserted into the surface, mixed with nutrients, and then injected upstream of the contaminated plume to bio-stimulate contaminant breakdown ex situ by native species.

a) **Land farming**: Land farming is a basic process that requires the removal of polluted substances, spreading it all over the surface, and manually tilling it periodically until pollutants have decomposed. The goal is to encourage native bio-derivative bacteria, thus making it easy to degrade pollutants.
b) **Composting**: Composting is the process of mixing contaminated soil/substances with non-hazardous organic waste. The availability of these organic materials encourages the growth of a diverse microbial community while also raising the temperature.
c) **Bio-piles**: Landfarming and composting are combined in bio-piles. Engineered cells are similar to aerated compost piles in terms of production. Surface contamination is usually treated with this method.
d) **Bioreactors**: Ex situ treatment of contaminated water released from a polluted location is done in reactors that are slurry, or aqueous reactors. In reactor bioremediation, contaminated water is processed through a designed containment system. A slurry bioreactor is a containment tank and device that is used to generate a three-phase (solid, liquid, and gas) mixing environment to speed up bioremediation.

14.7 NEW TECHNIQUES ON BIOREMEDIATION

i) **Genetic engineering approaches for microbial remediation**: The novel field of genetic engineering to generate bioremediation-specific species has a lot of prospects. The use of genetically modified microbes to impact their potential to use specific pollutants like pesticides and hydrocarbons is gaining popularity. By detecting and identifying degradative genes and converting them into appropriate hosts through a suitable vector in a contained environment, recombinant DNA techniques investigate an organism's property to degrade a xenobiotic (Omokhagbor Adams et al. 2015). *Bacillus infernos*, which can withstand a very high temperature, and *Methanococcus jannaschii*, which can withstand pressures that are as high as 230 atm as well as high temperatures, are two further examples. *Dehalococcoides* have the potential ability to de-chlorinate tri, tetra, and pentachloro dioxin (Srivastava et al. 2014).
ii) **Mycoremediation for microbial bioremediation**: Mycoremediation is a type of bioremediation that involves the use of fungus to decontaminate an area. Paul Stamets created the term "mycoremediation," which refers to the use of fungal mycelia in bioremediation. Decomposition, which is by mycelium, is one of the most novel functions of fungi in the environment. Acids and extracellular enzymes secreted by the mycelium degrade cellulose and lignin, the two major building elements required for plant fiber. These are molecules with lengthy carbon and hydrogen chains that are structurally comparable to numerous organic contaminants.
iii) **Bioremediation of chromium-contaminated soils**: The remediation of metal and organic polluted soils, water, and sediments has been intensively

researched. There is a comparison on the benefits of employing bioremediation vs other techniques. Each of the strategies outlined contains benefits and drawbacks. Bioremediation has been widely employed in comparison to all other approaches since it is cost-effective and does not generate more waste in the environment (Ramasamy, Kamaludeen, and Banu 2007).

14.8 FACTORS TO MONITOR MICROBIAL BIOREMEDIATION

The Oxidation Reduction Potential, or redox, in soil, groundwater or aquatic environment, for example oil spills (Figure 14.4), as well as oxygen content, pH, electron acceptor/donor concentrations, temperature, and breakdown product concentrations, is used to monitor the bioremediation process indirectly (e.g. carbon dioxide).

i) **Advantages of microbial bioremediation**: Since biological remediation is indeed a natural process, the public views it as a suitable waste method of treatment for polluted environmental materials such as water, soil, etc. The number of bacteria effective in decomposing the pollutant rises when it is available; when the pollutant is removed, the biodegradative population declines. Water, carbon dioxide, and cell biomass are some examples of treating by-products that are normally non-toxic.
   a. Bioremediation can theoretically be used to destroy a broad range of pollutants. Many chemicals that are considered dangerous by law can be changed into non-hazardous products.
   b. Instead of moving pollutants from one medium to another medium, such as from water to land or air, target pollutants can be destroyed.

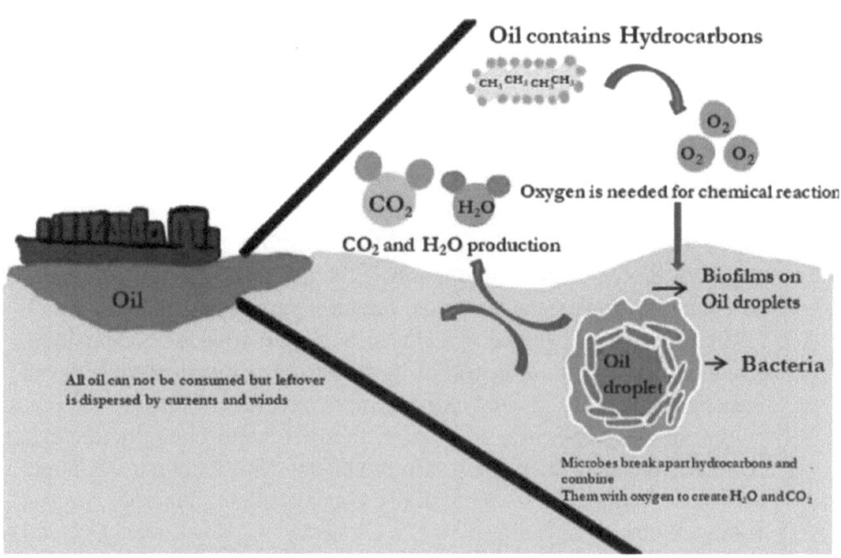

FIGURE 14.4 Graphical representation of oil spills in an aquatic environment.

c. Bioremediation can frequently be done onsite, with little or no disturbance to routine operations.
d. Bioremediation has the potential to be less costly than other different technologies for hazardous waste clean-up.

ii) Disadvantages of microbial bioremediation

Bioremediation is only possible with biodegradable chemicals. Not all chemicals are capable of complete and rapid breakdown.

a. Some people are concerned that biodegradation products would be more persistent or harmful than the parent molecule.
b. Biological processes are often very specialized. Important site concerns include the presence of biologically capable bacterial diversity, appropriate environmental developmental stage, and optimal concentrations of nutrients and contaminants.
c. It's difficult to extrapolate from lab and small-scale experiments to large field operations.
d. More study is needed on microbial bioremediation strategies, which are ideal for areas with complex combinations of contaminants that are not equally disseminated in the surroundings.
e. Bioremediation process may require more time than other treatment alternatives such as soil excavation and removal or incineration.
f. Acceptable performance standards for bioremediation remain a source of regulatory confusion. There being no universally accepted concept of "clean," measuring bioremediation performance is challenging, and bioremediation treatments have no acceptable goals.

14.9 CONCLUSION

Bioremediation is significantly cost-effective than other hazardous waste cleaning solutions. It has many economic and efficient advantages that can be used in places without excavation. The most important step in bioremediation is the characterization of the site/place that aids in the selection of the appropriate and feasible microbiological technique (ex situ or in situ). Because of the increased expenditures associated with excavation and shipping, ex situ bioremediation procedures are typically more expensive. Nonetheless, they can indeed be utilized in a controlled manner to remediate a wide spectrum of contaminants. In situ bioremediation techniques, on the other hand, do not involve any additional costs due to excavation; the cost of on-site equipment installation can also be negligible in many cases. As a result, the expense of treatment is not the most important factor when selecting a bioremediation strategy for an in situ technique. During the determination of the important and appropriate method(s) to treat a contaminated environment successfully, geographical attributes of the contaminated site, such as type of soil, pollutant depth and type, site concerning human habitat, functional properties, etc., should also be considered.

Acknowledgement: JB acknowledges the financial assistance (via research projects) from DBT Govt. of India (sanction no. BT/PR25026/NER/95/963/2017).

Conflict of Interest: The authors declare that they have no conflict of interest.

REFERENCES

Azubuike, Christopher Chibueze, Chioma Blaise Chikere, and Gideon Chijioke Okpokwasili. 2016. "Bioremediation Techniques–Classification Based on Site of Application: Principles, Advantages, Limitations and Prospects." *World Journal of Microbiology and Biotechnology* 32 (11): 180.

Blázquez, G., F. Hernáinz, M. Calero, M. A. Martín-Lara, and G. Tenorio. 2009. "The Effect of PH on the Biosorption of Cr (III) and Cr (VI) with Olive Stone." *Chemical Engineering Journal* 148: 473–479.

Cairney, Thomas. 1993. *Contaminated Land: Problems and Solutions*. London: Blackie Academic & Professional.

Goyal, N., S. C. Jain, and U. C. Banerjee. 2003. "Comparative Studies on the Microbial Adsorption of Heavy Metals." *Advances in Environmental Research* 7 (2): 311–319.

Han, Xu, and Ji-Dong Gu. 2004. "Sorption and Transformation of Toxic Metals by Microorganisms." *Environmental Microbiolog* 2: 153–175.

Hinchee, Robert E., John Thomas Wilson, and Douglas C. Downey. 1995. "Intrinsic Bioremediation." *International in Situ and On-Site Bioreclamation Symposium* 3: 266.

Hiraishi, Akira. 2008. "Biodiversity of Dehalorespiring Bacteria with Special Emphasis on Polychlorinated Biphenyl/Dioxin Dechlorinators." *Microbes and Environments* 23 (1): 1–12.

Holliger, Christof, and Alexander J. B. Zehender. 1996. "Anaerobic Biodegradation of Hydrocarbons." *Current Opinion in Biotechnology* 7: 326–330.

Ingham, Colin J., Ad Sprenkels, Johan Bomer, Douwe Molenaar, Albert Van Den Berg, Johan E. T. Van Hylckama Vlieg, and Willem M. De Vos. 2007. "The Micro-Petri Dish, a Million-Well Growth Chip for the Culture and High-Throughput Screening of Microorganisms." *Proceedings of the National Academy of Sciences of the United States of America* 104 (46): 18217–18222.

Kensa, V. Mary. 2011. "Bioremediation-An Overview." *Journal of Industrial Pollution Control* 27 (2): 161–168.

Lăzăroaie, Mihaela Marilena. 2010. "Multiple Responses of Gram-Positive and Gram-Negative Bacteria to Mixture of Hydrocarbons." *Brazilian Journal of Microbiology* 41 (3): 649–667.

Maphosa, Farai, Shakti H. Lieten, Inez Dinkla, Alfons J. Stams, Hauke Smidt, and Donna E. Fennell. 2012. "Ecogenomics of Microbial Communities in Bioremediation of Chlorinated Contaminated Sites." *Frontiers in Microbiology* 3: 1–14.

Margesin, R., A. Zimmerbauer, and F. Schinner. 2000. "Monitoring of Bioremediation by Soil Biological Activities." *Chemosphere* 40 (4): 339–346.

Mukherjee, Ashis K., and Kishore Das. 2005. "Correlation between Diverse Cyclic Lipopeptides Production and Regulation of Growth and Substrate Utilization by Bacillus Subtilis Strains in a Particular Habitat." *FEMS Microbiology Ecology* 54 (3): 479–489.

Narancic, Tanja, Lidija Djokic, Shane T. Kenny, Kevin E. O'Connor, Vanja Radulovic, Jasmina Nikodinovic-Runic, and Branka Vasiljevic. 2012. "Metabolic Versatility of Gram-Positive Microbial Isolates from Contaminated River Sediments." *Journal of Hazardous Materials*: 215–216.

Nweke, C. O., C. S. Alisi, J. C. Okolo, and C. E. Nwanyanwu. 2007. "Toxicity of Zinc to Heterotrophic Bacteria from a Tropical River Sediment." *Applied Ecology and Environmental Research* 5 (1): 123–132.

Omokhagbor Adams, Godleads, Prekeyi Tawari Fufeyin, Samson Eruke Okoro, and Igelenyah Ehinomen. 2015. "Bioremediation, Biostimulation and Bioaugmention: A Review." *International Journal of Environmental Bioremediation & Biodegradation* 3 (1): 28–39.

Pandey, Gunjan, and Rakesh K. Jain. 2002. "Bacterial Chemotaxis Toward Environmental Pollutants: Role in Bioremediation." *Applied and Environmental Microbiology* 68 (12): 5789–5795.

Pazirandeh, M., L. A. Chrisey, J. M. Mauro, J. R. Campbell, and B. P. Gaber. 1995. "Expression of the Neurospora Crassa Metallothionein Gene in Escherichia Coli and Its Effect on Heavy-Metal Uptake." *Applied Microbiology and Biotechnology* 43 (6): 1112–1117.

Pieper, Dietmar H., and Walter Reineke. 2000. "Engineering Bacteria for Bioremediation of Persistent Organochlorine." *Current Opinion in Biotechnology* 11: 262–270.

Ramasamy, K., Kamaludeen, and Sara Parwin Banu. 2007. "Bioremediation of Metals: Microbial Processes and Techniques". In Shree N. Singh, Rudra D. Tripathi (Eds.), *Environmental Bioremediation Technologies* (pp.173–187). Berlin and Heilderberg: Springer.

Robles-González, Ireri V., Fabio Fava, and Héctor M. Poggi-Varaldo. 2008. "A Review on Slurry Bioreactors for Bioremediation of Soils and Sediments." *Microbial Cell Factories* 7 (5): 1–16.

Roy, Madhumita, Ashok K. Giri, Sourav Dutta, and Pritam Mukherjee. 2015. "Integrated Phytobial Remediation for Sustainable Management of Arsenic in Soil and Water." *Environment International* 75: 180–198.

Shukla, O. P., U. N. Rai, and Smita Dubey. 2009. "Involvement and Interaction of Microbial Communities in the Transformation and Stabilization of Chromium During the Composting of Tannery Effluent Treated Biomass of Vallisneria Spiralis L." *Bioresource Technology* 100 (7): 2198–2203.

Srivastava, J., R. Naraian, S. J. S. Kalra, and H. Chandra. 2014. "Advances in Microbial Bioremediation and the Factors Influencing the Process." *International Journal of Environmental Science and Technology* 11 (6): 1787–1800.

Venosa, Albert D., Kenneth Lee, Makram T. Suidan, Susana Garcia-Blanco, Susan Cobanli, Moustafa Moteleb, John R. Haines, Gilles Tremblay, and Melynda Hazelwood. 2002. "Bioremediation and Biorestoration of a Crude Oil-Contaminated Freshwater Wetland on the St. Lawrence River." *Bioremediation Journal* 6 (3): 261–281.

Watanabe, K., K. Watanabe, Y. Kodama, K. Syutsubo, and S. Harayama. 2000. "Molecular Characterization of Bacterial Populations in Petroleum-Contaminated Groundwater Discharged from Underground Crude Oil Storage Cavities." *Applied and Environmental Microbiology* 66 (11): 4803–4809.

15 Mitigation of Metal-Induced Stress in Plants with the Help of Microbial Association

Pratik Talukder, Anushka Ghosh, Arpan Banerjee, Subhabrata Paul

15.1 INTRODUCTION

Plants are constantly exposed to a variety of abiotic factors, and most of these abiotic factors are toxic or harmful for plants, which negatively impact on their growth, development, and productivity enormously (Ding et al., 2018). Such types of abiotic factors cause stress to the plants and, hence, are called abiotic stress. These stresses include salinity of the soil and water, drought, low or high availability of nutrients, soil acidity, drastic change in light intensity, exposure to heavy metal toxicity, etc. Now it totally depends upon the stressor (the material which causes stress) how intensely it will affect the plant's physiological activities. Plants, on the other hand, have some defence mechanisms against the adverse effects of the stressors. These combat mechanisms include early senescence, expression of heat shock proteins, avoidance, formation of enzymes having antioxidant activity, accumulation of osmolytes, etc. (Cramer et al., 2011). But the stressors can nullify these defence mechanisms and, hence, can cause detrimental effects on the plant's physiological activities. The increase in the level of stress-combating mechanisms has been observed widely in the presence of plant growth-promoting (PGP) microorganisms, and automatically many PGP microorganisms have the ability to increase the effects of antioxidant enzymes, osmolyte accumulation, and also the expression of certain proteins that will elevate the combat mechanisms (Koyro et al., 2012). Microbes have the potential to change the soil parameters by forming mycorrhizal association. Repeated exposure to stressors will help in the formation of strains of microorganisms that are resistant to it and can sustain in the presence of stressors. Abiotic stressors like salinity, drought, etc. are often chronic in nature and, hence, will provide microorganisms with altered metabolic activities and help in alleviating stress. Hence, microbial community, as a whole, helps in improving the defence mechanisms of plants. Under adverse effects of stressors, microorganisms help in the formation of biomolecular derivatives like microbial exopolysaccharides, the formation of 1-aminocyclopropane-1-carboxylate deaminase (ACC deaminase) enzyme, siderophores, etc. ACC deaminase production occurs in the presence of PGP microbes, which helps in lowering the ethylene level in plants. Ethylene is a gaseous plant growth regulator. Ethylene is synthesized under both biotic and abiotic

DOI: 10.1201/9781003297901-15

stresses and has adverse effect on all the stages of growth and development. This specific enzyme is responsible for cleaving the precursor of ethylene to ACC. ACC deaminase helps in the growth and development of plants. As a whole, ACC deaminase production, by lowering ethylene levels, helps in the alleviation of stress-induced damage (Kumar et al., 2018). Microbial exopolysaccharide is an important carbohydrate. They may be either homo or heteropolymer, depending on the type of organism synthesizing it. There are many biologically beneficial functions of exopolysaccharide (EPS). EPS has the ability to hold water effectively, and hence it helps in holding water in water-deficit soils, thereby promoting the growth of plants. Also, EPS has a very strong affinity towards metal ions like sodium; hence, not much problem will be created if the plants are growing in saline soil (Minocha et al., 2014). EPS also helps in microbial colonization by providing adherence to microorganisms. It also has some reactive sites where they can trap micronutrients and can enhance the combating mechanisms. Therefore, the plant's association with its microbiota helps the plant by providing a sustainable condition which can control the stressors.

15.2 TYPES OF STRESS IN PLANTS

In plants, mainly two types of stress occur, which fall under the category of environmental stress. They are biotic stress and abiotic stress. Biotic stress is the type of stress which is due to the invasion of microorganisms like viruses, nematodes, bacteria, insects, etc. These biotic stresses cause various types of plant diseases, which ultimately decrease the plant growth rate and productivity, sometimes, in severe cases, causing the death of the plant (Saha et al., 2019). There are various ways to overcome biotic stresses, namely by forming a genetically modified organism, overproducing some enzymes, or understanding the genetic makeup that can help in developing resistance against some diseases. Abiotic stress is mainly environmental in nature, and it also can cause a massive reduction in productivity, affecting the plant's health drastically. These abiotic stresses include the production of osmolytes, increased exposure to low and high light intensity, drought, soil salinity, extreme exposure to temperature, etc. So it is important to recognize the genes that can respond against abiotic stresses in order to know their stress response mechanisms.

15.2.1 Types of Abiotic Stress in Plants

15.2.1.1 Cold-Induced Stress

It is the most important abiotic stress that plants encounter. As they are sessile in nature, they have developed some defence mechanisms in order to protect themselves from such environmental stresses. In a chilling environment, the plant needs some chilling tolerance methods called acclimation. But most of the plants are not acclimatized. Several signal transduction pathways are present by which cold stresses are reduced, like ABA, calcium ion, protein phosphate, protein kinase synthesis, etc. (Suzuki et al., 2014).

15.2.1.2 Salt-Induced Stress

Saline soil abruptly decreases the productivity of plants in two ways, namely by affecting the osmotic balance and by causing ion toxicity. In saline soil, due to the presence of a huge number of ions, the osmotic pressure in plant cells increases from the normal osmotic pressure of the cells, thereby affecting the entire metabolic activity of the plant (Mishra et al., 2017). It also hampers the plant's ability to absorb water and minerals like sodium and calcium from soil. These are the most common primary effects, but some secondary effects like decreased cytoplasmic metabolism and reduced cell expansion and membrane function are also observed, which limits the plant productivity.

15.2.1.3 Toxin-Induced Stress

Due to the rapid increase in human population, the demand for more crop production is there; hence, crop yield must be high. So farmers use a lot of chemical fertilizers, which add to the toxicity of the soil. Also, due to irrigation of sewage wastewater and rapid industrialization, the situation is deteriorating faster. Hence all these processes contribute to the toxicity of the soil, which in turn causes a negative impact on plant physiology (Cvjetko et al., 2014).

15.2.1.4 Heat-Induced Stress

Due to enhanced global warming, the earth's temperature is increasing, and thereby plant productivity is reducing tremendously day by day. When encountering heat stress, seed germination rate and photosynthetic efficiency decrease and automatically affect the plant growth abruptly.

15.2.1.5 Drought-Induced Stress

Due to the pulsating nature of the environment, drought is much more common in some areas. Due to the unavailability of water, growth arrest of plant cells takes place, reduces the growth of shoots, and also decreases their metabolic activities, leading to the premature death of some plants. Sometimes during drought, some protective compounds are also synthesized by plants to maintain osmotic balance (Mishra et al., 2017).

15.2.1.6 Metal-Induced Stress

Metal-induced stress is the type of environmental abiotic stress that is caused by exposure of plants to heavy metals, thereby causing ROS (reactive oxygen species) accumulation in the mitochondria and also to various parts of the cell. As mitochondria is involved in redox reactions, exposure to heavy metals can upregulate the overall stress response and, thereby, can engage in initiating enhanced signalling cascade (Sytar et al., 2013).

Heavy metals such as iron (Fe), manganese (Mn), copper (Cu), nickel (Ni), cobalt (Co), cadmium (Cd), zinc (Zn), mercury (Hg), and arsenic (Ar) are being accumulated in the soil for a prolonged period of time through sewage disposals. Heavy metals actually serve as sources of nutrients for plants that are usually required in a very little amount, but prolonged exposure to these toxic metals will affect different physiological functions of the plants (Hossain and Komatsu, 2013).

But among all other heavy metals, mainly the four metals, namely arsenic, cadmium, lead, and mercury, are proved to be highly toxic to plants. Heavy metals can cause groundwater contamination and so have the ability to accumulate in the soil for prolonged periods and thereby cause worldwide problems. They also cause intra-chromosomal mutations, inhibit phosphorylation and ATP synthesis, generate ROS mechanisms, and take a toll on transpiration, growth development, and other physiological activities (Seneviratne et al., 2019). They can also alter cell membrane permeability and cause phytotoxicity. They can also enhance the rate of lipid peroxidation, cause DNA damage, and lead to decreased ATP production. They can also induce alterations in enzymatic activities, cause leaf epinasty and chlorosis, cause harm to pollen germination and tube growth, and cause disruption in the metabolism of plants (Wallace et al., 2020).

Plants have very different mechanisms to fight against stressors, and plants also help in maintaining the homeostasis of essential metals. The first line of defence that a plant has is one that helps in reducing the uptake of metals when heavy metal toxicity is encountered. Many plants have a unique way of guarding cells from heavy metal exposure. They compartmentalize these ions, thereby providing no access to the sensitive tissue or cells of the plants. During the time of metal toxicity, oxidative stress is very high, and the production of stress-related proteins, hormones, etc. occurs, and there is also a very high chance of increasing the levels of heat shock proteins (Hasan et al., 2017).

15.3 PLANT MICROBIAL ASSOCIATION

Plants, through roots, with the help of chemical signalling, actively communicate with the microbes present in the soil and thereby form a plethora of beneficial microorganisms called plant growth-promoting bacteria (PGPB) (Schirawski and Perlin, 2018). This association is known as plant microbial interactions or association. The major types of plant microbe interactions are as follows:

- Mutualism: It's the most important type of plant microbial interaction, where both of the species are metabolically dependent upon each other, and both of them mutually help in their survival (Divya and Deepak Kumar, 2011). It is also under the positive type of interaction, where no one will be harmed.
- Competition: It's again one of the most important type of interactions, where both the individuals compete for the same product. Here neither of the individuals will get the favour. Hence it's a negative interaction.
- Parasitism: This type of interaction between plants and microbes is symbiotic and obligatory. The microbe draws nutrients from the plant and sometimes can cause harm to the plant or sometimes may not be harmful to the plant. This type of relationship can harm the host organism, thereby making it a negative type of interaction for the host and a positive type of interaction for the microbes (Čapek et al., 2018).
- Commensalism: This interaction is also more or less similar to that of parasitism, where an organism, like microbes, is metabolically dependent upon the host organism, but it doesn't cause any harm to the host organism. It's a positive type of interaction (Sahu et al., 2020).

In the presence of plant growth promoter (PGP), the combat mechanisms against the stressors are alleviated alarmingly by increasing the effectiveness of the antioxidant enzymes, helping in the early establishment of juvenile seedlings, and helping in the formation of 1-aminocyclopropane-1-carboxylate deaminase (ACC deaminase) by degrading the ethylene hormone. Microbial exopolysaccharide also acts as an effective tool in trapping water so that even in drought conditions the plants can grow. They also help in maintaining plant growth and developmental processes by the following mechanisms: volatilization, efflux, impermeability towards heavy metals, etc. (DalCorso, 2012). Therefore, the plant's associated microbes help in providing a sustainable role in controlling the stressors.

15.4 DIFFERENT MECHANISMS INVOLVED IN BIOTIC STRESS PERCEPTION IN PLANTS

Abiotic stress in plants causes huge problems for plant growth and plant existence. This type of stress is also caused by heavy metals, metalloids, and some essential metals in an excess amount in plants. Based upon the requirement in the plant, the elements are divided into main two categories: macroelements (required in a very minute amount, less than microelements) and microelements (required more than macroelements) (Rajput et al., 2021). These elements include metals, metalloids, and nonmetals. These elements are required in a very minute amount in a plant to maintain the whole processes, mechanisms, different growth-regulated factors and homeostasis in a plant (Rhaman et al., 2020). But the excess deposition in a plant causes huge problems and inhibits the vital mechanisms in a plant. An excess amount of heavy metal-induced abiotic stress in plants has a negative impact on the vital pathways like energy production mechanisms and chlorophyll synthesis in leaves, causes discolouration of plant leaves, shoots, and tissues, causes cell death, and destroys the transportation of food, water, and nutrients in plants by hampering the plant's transportation system (Sharma et al., 2021). Heavy metals, for example cadmium (Cd), chromium (Cr), cobalt (Co), arsenic (Ar), nickel (Ni), and lead (Pb); microelements like iron (Fe), chlorine (Cl), bromide (Br), manganese (Mn), zinc (Zn), copper (Cu); and metalloids like silicon, tellurium, selenium, and germanium, play an important role as they are essential for plants' reproductive growth and also help in food production pathways (Wu et al., 2020). Many abiotic stressors can block the calcium channels, which directly has a negative impact on plant transport mechanism and water uptake system in roots (Ahammed et al., 2020). Deposition or uptake of these elements in excess amounts in the plant can destroy the plant's growth, metabolism, and physicochemical properties by inhibiting the various transition elements and enzymes directly or indirectly (Sachdev et al., 2021). A huge amount of heavy metal uptake in a plant can cause dry leaves, early falling of leaves, several adverse effects on seed germination, and, in severe cases, even complete death of plants (Pandey, 2020).

15.4.1 Role of Microelements in Inducing Plant Stress

Microelements are required at a very minute level for the growth of plant cells and tissues. Boron, chlorine, copper, iron, manganese, molybdenum, and zinc are mainly the microelements that are required (Souza et al., 2020). But in an abnormal situation where the levels of these elements go high in a plant, it will show various adverse effects on plant growth and induce plant stress (Shafi and Zahoor, 2020).

Iron (Fe): A high level of iron concentration in a plant is very deleterious for the plant's health. It increases the production of oxidized polyphenols, which are responsible for leaf bronzing (chlorosis), and also inhibits the root oxidation power (Rodrigues et al., 2020). After that, the leaves turn from orange-yellow to brown and die early (Diedhiou et al., 2020). The total root system gets damaged and turned dry and black in colour (He et al., 2020). An excess amount of Fe can shunt the growth of the plant (Zaid et al., 2020).

Chlorine (Cl): A very minute amount of chlorine is required in a plant, to act as a growth-enhancing factor. But excessive levels of it cause burned leaves. It blocks the vein of the leaf and causes scorched leaves that are smaller than normal leaves and also drop early (Du et al., 2017).

Boron (Br): A high concentration of boron in plants causes tissue death (Sharafi and Raina, 2021). High deposition of boron in leaves causes dryness in leaves and turns them yellow. Too much boron is the main cause of yellowing or browning of foliage with necrotic spots (Pereira et al., 2021). An optimum level of boron is very important for the growth of the shoot.

Manganese (Mn): Manganese is one of the most abundant elements in soil, but neutral pH and alkaline pH control the excessive uptake of Mn (Liu et al., 2021). Like boron, excessive manganese levels also cause necrotic spots and chlorosis in the leaves (Scattolin et al., 2021).

Zinc (Zn): A high concentration of zinc causes toxicity in plants. The common symptoms are the death of leaf tips, shoot shunting, curling, and rolling of newborn leaves (Mazaheri-Tirani and Dayani, 2020). Sometimes it also causes chlorosis. Other toxicities like inhibiting reproductive growth and various adverse effects on germination, morphology, and physiology of plants are also observed (Amuthavalli et al., 2021).

Copper (Cu): Minute level of excess copper is very toxic for the plant. Just a 5 μM level of Cu causes severe growth reduction, and 20 μM of Cu can completely inhibit root growth and cause death in plants (Listiani and Yuniati, 2021). It also causes chlorosis in younger leaves. Excess disposition of Cu causes shoots' death. Cu inhibits the overall shoot growth, biomass, and water-holding capacity due to the high production and accumulation of thiobarbituric acid reactive substance (TBARS) in the plant (Rizwan et al., 2021). A high amount of Cu deposition increases the oxidative stress in the plant, which directly perturbs the metabolic process in the plant.

Molybdenum (Mo): It is very essential for plant function because it plays a vital role in converting nitrate into amino acids in the plant. But a high level of Mo is also very toxic for plants. A higher concentration of Mo in a mature leaf can minimize the presence of nitrates and nitrites in plant tissue (Scattolin et al., 2021). Excess deposition of Mo in leaves turns it into a purple colour because of anthocyanin accumulation.

15.4.2 Role of Macroelements in Inducing Plant Stress

Macroelements are very crucial for a plant's growth. Macroelements are divided into two categories: primary macronutrients (nitrogen, phosphorus, potassium) and secondary macronutrients (calcium, magnesium, and sulphur) (Sung et al., 2021). In a plant's daily nutrition need, the concentration of primary macronutrients is higher than the concentration of secondary macromolecules.

Primary macronutrients introduce stress in plants. Primary macronutrients play an important role in plants' diets, but too much exposure to all these elements can cause tremendous harmful effects on them.

Nitrogen (N): Nitrogen is a very essential element in plants, though too much of it causes severe and adverse effects on the plant. A high level of nitrogen can completely kill the plant. A high level of nitrogen concentration is required in the vegetative phase of growth in a plant (as in this stage development of the stems, roots, and foliage occurs) (Min et al., 2021). But the same level of nitrogen uptake will heavily hamper the flowering stage in the plant. In this stage, plants shift their energy consumption from the development of stems, roots, and leaves to the development of buds, flowers, and fruits. During this phase, high deposition of nitrogen has an adverse effect on these developments and inversely affects the growth. Another deadly effect of excess nitrogen deposition in plant leaves is discolouration of the leaves (block chlorophyll synthesis), and they turn into extremely dark green leaves (Menegatti et al., 2021). Extreme nitrogen deposition can burn the leaf tips and turn them brown. In some leaves, nitrogen blocks the uptake of other nutrients, and yellowish spots occur on the leaf surface.

Phosphorus (P): Phosphorus is one of the most essential elements for a plant's normal function. Phosphorus fertilizer is necessary for crop growth and function (Devi and Kumar, 2020). Necrotic symptoms in mesophyll cells occur due to the heavy accumulation of phosphorus in the leaf (Martın-Rodrıguez et al., 2021). Leading to lipid peroxidase, phosphorus decreases Rubisco activation rate and inhibits the photosynthesis mechanism (Pardo-Hernández et al., 2021). A huge interveinal death of leaf tissue is a severe symptom of phosphorus toxicity.

Potassium (K): Potassium is a very critical nutrient for a plant; it regulates the movement of water, nutrient, water vapour and exchange of oxygen and carbon dioxide (De Conti et al., 2020). It also controls the opening and closing of stomata. But an excess amount of K is also harmful to plants. Excess deposition of potassium in the root region prevents the absorption of other nutrients from the soil. Extremely high uptake of potassium blocks the uptake of phosphorus and enhances the algae growth rate and kills the plant (Dad et al., 2021). It also reduces the leaf growth rate and reduces the leaves' size and shape. Potassium deposition in a leaf causes discolouration and turns it yellowish (Wasim et al., 2021). Also, it burns the tip of the leaf, and splitting of leaf occurs.

Secondary macroelements induce stress in plants. Secondary macroelements are very crucial in plant defence. These elements help to build a defence mechanism against herbivores and pathogens (Devi and Kumar, 2020). An excess amount of these elements can have various toxic effects on the plant.

Calcium (Ca): It is a critical secondary macroelement that plays an important role in cell wall and cell membrane synthesis. But a high level of calcium competes and blocks potassium and magnesium intake in the plant. High deposition of calcium inhibits boron intake also (De Conti et al., 2020). In young leaves, chlorosis can occur due to the deposition of calcium. In the tip of a new leaf, brown and black scorching occurs. Also, it inhibits the root growth and forms short and thickened roots. Sometimes Mg and Br deficiency occurs in the plant due to excess intake of calcium (El-Beltagi et al., 2020).

Magnesium (Mg): Magnesium is one of the most vital elements in the plant that has various important roles. Yellow patches occur in the centre of young leaves, and old leaves are discoloured due to the high concentration of Mg (De Conti et al., 2020). Excess magnesium deposition in the leaf causes curling of the leaf. High levels of Mg intake can compete and block calcium intake and cause calcium deficiency. Extremely high levels of magnesium in plants show symptoms of excess salt intake (De Conti et al., 2020). It also inhibits growth regulatory pathways and results in stunted growth. Due to magnesium deposition, some plants display dark-coloured vegetation.

Sulphur (S): Excess sulphur causes chlorosis in the whole plant and creates patches of yellow colour (Devi and Kumar, 2020). Young leaves are curled and reduced in size. Shoot and root growth are completely inhibited by excess sulphur. Fruits and flowers will be deformed due to excess sulphur intake in a plant (De Conti et al., 2020). Sometimes necrosis occurs in the main veins of leaves.

Due to exposure to heavy metals, plants experience oxidative stress that leads to tissue damage, inhibition of growth, an early drop of young leaves, fruits, etc. Accumulations of heavy metal ions in plants ruin the cellular ionic homeostasis. Uncontrolled anthropogenic activities induce soil contamination with toxic heavy metals (mainly cadmium, chromium, lead, arsenic, cobalt, and nickel). One of the common dangers of heavy metal contamination in plants is lipid peroxidation (Khalilzadeh et al., 2020). It is the prime cause of bio-membrane deterioration in plants. For this oxidative stress, the polyunsaturated fatty acid of the membrane decomposes into malondialdehyde (MDA).

Cadmium (Cd): The threshold limit of Cd in agricultural soil is 100mg/kg. A minute excess of Cd is dangerous for plants. Extreme uptake of Cd in a plant causes severe injuries like delayed photosynthesis by the reduction in uptake of water and nutrients (Tahir et al., 2020). A high level of deposition of cadmium causes chlorosis (browning of leaves), growth inhibition, and, at last, death of the whole plant occurs. High thiobarbituric acid reactive substances (TBARS) accumulate in plants due to cadmium stress in plants. More than twenty proteins are inhibited due to deposition of higher amount of Cd, which are involved in detoxification of plants, activation of antioxidants, water consumption, germination, and protein biosynthesis in plants (Devi and Kumar, 2020).

Chromium (Cr): Chromium is one of the harmful heavy metals having a very serious impact on plants. The Environmental Protection Agency recommends that a deposition of 1.07–7.80 mg/l of Cr (VI) is most harmful for plants (Khalilzadeh et al., 2020). It totally stops the growth of shoots, leaves, and roots in a plant. It also causes chlorosis in younger leaves (by complete inhibition of biosynthesis of chlorophyll) and stops the photosynthesis mechanism. Excess amounts of Cr have a deleterious impact on the water uptake system and causes nutrient imbalance in plants. By generating reactive oxygen species (ROS), it alters the conformation of various enzymes and, through it, inhibit the various biochemical pathways in plants (Devi and Kumar, 2020). Also, it reduces the germination process and overall growth of a plant. Inhibiting the excess amount of Cr also inhibits the thylakoid activity and thus perturbs the food production process.

Lead (Pb): Pb is the most abundant heavy metal and the most dangerous for plants. It promotes oxidative stress to the plant by directly enhancing the formation of reactive oxygen species (Devi and Kumar, 2020). High exposure to Pd causes photosynthesis inhibition, water imbalance, and also deteriorates the enzyme activity. Reacting with the sulfhydryl group of cellular enzymes, it alters the cellular enzymatic activities. It also reduces the membrane permeability. It has a hugely adverse effect on germination; a high concentration of Pb can inhibit the germination process completely. It also increases the dry weight in plants by blocking the water transport system. Excess of it inhibits shoot growth also. A high concentration of Pb in plants can kill them (Musa, 2020).

Arsenic (As): As has a major role in inhibiting the phosphorylation process in plants. Although some plants have genes which codes proteins which can confer arsenic tolerance, extreme deposition of arsenic causes numerous damages in the plant (Devi and Kumar, 2020). Excess deposition of it ensures that the energy production process is stopped. Inorganic arsenite (derivatives of arsenic) is more dangerous and is the main reason for the inhibition of phosphorylation in plants. It also produces ROS in the plant. It also affects the overall photosynthesis process (Devi and Kumar, 2020).

Cobalt (Co): Excess amount of cobalt contamination in soil occurs due to the burning of fuels (Souza et al., 2020). Some phyto-toxicity studies show that the excess deposition of Co in the plant inhibits the growth of shoots. It restricts the transportation of iron, chlorophyll, and various proteins in leaves. It also restricts the uptake of P, Zn, Mn, S, Cu, which are elements that play a vital role in plant growth and various physiological and morphological mechanisms of plants (Devi and Kumar, 2020). It acts as an indirect plant growth inhibitor by aerial parts of plants.

Nickel (Ni): The normal range of Ni in the soil is 100–1000 mg/kg; when that concentration is around 200–2600 mg/kg, it affects plants. Various physiological alterations occur due to an excess amount of nickel deposition in plant parts; it also causes necrosis and chlorosis in plants (Souza et al., 2020). It reduces nutrient uptake and leads to nutrient imbalance and halts growth of the plant parts (shoots, roots). It also affects the cell membrane function, and cell death occurs due to improper energy and nutrient transport. Ni ions block the H-ATPase activities and lipid compositions (e.g. inhibition of H-ATPase activity in *Oryza sativa* shoots and its plasma membrane) (Khalilzadeh et al., 2020). Excess deposition of Ni also disturbs the cytoplasmic ion balance (e.g. potassium ions) and disturbs the water uptake in plants.

15.4.3 Role of Metalloids in Inducing Plant Stress

Metalloids are also capable of inducing abiotic stress in plants. Mainly, four metalloids have very adverse effects on plants; they are silicon, tellurium, selenium, and germanium. Though silicon and selenium play an important role in plant growth when present in a tiny amount, metalloids like tellurium and germanium are heavily toxic for plants even when present in minute amounts (Angulo-Bejarano et al., 2021). Due to heavy industrial and urban waste exposure, plants are contaminated by these toxic metalloids. Antimony and germanium have various adverse effects on plants, heavily damage the seed germination process, destroy the whole physiological system, and increase the dry weight due to dehydration. These metalloids block the water distribution mechanism also. Some vital antioxidants like aldose reductase, peroxiredoxin, glyoxalase I, UlpI protease, DnaK-type molecular chaperone, and receptor-like kinase get damaged due to the excess deposition of silicon and selenium in a plant (Angulo-Bejarano et al., 2021). On the other hand, excess silicon deposition can cause tissue damage in leaves. Excess amounts of these heavy metalloids in plants, especially tellurium, selenium, and germanium can downregulate the α-amylase and enolase activity in the plant and, by inhibiting the thylakoid activity, also disturbs the food production process. Very toxic metalloids like antimony and germanium also inhibit the activity of some proteins that are related to the plant's response against these abiotic stresses (Kesawat et al., 2020). High deposition of selenium in the plant can raise hydrogen peroxide (H_2O_2) level. Metalloid-induced stress enhances the reduction of thylakoid, photosynthetic pigments, and O_2 evolution in leaves. Tellurium deposition inhibits the carbon metabolism and disturbs the electron transport chain both in chloroplast and mitochondria (Souza et al., 2020).

15.5 MITIGATION OF METAL-INDUCED STRESS AND TOXICITY IN PLANTS

Like any other organisms, plants also possess some specific tolerance mechanism that helps in an unfavourable stress condition. Various cascades of signalling pathways are stimulated in these conditions such as calcium-calmodulin pathways, reactive oxygen species pathways, and various other hormonal pathways. Transcription factors are produced to act against stress. The signalling pathways are needed to maintain metal homeostasis in plants. Some of them target a specific metal ion, separating it from others and restricting it in a specific place inside the cell and thus preventing it from harming the entire cell. Plants change their pH levels to regulate the uptake of metals according to their requirements. As a result, the metal ions are being stored in the apoplast and cannot go inside the cells. As an example, sorghum secretes malate (Lin et al., 2021), and the tomato plant secretes oxalate from roots to withstand toxic metals. Similarly, wheat plants secrete malic acid against aluminum. Another molecule is nitric oxide (NO), which is very efficient in regulating hormones and ROS levels (Huo et al., 2021; Štolfa et al., 2015). It can stimulate the production of ATPase and proton phosphatase (PPase) in tomato plants, making them more tolerant by controlling some biochemical changes. The plant cell wall has callose, which does not let the heavy metals pass through it (Sarma et al., 2018). Pectin, a component of the cell wall, also helps in the same. Sometimes plants transport metals to various organs just to stop the local accumulation of them. Two cysteine-rich polypeptides are found, which are phytochelatins and metallothioneins, which bind to metal ions using their thiol group. HSP70 and HSP25 are two heat shock proteins that are also found to mitigate heavy metal toxicity in plants.

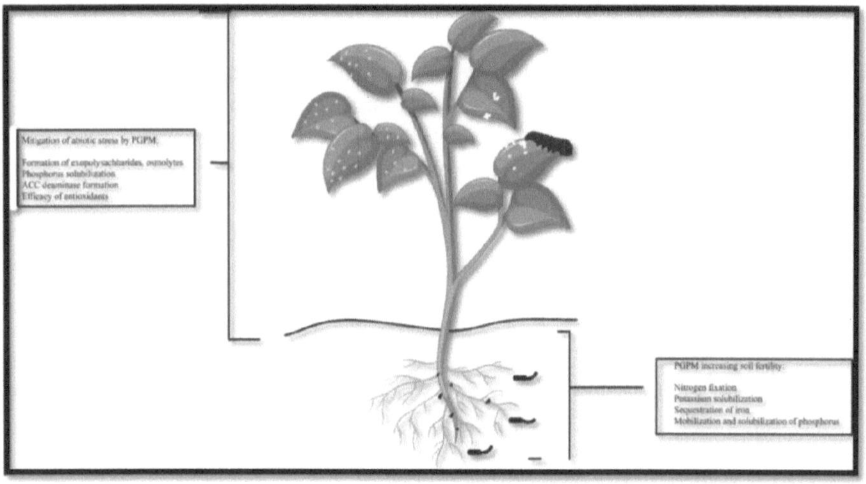

FIGURE 15.1 Mitigation of abiotic stress by PGPM in plants.

15.6 ROLE OF MICROBIAL ASSOCIATION IN MITIGATION OF METAL-INDUCED ABIOTIC STRESS IN PLANTS

Several types of microbes are seen to have association with different plant species. These kinds of microbial associations can be both harmful and helpful for plants. These interactions interestingly influence the overall growth and development of the plants (Kumar et al., 2018). For the past few decades, a huge progress has been made in the field of study on the molecular basis of these interactions. Plants face various abiotic factors on a daily basis that can be disastrous to plants, affecting their growth, productivity, etc. All of these various abiotic stresses are collectively known as abiotic stress. Both the microbial population and the plant growth can experience a negative impact of abiotic stresses. Plants have various defence mechanisms to avoid the negative impacts of abiotic stress factors. Some mechanisms are triggered in the presence of some plant growth-promoting (PGP) microorganisms (Lata et al., 2018). Many PGP microorganisms help in secreting some antioxidant enzymes, expression of some genes that can respond against stress, and osmolytes accumulation. Consistent industrialization and modern agricultural practices are contributing to heavy metal accumulation in soil. Heavy metals are those elements with a high density of more than 4g/cc which are not degradable, and also they are harmful even at a low concentration. The artificial methods to remove these metals are costly and also destructive to the soil. So phytoremediation is a useful alternative that involves the plants and their associated microbes. Normally the metal-induced stress affects the plants mostly during their germination period, due to which even if the seed somehow manages to germinate, the plant will have poor physiological and morphological characteristics (Egamberdieva et al., 2017). Stress can also hamper the signalling cascades of the associated microorganism that will lead to decreased colonies. But some metal stress-tolerant microbes like rhizobacteria, mycorrhiza, and firmicutes have the ability to stimulate plant growth and development under stressed environment (Mishra et al., 2017). These microbes use various mechanisms to maintain plant growth and development such as volatilization, efflux, enzyme detoxification, sequestration of extracellular polymeric substances (EPS), impermeability towards heavy metals, etc. (Ahmad and Prasad, 2011). The PGP microbes can also stimulate plant growth by decreasing the ethylene level, increasing the production of indole acetic acid (IAA) and aminocyclopropane-1-carboxylic acid (ACC). Besides all these, phosphorus solubilization, nutrient mobilization, and nitrogen fixation can also help in the removal of heavy metals. Both living and non-living biomass of microbes are used in this removal technique (Ahemad, 2019). Some specific microbes which are capable of removing heavy metals are proteobacteria, actinobacteria, firmicutes, enterobacter, and klebsiella. These can effectively decrease the concentration of Mn, Pb, As, Cd, Zn, etc.

15.6.1 Role of Bacteria in Mitigating Metal-Induced Stress in Plants

Heavy metals are seen to impart adverse effects on bacterial association with plants. Several studies have shown that sometimes microbes adapt themselves to the changing adverse environment. In some cases, it has been seen that inoculating maize with four different strains of bacteria led to increased accumulation of Zn, Cu, Cr, etc. Another study showed that inoculating willow trees with *Streptomyces* AR17 had increased Zn absorption, whereas inoculating with *Pseudomonas* PR04 and *Streptomyces* AR36 had led to decreased absorption of Zn. It suggested that different inoculations stimulated different gene expressions in the same tree, causing different concentrations of metal absorption.

Plant growth-promoting bacteria (PGPB) leads this pathway of removal of heavy metals by releasing plant growth hormones, synthesizing ACC deaminase, solubilization of nutrients, etc. Usage of rhizospheric soil is well known in inoculating legume plants. They have a high concentration of plant growth hormones, fatty acids, organic acids, various amino acids, nucleotides, vitamins, phenol, etc. released from plant roots by secretion and exudation methods. Such rhizobacteria are classified into four categories based on their nearness to the roots: a) bacteria living near the roots in the soil, b) bacteria living inside the root tissues, c) bacteria forming colonies on the root surface (rhizoplane), and d) bacteria residing inside the root nodules, a specialized structure that includes the well-known legume associated rhizobium bacteria (Gopalakrishnan et al., 2015). These rhizobium bacteria are called intracellular plant growth-promoting rhizobacteria (iPGPR).

Mechanisms adapted by rhizobacteria against metal-induced stress to stimulate plant growth promotion mechanisms are as follows:

Nitrogen fixation: Nitrogen is an important macroelement required for amino acids synthesis, nucleotide formation, enzymes, and chlorophyll production. Plants cannot absorb nitrogen (N) from the atmosphere. The rhizobium bacterium is known as a biological nitrogen fixer. So rhizobia are used as inoculum in legume production. It is a type of bacteria which establishes a symbiotic association with plants. Nitrogenase, the major type of enzyme in nitrogen fixation, consists of two components: dinitrogenase reductase which is the iron protein, and dinitrogenase that is the cofactor (Wang et al., 2018). Clusters of several genes contribute to nitrogen fixation, nodulation, glutamate synthesis, lipopolysaccharides (LPS), and exopolysaccharides synthesis (EPS). The genes responsible for symbiosis and pathogenicity in rhizobium also stimulate nodulation.

Phosphate solubilizer: Phosphate, being the major limiting element for plants, exists in very little amount in the soil, and only 0.1% is available for the plants to absorb. Rhizobia, consisting of *R. leguminosarum*, *Bradyrhizobium sp.*, *M. Mediterranean*, are important phosphate solubilizers (Tiwari and Lata, 2018).

Phytohormone production: *Rhizobium* sp. can also stimulate the production of growth hormones like cytokinin, gibberellin, indole acetic acid (IAA), etc.

ACC deaminase production: ACC deaminase enzyme hydrolyses ACC into alpha-ketobutyrate and ammonia. As a consequence, it helps in decreasing the ethylene level during the stress period. PGPR forms colonies near the seed or roots, and because of their high tryptophan availability, they can secrete IAA, which can further stimulate the synthesis of ACC synthase that produces ACC. Heavy metal contaminated soil also leads to decreased amount of iron (Fe) availability. Plants cannot take up sufficient Fe as their phytosiderophores have less affinity towards Fe. It is observed that plants can use bacterial ferric siderophore complexes to do so. Hence bacterial association helps the plants in maintaining a stable concentration of Fe (Glick, 2010).

Here are some examples of PGPR responding against metal-induced stress condition: *R. leguminosarum* performs sequestration of Cd; *Bradyrhizobium* RM8 increases growth performances in green grams against Ni and Zn; in maize, *R. leguminosarum* induces growth and biomass production against Pb; similarly, *Rhizobium* RL9 helps in nodulation, plant growth, leghaemoglobin, and chlorophyll production against Pb, Ni in lentils (Yadav et al., 2021).

Another important PGPB is cyanobacteria. Cyanobacteria are also known as blue-green algae. They are photosynthetic, oxygenic bacteria and one of the most diversified prokaryotic groups (Singh, 2014). They can form symbiotic associations with plant hosts, and they provide the plant host with nitrogen as they are capable of nitrogen fixations. Many studies have been performed on them that prove that they have some metal tolerance mechanisms also.

Extracellular polysaccharides or extracellular mantle barrier: They secret some extracellular polymeric substances (EPS) that help cyanobacteria in biomineralization of calcium (Ca) and magnesium (Mg) carbonate that further forms stromatolites. They can absorb Fe atoms for future requirements, and, thus, by increasing the availability of Fe they can help the host plant. It can also protect from CeO_2- and TiO_2-causing oxidative stress (Yadav et al., 2021).

Metallothionein or efflux ATPase operation: Cyanobacteria can synthesize metallothionein (MTs), a metal (As, Cd, Hg, Cu, Zn, etc.) binding and chelating protein. It is rich in cysteine. This MT protein is encoded by the gene smtA, which is stimulated by a high concentration of Pb, Hg, Cd, Ni, Co, Zn, etc. The smtA gene is regulated by smtB, a metal repressor gene. Another homologous gene ZiaR, a Zn repressor, regulates the synthesis of ZiaA which in turn helps to maintain a proper balance of Zn, Ni, Co, etc. Another gene for maintaining Mn homeostasis, mntC, is found in cyanobacteria.

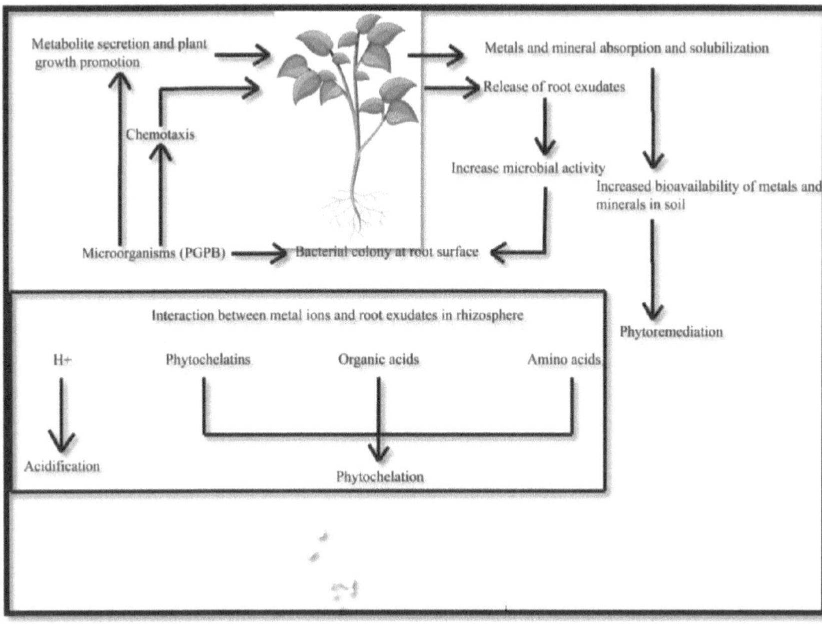

FIGURE 15.2 Schematic representation of mechanism of mitigating metal stress by PGPB.

Transport systems and redox enzymes against arsenic (As): As can remain as arsenite (III) or arsenate (V). Arsenite sometimes binds with the thiol group of some proteins, leading to a change in their activity, and also it can reduce the level of glutathione, an antioxidant. On the other hand, arsenate can affect various biochemical processes (Cassier-Chauvat and Chauvat, 2015). Several genes are involved in the resistance mechanism of As. One of those genes is ArsB, which reduces arsenate to arsenite conversion, ArsC (another arsenate reductase), etc.

15.6.2 Role of Fungi in Mitigating Metal-Induced Stress in Plants

Plants are often seen to be in a symbiotic relationship with arbuscular mycorrhizal fungi (AMF) where the photosynthetic member is a plant that provides food to AMFs, and on the other hand fungi helps the plant in absorbing nutrients, protects it from heavy metal toxicity, increases the biomass accumulation, etc. AMFs are considered important bioagents as they can help in decreasing the adverse effects of HMs in plants (Dhalaria et al., 2020). Normally it is known that HMs get inactivated in the fungal hyphae or in their cell wall or vacuoles, thus they decrease the metal toxicity in plants. Various mechanisms are adapted by them, such as increment in antioxidant properties of plants, binding the toxic metals to their cell wall, chelation of HMs, etc. AMFs can bind with Cd and Zn in the cell wall of hyphae, restricting their uptake by the plant (Begum et al., 2019).

The HMs like Cd, As, Pb, etc. are not that much necessary for a plant's normal growth, whereas increased concentration affects normal plant development. Vesicular arbuscular mycorrhiza (VAM) is a subtype of AMF; it helps to fight against these metal-causing stresses. Immobilization of metals, chelation, and precipitation of phosphate granules are some common mechanisms used by them (Javaid and Bajwa, 2011). But VAM can also bind the HMs with the help of the released insoluble glycoprotein glomalin. It is already observed that glomalin can absorb up to 1.12mg Pb, 4.3mg Cu, etc. (Miransari, 2017). In addition, fungal hyphae also act as a sink for the HMs because of their large surface area. They passively absorb these metals. The mycorrhizal association also helps in better uptake and immobilization of lead. VAM can supply mycorrhizal buffer against Cd-causing stress. In maize, metals are captured in the parenchyma cells near the fungal association. VAM also helps in decreasing the concentration of As and Ni so that the availability of phosphorous increases to the plants (Miransari, 2017).

Like the VAM, the ectomycorrhizal fungi (EMF) also respond against HM stresses on plants. They also follow these certain mechanisms to help in this stressed situation:

Cellular mechanism: Some HMs are considered beneficial to the plants when those are present in optimum concentrations like Mn, Zn, Cu, etc. But in excess amounts they are harmful; so EMF adapts some cellular mechanisms like subcellular sequestration of these HMs to biofilter them. They also stimulate the absorption of some beneficial HMs and separate the toxic ones in to vacuoles that lead to intracellular detoxification. By absorbing more phosphorous, they help in the formation of the Mn-P complex and thereby decrease the Mn toxicity. Against the harmful, dangerous HMs, they adapt several strategies, such as chelation of metals, attaching them to the EPS, decreasing the intake of metals by ion pumping methods, repairing the biomolecules damaged by metals, and transporting metals to other subcellular spaces other than cytosol (Colpaert et al., 2011).

Molecular mechanism: EMFs can control the rate of transcription of some metal carrying protein at the molecular level. It also regulates the amount of chelating molecules and also the transcriptomes of the host plants. In *H. cylindrosporum*, HcZnT1 gene was first observed to control the Zn homeostasis. One of the most common genes in *Arabidopsis* AtMTP11 was found to regulate the Mn toxicity level by the Golgi-mediated exocytosis mechanism. In H. *cylindrosporum*, tolerance against Cu is controlled by the expression of the Pim1 gene. The growth of eucalyptus under Ni stressed condition was improved by Ni-tolerant gene encoded by *P. albus*. Other examples of Cd-tolerant genes were TeMT, metallothionein, and TmelPCS, a phytochelatin (Khullar et al., 2018).

Physiological mechanism: Due to the functions of EMF, various changes occur at the physiological level of the fungi as well as the host plant. Some physiological changes are changed nutritional uptakes, altered photosynthetic rates and photosynthate supplies, concentration changes in the case of free amino acids, fatty acids, increment in osmolarity, etc. *P. canescens* and *P. involutus* colonies show increased phosphorous uptakes in roots and higher Ca^{2+} and K^{+} influx rates (Khullar et al., 2018). These colonies show increased uptake of Zn, P, Fe, Ca that helps in increasing the photosynthetic rate, the concentration of nutrients, detoxification ability, and defence mechanisms. EMF-induced physiological changes are directly related to the tolerance against non-nutritional HMs like Cd, Ni, Pb, Hg, etc. EMFs also release formic acid, oxalic acid, succinic acid, and malic acid that help to reduce the toxic metals' availability in the rhizosphere zone.

15.6.3 Role of Other Microbes in Mitigating Metal-Induced Stress in Plants

There are other microbes that help in reducing the effect of metal toxicity in the host plants (Schalk et al., 2011). Some bacteria secrete siderophores that act as chelators. *Candida tropicalis* is a type of yeast that is observed to have some ability to process metals. Glutathione and some other non-protein thiols also have important roles in metal tolerance (Rehman and Anjum, 2011). *C. albicans* and *C. tropicalis* both have the highest resistance against Pb, Hg, Cd, selenite, and arsenate ions. Heavy metal tolerant (HMT) microbes have the ability to alter the availability of HMs in the rhizosphere, thereby promoting plant growth. Excess secretion of some organic acids by the growth-promoting microbes helps in the solubilization of Cd, Pb, etc. Wood rotting fungi secrets oxalate crystals which convert the metals into less toxic forms. Microbes-induced redox reactions also help in this. Membrane porin-cytochrome complexes, outer membranous c-type cytochromes, do have a role in reducing metal accumulation (Singh et al., 2016).

15.7 CONCLUSION

Plant-associated microorganisms play a pivotal role in helping the plants cope against the stressors and combat them. The habitats or the plants that have been under constant exposure to stressors may have become resilient and can withstand the negative effect of stressors due to the resistance offered by the PGPs. The most important of all other stressors is heavy metal–induced stress. Depending on the requirements of metals by plants, they can be grouped into macro and microelements, but too much exposure to heavy metals can cause harm to the plants themselves, like burned or scooted leaves, necrotic spots, inhibited growth, and other physiological activities. Another compound that can harm plants is metalloid. They cause negative impact in plants by enhancing toxicity at physiological levels and also down regulate nutrient uptake and alter photosynthetic efficiency. But plants, on the other hand, have evolved mechanisms to combat stress so that they can thrive in stressful conditions. Here the PGP plays an important role by enhancing water storage, by degrading ethylene and helping in the formation of ACC deaminase, and stimulating ROS pathway, calcium calmodulin pathways, etc. Some free-living and symbiotic nitrogen-fixing bacterial species also help in providing protection against stressors, helping in phosphate solubilization, fixing atmospheric N_2, etc. Also, some microorganisms form decreasing stressor effects. Heavy metal–tolerant microbes help by diminishing the availability of heavy metals to the plant. Growth-promoting microbes have bioleaching properties also. Hence, without microorganisms, it is very difficult for the plant system to grow in the presence of stressors, and hence the plants have evolved a unique strategy of microbial association not only to resist heavy metal toxicity but also to thrive under metal-induced toxic conditions.

REFERENCES

Ahammed, G.J., Li, X., Liu, A. and Chen, S. 2020. Brassinosteroids in plant tolerance to abiotic stress. *Journal of Plant Growth Regulation*, 39(4):1451–1464.

Ahemad, M. 2019. Remediation of metalliferous soils through the heavy metal resistant plant growth promoting bacteria: paradigms and prospects. *Arabian Journal of Chemistry*, 12(7):1365–1377.

Ahmad, P. and Prasad, M.N.V. 2011. Environmental adaptations and stress tolerance of plants in the era of climate change. Springer Science.

Amuthavalli, P., Hwang, J.S., Dahms, H.U., Wang, L., Anitha, J., Vasanthakumaran, M., Gandhi, A.D., Murugan, K., Subramaniam, J., Paulpandi, M. and Chandramohan, B. 2021. Zinc oxide nanoparticles using plant Lawsonia inermis and their mosquitocidal, antimicrobial, anticancer applications showing moderate side effects. *Scientific Reports*, 11(1):1–13.

Angulo-Bejarano, P.I., Puente-Rivera, J. and Cruz-Ortega, R. 2021. Metal and metalloid toxicity in plants: an overview on molecular aspects. *Plants*, 10(4):635.

Arshad J. 2011. Importance of arbuscular mycorrhizal fungi in phytoremediation of heavy metal contaminated soils. In Biomanagement of metal-contaminated soils, eds. M.S. Khan, A. Zaidi, R. Goel and J. Musarrat, 125–141. Springer.

Begum, N., Qin, C., Ahanger, M.A., Raza, S., Khan, M.I., Ashraf, M., Ahmed, N. and Zhang, L. 2019. Role of arbuscular mycorrhizal fungi in plant growth regulation: implications in abiotic stress tolerance. *Frontiers in Plant Science*, 10:1068.

Čapek, P., Manzoni, S., Kaštovská, E., Wild, B., Diáková, K., Bárta, J., Schnecker, J., Biasi, C., Martikainen, P.J., Alves, R.J.E. and Guggenberger, G. 2018. A plant–microbe interaction framework explaining nutrient effects on primary production. *Nature Ecology & Evolution*, 2(10):1588–1596.

Cassier-Chauvat, C. and Chauvat, F. 2015. Responses to oxidative and heavy metal stresses in cyanobacteria: recent advances. *International Journal of Molecular Sciences*, 16(1):871–886.

Colpaert, J.V., Wevers, J.H., Krznaric, E. and Adriaensen, K. 2011. How metal-tolerant ecotypes of ectomycorrhizal fungi protect plants from heavy metal pollution. *Annals of Forest Science*, 68(1):17–24.

Cramer, G.R., Urano, K., Delrot, S., Pezzotti, M. and Shinozaki, K. 2011. Effects of abiotic stress on plants: a systems biology perspective. *BMC Plant Biology*, 11(1):1–14.

Cvjetko, P., Zovko, M. and Balen, B. 2014. Proteomics of heavy metal toxicity in plants. *Archives of Industrial Hygiene and Toxicology*, 65(1):1–18.

Dad, F.P., Khan, W.U.D., Tanveer, M., Ramzani, P.M.A., Shaukat, R. and Muktadir, A. 2021. Influence of Iron-Enriched Biochar on Cd Sorption, Its Ionic Concentration and Redox Regulation of Radish under Cadmium Toxicity. *Agriculture*, 11(1):1.

DalCorso, G. 2012. Heavy metal toxicity in plants. In Plants and heavy metals, ed. Antonella Furini. Springer.

De Conti, L., Cesco, S., Mimmo, T., Pii, Y., Valentinuzzi, F., Melo, G.W., Ceretta, C.A., Trentin, E., Marques, A.C. and Brunetto, G. 2020. Iron fertilization to enhance tolerance mechanisms to copper toxicity of ryegrass plants used as cover crop in vineyards. *Chemosphere*, 243:125298.

Devi, P. and Kumar, P., 2020. Effect of bioremediation on internodal length and leaf area of maize plant cultivated in contaminated soil with chromium metal. *Journal of Pharmacognosy and Phytochemistry*, 9(4):1408–1413.

Dhalaria, R., Kumar, D., Kumar, H., Nepovimova, E., Kuča, K., Torequl Islam, M. and Verma, R. 2020. Arbuscular mycorrhizal fungi as potential agents in ameliorating heavy metal stress in plants. *Agronomy*, 10(6):815.

Diedhiou, S., Goudiaby, A.O.K., Sagna, Y.P., Diatta, Y., Diallo, M.D. and Ndoye, I. 2020. Effect of iron toxicity on rice growth in sulfato-ferruginous lowland of south senegal. *American Journal of Agriculture and Forestry*, 8(1):9–14.

Ding, L., Lu, Z., Gao, L., Guo, S. and Shen, Q. 2018. Is nitrogen a key determinant of water transport and photosynthesis in higher plants upon drought stress? *Frontiers in Plant Science*, 9:1143. doi: 10.3389/fpls.2018.01143.

Divya, B. and Deepak Kumar, M. 2011. Plant–microbe interaction with enhanced bioremediation. *Research Journal of Biotechnology*, 6:4.

Du, Y., Lv, X.T., Wu, Q.Y., Zhang, D.Y., Zhou, Y.T., Peng, L. and Hu, H.Y. 2017. Formation and control of disinfection byproducts and toxicity during reclaimed water chlorination: a review. *Journal of Environmental Sciences*, 58:51–63.

Egamberdieva, D., Wirth, S.J., Alqarawi, A.A., Abd_Allah, E.F. and Hashem, A. 2017. Phytohormones and beneficial microbes: essential components for plants to balance stress and fitness. *Frontiers in Microbiology*, 8:2104.

El-Beltagi, H.S., Sofy, M.R., Aldaej, M.I. and Mohamed, H.I. 2020. Silicon alleviates copper toxicity in flax plants by up-regulating antioxidant defense and secondary metabolites and decreasing oxidative damage. *Sustainability*, 12(11):4732.

Glick, B.R. 2010. Using soil bacteria to facilitate phytoremediation. *Biotechnology Advances*, 28(3):367–374.

Gopalakrishnan, S., Sathya, A., Vijayabharathi, R., Varshney, R.K., Gowda, C.L. and Krishnamurthy, L. 2015. Plant growth promoting rhizobia: challenges and opportunities. *Biotech*, 5(4):355–377.

Hasan, M., Cheng, Y., Kanwar, M.K., Chu, X.Y., Ahammed, G.J. and Qi, Z.Y. 2017. Responses of plant proteins to heavy metal stress—a review. *Frontiers in Plant Science*, 8:1492.

He, L., Yue, Z., Chen, C., Li, C., Li, J. and Sun, Z. 2020. Enhancing iron uptake and alleviating iron toxicity in wheat by plant growth-promoting bacteria: theories and practices. *International Journal of Agriculture and Biology*, 23(1):190–196.

Hossain, Z. and Komatsu, S. 2013. Contribution of proteomic studies towards understanding plant heavy metal stress response. *Frontiers in Plant Science*, 3:310.

Huo, Y., Kang, J.P., Ahn, J.C., Kim, Y.J., Piao, C.H., Yang, D.U. and Yang, D.C. 2021. Siderophore-producing rhizobacteria reduce heavy metal-induced oxidative stress in *Panax ginseng* Meyer. *Journal of Ginseng Research*, 45(2):218–227.

Javaid, A. and Bajwa, R. 2011. Effect of effective microorganism application on crop growth, yield, and nutrition in Vigna radiata (L.) Wilczek in different soil amendment systems. *Communications in Soil Science and Plant Analysis*, 42(17):2112–2121.

Kesawat, M.S., Shivaraj, S.M., Kim, D.K., Kumar, M., Hahn, B.S. and Deshmukh, R. 2020. Metalloids and their role in the biological system. metalloids in plants: advances and future prospects, eds. R Deshmukh, D.K. Tripathi, G. Guerriero, 1–17. John Wiley & Sons.

Khalilzadeh, R., Pirzad, A., Sepehr, E., Khan, S. and Anwar, S. 2020. Long-term effect of heavy metal–polluted wastewater irrigation on physiological and ecological parameters of Salicornia europaea L. *Journal of Soil Science and Plant Nutrition*, 20(3):1574–1587.

Khullar, S. and Reddy, M.S. 2018. Ectomycorrhizal fungi and its role in metal homeostasis through metallothionein and glutathione mechanisms. *Current Biotechnology*, 7(3):231–241.

Koyro, H.W., Ahmad, P. and Geissler, N. 2012. Abiotic stress responses in plants: an overview. In Environmental adaptations and stress tolerance of plants in the era of climate change, ed. P. Ahmad and M.N.V. Prasad, 1–28. Springer.

Kumar, A. and Verma, J.P. 2018. Does plant—microbe interaction confer stress tolerance in plants: a review? *Microbiological Research*, 207:41–52.

Kumar, S., Sachdeva, S., Bhat, K.V. and Vats, S. 2018. Plant responses to drought stress: physiological, biochemical and molecular basis. In: Biotic and abiotic stress tolerance in plants, ed. S. Vats. Springer. doi: 10.1007/978-981-10-9029-5_1

Lata, R., Chowdhury, S., Gond, S.K. and White Jr, J.F. 2018. Induction of abiotic stress tolerance in plants by endophytic microbes. *Letters in Applied Microbiology*, 66(4):268–276.

Listiani, S. and Yuniati, R. 2021, January. The effect of mycorrhizae on the growth of *Paraserianthes falcataria* L. (Nielsen) in an artificial growth medium containing copper and cadmium. *Journal of Physics: Conference Series*, 1725(1):012052.

Liu, K., Xu, J., Dai, C., Li, C., Li, Y., Ma, J. and Yu, F. 2021. Exogenously applied oxalic acid assists in the phytoremediation of Mn by Polygonum pubescens Blume cultivated in three Mn-contaminated soils. *Frontiers of Environmental Science & Engineering*, 15(5):1–13.

Martın-Rodrıguez, J.A., Ariani, A., Leija, A., Elizondo, A., Fuentes, S.I., Ramirez, M., Gepts, P., Hernandez, G. and Formey, D. 2021. Differentially regulate plant tolerance to aluminum toxicity. *The Plant Journal*, 105:1521–1533.

Mazaheri-Tirani, M. and Dayani, S. 2020. In vitro effect of zinc oxide nanoparticles on *Nicotiana tabacum* callus compared to ZnO micro particles and zinc sulfate ($ZnSO_4$). *Plant Cell, Tissue and Organ Culture (PCTOC)*, 140(2):279–289.

Menegatti, R.D., Souza, A.D.G. and Bianchi, V.J. 2021. Nutritional efficiency for nitrogen, phosphorus and potassium in peach rootstocks. *Journal of Plant Nutrition*, 44(2):228–237.

Min, J., Sun, H., Kronzucker, H.J., Wang, Y. and Shi, W. 2021. Comprehensive assessment of the effects of nitrification inhibitor application on reactive nitrogen loss in intensive vegetable production systems. *Agriculture, Ecosystems & Environment*, 307:107227.

Minocha, R., Majumdar, R. and Minocha, S.C. 2014. Polyamines and abiotic stress in plants: a complex relationship. *Frontiers in Plant Science*, 5:175.

Miransari, M. 2017. Arbuscular mycorrhizal fungi and heavy metal tolerance in plants. In Arbuscular mycorrhizas and stress tolerance of plants, eds. Q.S. Wu, 147–161. Springer.

Mishra, J., Singh, R. and Arora, N.K. 2017. Alleviation of heavy metal stress in plants and remediation of soil by rhizosphere microorganisms. *Frontiers in Microbiology* 8:1706.

Musa, K.A.R. 2020. Expression patterns of oxidative stress-related genes of cucurbita pepo and relation to cellular H_2O_2 under short-term heavy metal stress. *Iğdır Üniversitesi Fen Bilimleri Enstitüsü Dergisi*, 10(4):2952–2961.

Pandey, S.N. 2020. Role of micronutrients in biochemical responses of crops under abiotic stresses. In Sustainable agriculture in the era of climate change, eds. R. Roychowdhury, S. Choudhury, M. Hasanuzzaman and S. Srivastava, 93–112. Springer.

Pardo-Hernández, M., López-Delacalle, M., Martí-Guillen, J.M., Martínez-Lorente, S.E. and Rivero, R.M. 2021. ROS and NO phytomelatonin-induced signaling mechanisms under metal toxicity in plants: a review. *Antioxidants*, 10(5):775.

Pereira, G.L., Siqueira, J.A., Batista-Silva, W., Cardoso, F.B., Nunes-Nesi, A. and Araújo, W.L. 2021. Boron: more than an essential element for land plants? *Frontiers in Plant Science*, 11:2234.

Rajput, V.D., Minkina, T., Kumari, A., Singh, V.K., Verma, K.K., Mandzhieva, S., Sushkova, S., Srivastava, S. and Keswani, C. 2021. Coping with the challenges of abiotic stress in plants: new dimensions in the field application of nanoparticles. *Plants*, 10(6):1221.

Rehman, A. and Anjum, M.S. 2011. Multiple metal tolerance and biosorption of cadmium by Candida *tropicalis* isolated from industrial effluents: glutathione as etoxifying agent. *Environmental Monitoring and Assessment*, 174(1):585–595.

Rhaman, M.S., Imran, S., Rauf, F., Khatun, M., Baskin, C.C., Murata, Y. and Hasanuzzaman, M. 2020. Seed priming with phytohormones: an effective approach for the mitigation of abiotic stress. *Plants*, 10(1):37.

Rizwan, M.S., Imtiaz, M., Zhu, J., Yousaf, B., Hussain, M., Ali, L., Ditta, A., Ihsan, M.Z., Huang, G., Ashraf, M. and Hu, H. 2021. Immobilization of Pb and Cu by organic and inorganic amendments in contaminated soil. *Geoderma*, 385:114803.

Rodrigues, W.D.S., Pereira, Y.C., de Souza, A.L.M., Batista, B.L. and Lobato, A.K.D.S. 2020. Alleviation of oxidative stress induced by 24-epibrassinolide in soybean plants exposed to different manganese supplies: UpRegulation of antioxidant enzymes and maintenance of photosynthetic pigments. *Journal of Plant Growth Regulation*, 39(4):1425–1440.

Sachdev, S., Ansari, S.A., Ansari, M.I., Fujita, M. and Hasanuzzaman, M. 2021. Abiotic stress and reactive oxygen species: generation, signaling, and defense mechanisms. *Antioxidants*, 10(2):277.

Saha, B., Chowardhara, B., Kar, S., Devi, S.S., Awasthi, J.P., Moulick, D., Tanti, B. and Panda, S.K. 2019. Advances in heavy metal-induced stress alleviation with respect to exogenous amendments in crop plants. In Priming and pretreatment of seeds and seedlings: Implication in plant stress tolerance and enhancing productivity in crop plants, eds. M. Hasanuzzaman and V. Fotopoulos, 313–328. Springer.

Sahu, J., Vaishnav, A. and Singh, H.B. 2020. Insights in plant-microbe interaction through genomics approach (part 1). *Current Genomics*, 21(3):155–156.

Sarma, R.K., Gowtham, I., Bharadwaj, R.K.B., Hema, J. and Sathishkumar, R. 2018. Recent advances in metal induced stress tolerance in plants: possibilities and challenges. In Plants under metal and metalloid stress, eds. M. Hasanuzzaman, K. Nahar and M. Fujita, 1–28. Springer.

Scattolin, M., Peuble, S., Pereira, F., Paran, F., Moutte, J., Menad, N. and Faure, O. 2021. Aided-phytostabilization of steel slag dumps: The key-role of pH adjustment in decreasing chromium toxicity and improving manganese, phosphorus and zinc phytoavailability. *Journal of Hazardous Materials*, 405:124225.

Schalk, I.J., Hannauer, M. and Braud, A. 2011. New roles for bacterial siderophores in metal transport and tolerance. *Environmental Microbiology*, 13(11):2844–2854.

Schirawski, J. and Perlin, M.H. 2018. Plant–microbe interaction 2017–the good, the bad and the diverse. *International Journal of Molecular Sciences*, 19:1374. doi: 10.3390/ijms19051374

Seneviratne, M., Rajakaruna, N., Rizwan, M., Madawala, H.M.S.P., Ok, Y.S. and Vithanage, M. 2019. Heavy metal-induced oxidative stress on seed germination and seedling development: a critical review. *Environmental Geochemistry and Health*, 41(4):1813–1831.

Shafi, A. and Zahoor, I. 2020. Abiotic and biotic stress-induced alterations in the micronutrient status of plants. In plant micronutrients, eds. T. Aftab and K.R. Hakeem, 285–310. Springer.

Sharafi, Y. and Raina, M. 2021. Effect of boron on pollen attributes in different cultivars of Malus domestica L. *National Academy Science Letters*, 44(3):189–194.

Sharma, P., Bakshi, P., Kapoor, D., Arora, P., Kour, J., Kaur, R., Sharma, A., Mir, B.A. and Bhardwaj, R. 2021. Thiol assisted metal tolerance in plants. In *Handbook of assisted and amendment: enhanced sustainable remediation technology* (pp. 389–401). John Wiley & Sons, Ltd, USA.

Singh, S. 2014. A review on possible elicitor molecules of cyanobacteria: their role in improving plant growth and providing tolerance against biotic or abiotic stress. *Journal of Applied Microbiology*, 117(5):1221–1244.

Singh, S., Parihar, P., Singh, R., Singh, V.P. and Prasad, S.M. 2016. Heavy metal tolerance in plants: role of transcriptomics, proteomics, metabolomics, and ionomics. *Frontiers in Plant Science*, 6:1143.

Souza, S.C., Souza, L.A., Schiavinato, M.A., de Oliveira Silva, F.M. and de Andrade, S.A. 2020. Zinc toxicity in seedlings of three trees from the Fabaceae associated with arbuscular mycorrhizal fungi. *Ecotoxicology and Environmental Safety*, 195:110450.

Štolfa, I., Pfeiffer, T.Ž., Špoljarić, D., Teklić, T. and Lončarić, Z. 2015. Heavy metal-induced oxidative stress in plants: response of the antioxidative system. In Reactive oxygen species and oxidative damage in plants under stress, eds. D.K. Gupta, J.M. Palma and F.J. Corpas, 127–163. Springer.

Sung, J., Lee, Y. and Lee, S. 2021. Primary metabolic responses in the leaves and roots of bell pepper plants subjected to microelements-deficient conditions. *Korean Journal of Agricultural Science*, 48(1):179–189.

Suzuki, N., Rivero, R.M., Shulaev, V., Blumwald, E. and Mittler, R. 2014. Abiotic and biotic stress combinations. *New Phytologist*, 203(1):32–43.

Sytar, O., Kumar, A., Latowski, D., Kuczynska, P., Strzałka, K. and Prasad, M.N.V. 2013. Heavy metal-induced oxidative damage, defense reactions, and detoxification mechanisms in plants. *Acta Physiologiae Plantarum*, 35(4):985–999.

Tahir, M., Rashid, M.S., Ali, Q. and Malik, A. 2020. Evaluation of genetic variability in wheat and maize under heavy metal and drought stress. *Genetics and Molecular Research*, 19(3).

Tiwari, S. and Lata, C. 2018. Heavy metal stress, signaling, and tolerance due to plant-associated microbes: an overview. *Frontiers in Plant Science*, 9:452.

Wallace, D.R., Taalab, Y.M., Heinze, S., Tariba Lovaković, B., Pizent, A., Renieri, E., Tsatsakis, A., Farooqi, A.A., Javorac, D., Andjelkovic, M. and Bulat, Z. 2020. Toxic-metal-induced alteration in miRNA expression profile as a proposed mechanism for disease development. *Cells*, 9(4):901.

Wang, Q., Liu, J. and Zhu, H. 2018. Genetic and molecular mechanisms underlying symbiotic specificity in legume-rhizobium interactions. *Frontiers in Plant Science*, 9:313.

Wasim, H.M., Sarwar, G., Tahir, M.A., Aftab, M., Manzoor, M.Z., Zafar, A., Shehzad, I., Riaz, A., Niaz, A., Mufti, K.A. and Arif, M. 2021. Deleterious impact of sodium chloride (NaCl) toxicity on chemical properties of soil and ionic composition of rice. *Pakistan Journal of Agricultural Research*, 34(1):169–175.

Wu, S., Hu, C., Yang, X., Tan, Q., Yao, S., Zhou, Y., Wang, X. and Sun, X. 2020. Molybdenum induces alterations in the glycerolipidome that confer drought tolerance in wheat. *Journal of Experimental Botany*, 71(16):5074–5086.

Yadav, A.P.S., Dwivedi, V., Kumar, S., Kushwaha, A., Goswami, L. and Reddy, B.S. 2021. Cyanobacterial extracellular polymeric substances for heavy metal removal: A mini review. *Journal of Composites Science*, 5(1):1.

Zaid, A., Ahmad, B., Jaleel, H., Wani, S.H. and Hasanuzzaman, M. 2020. A critical review on iron toxicity and tolerance in plants: Role of exogenous. In Phytoprotectants, plant micronutrients, eds. T. Aftab and K.R. Hakeem, 83–99. Springer.

16 Potential Use of Algae and Bacteria for Bioremediation of Wastewater

A Critical Review

Dipannita Parial and Satarupa Dey

16.1 INTRODUCTION

Rapid industrialization, human activities, unsustainable agricultural practices all over the globe have resulted in the generation of a huge amount of wastewater, which is regularly released in water bodies. These effluents are mostly generated and discharged in the water systems without any proper treatment. As a result, these effluents containing high amounts of toxic metals, colorants, pharmaceutical products, antibiotic residues, pesticides, and suspended and dissolved organic components lead to the eutrophication of the water bodies, which is extremely harmful to the aquatic flora and fauna. Moreover, these pollutants are entering the food chain, and at higher trophic levels, they are causing more bioaccumulation of toxic materials, causing serious health issues.

In order to protect the environment and human health, there is an ever-increasing demand for sustainable wastewater treatment systems all over the world. Wastewater treatments are largely done in primary, secondary and tertiary stage following physical, chemical and biological procedures. In order to dispose wastewater safely, different methods are followed, which include processes like coagulation, flocculation, filtration, flotation, adsorption, photocatalysis and electrocatalysis. These methods are mainly aimed to remove solid particles, suspended particles, colloidal particles, organic matter and toxic metals. However, the main drawback of these processes is their intensive energy requirement, high cost and less environment-friendly nature; moreover, these processes tend to generate hazardous by-products and sludge, which results in secondary pollution. Thus, there is always a need for a treatment system which is cost-effective and eco-friendly in nature.

Biological wastewater treatments largely depend on the consortia of autotrophic and heterotrophic bacteria that help in the biodegradation of contaminants and can be an important step after pretreatment of the effluents. In most of the biological processes, indigenous microbes are used, which remove most of the pollutants,

DOI: 10.1201/9781003297901-16

heavy metals and nutrients such as phosphate/nitrate from wastewater and help in stabilizing the downstream treatment process. Biological treatment processes can also be carried out using microalgae as they show a high efficiency in removal of toxic metals, nutrients and pharmaceuticals. Microalgae also have a great capacity for uptake of nutrients such as nitrogen and phosphorus which are required for their growth. Microalgae help to maintain the dissolved oxygen and reduce pathogen and faecal bacteria in the wastewater and can treat wastewater in a single step and does not require multiple steps for fixing carbon, nitrogen and phosphorus contents. It can also convert CO_2 to chemical products, can produce biofuels and bio alcohol, reduce greenhouse gases and can be converted to biobased products. The symbiotic relationship between algae and bacteria is considered as the structural pillar of the ecosystem, and their consortia can be effectively used for the treatment of wastewater.

The current chapter deals with the use of different biological processes in the treatment of wastewater effluents. The sustainable, cost effective, eco-friendly nature of the biological treatment processes using microalgae, bacteria and their consortium was also discussed in detail. The application of the used algal biomass was also discussed. The future prospects and challenges faced by the biological treatment were also assessed.

16.2 USE OF ALGAE IN WASTEWATER TREATMENT

Algae are defined as a large and diverse polyphyletic group of predominantly aquatic and photoautotrophic organisms. It ranges from unicellular microalgae to multicellular forms like giant kelp and seaweeds. Regardless of their dissimilar cellular organization, almost all of them use solar energy to reduce inorganic nutrients to organic substance, thus producing biomass.

Algae play important roles in aquatic food chain, have several economic importance and are also regarded as a good source of renewable energy. The reasons why algae are applied in various fields include their rapid growth rate, ease of handling, requirement of only light, CO_2 and minerals to grow, ability to grow at different environmental conditions, valuable biochemical composition of algal cells, non-requirement of any land for cultivation, their role in CO_2 sequestration, higher carbon fixation rate than land plants, ability to evolve oxygen as a by-product and their ability to grow in both fresh and saline wastewater.

While growing in wastewaters, algae recover nutrients and use them to produce valuable algal biomass. Algal cells accumulate high amount of lipids and carbohydrates during their growth, which makes them suitable for biofuel production. Besides nutrients recovery, algae can also remove and/or bio transform toxic heavy metals and xenobiotic substances from wastewater. Moreover, algae minimize the requirement for mechanical aeration during wastewater treatment as oxygen released by their photosynthetic metabolism can be utilized by aerobic bacteria for decomposition of organic matter present in the wastewater. They also secrete secondary metabolites that inhibit further growth of pathogenic organisms. In addition, they influence the wastewater treatment process by acting as flocculants, thereby increasing the rate of sedimentation.

Phycoremediation is the precise term that represents the practice of using algae to treat wastewater. As algae exhibit different kinds of metabolism (autotrophic, mixotrophic and heterotrophic), they can be used to treat various types of waste streams, simultaneously fabricating valuable biomass. Algae synthesize various kinds of remarkable products such as pigments, proteins, polyunsaturated fatty acids, animal feed, biofuels, biofertilizer, etc., which are, in turn, used in different industries such as food, feed, cosmetic, pharmaceutical, nutraceutical, chemical, etc. and as real biorefinery alternatives. Therefore, the efficiency of this method lies in its dual function.

William Oswald and his fellow researchers first used algae to treat wastewater. He was also among the pioneer researchers to observe the collaborative interaction of bacteria and algae in treating wastewater (Oswald 1953). They proposed an inexpensive green technique to treat wastewater that is chiefly fueled by sunlight. Subsequently, several scientists treated various industrial and urban wastewaters with different algal strains, which have been discussed in this chapter.

16.2.1 Heavy Metal Bioremediation by Algae

The presence of heavy metals in wastewater pollutes natural water bodies like lakes, rivers and seas. Incorporation of these toxic metals into aquatic food chains may subsequently enter the human body, leading to several health issues such as kidney damage, reduced lung function, loss of bone mineral, nerve problems, cancer, etc. In spite of the presence of several conventional methods, use of algae in heavy metal removal offers an innovative technology which is more proficient, ecologically secure and inexpensive. Notably, *Scenedesmus* and *Chlorella* are regarded as hyper-adsorbents and hyper-accumulators, owing to their remarkable efficacy to remove heavy metals. Algae require heavy metals like zinc, molybdenum, manganese, iron, cobalt, copper and boron as trace elements for their growth and metabolism; however, other heavy metals like cadmium, chromium, lead, arsenic and mercury are harmful to them. Algae can also recover precious metals such as silver and gold and can also remove toxic radioactive elements from water.

Algae tolerate heavy metal toxicity by various mechanisms such as gene regulation, heavy metal immobilization, chelation, exclusion and by producing different enzymes that decrease metal toxicity. Algae controls heavy metal concentration in the cytoplasm by forming organometallic complexes and inducing the production of phytochelatins, a number of antioxidant enzymes and various non-enzymatic antioxidants that ultimately reduce heavy metal stress.

Algae remove heavy metals from wastewater by the processes known as biosorption and bioaccumulation. There are a number of reports of heavy metal removal from wastewater using algae, some of which are shown in Table 16.1. It shows that non-living algal biomass has been predominantly used to treat wastewater as live algae show restricted sorption due to poisoning of the living cells. Furthermore, the absorption process by live algal biomass is more complex as the live cells accumulate metal ions intracellularly, and the intracellular uptake, in turn, is affected by several factors like the growth phase of the algae used. In contrast, non-living or inactive algal biomass acts as an assemblage of polymers (cellulose, glycoproteins, pectins, sugars, etc.) and adsorbs metal ions at the extracellular level only and is considered as a better candidate for bioremediation. Table 16.1 also shows that the heavy metal removal efficiency varies with different algal strains and usually reaches to a

TABLE 16.1
Biosorption efficiency of different heavy metal ions using a number of macro and microalgal strains under optimal condition (#: seaweed/macroalgae)

Metal	Algal strain	Initial metal conc (mg/L)	Biomass (g/L)	Temp (°C)	Optimal pH	Time	Max. sorption (mg g^{-1})	References
As (III)	*Ulothrix cylindricum*	10	NA	20	6	60 min	67.2	Tuzen 2010
	Maugeotiagenuflexa	10	4	20	6	60 min	2.4	Sari 2011
	Mixture of green and blue-green algae	50	10	20	4	180 min	3.5	Sulaymon 2013
	Scenedesmus almeriensis	12	1	NA	10	180 min	5	Saavedra 2018
Al (III)	*Laminaria japonica*	NA	1	NA	5	30 h	75.27	Lee 2004
Au (III)	*Fucus vesiculosus#*	100	1	23	7	8 h	74.05	Mata 2009
Cd (II)	*Chlorella minutissima*	NA	4	28	6	20 min	303	Yang 2015
	Scenedesmus sp.	200	1.5	NA	6	NA	48.4	Jena 2015
	Lipid-extracted *Chlamydomonas* sp.	NA	1	30	8	60 min	23.3	Zheng 2016
	Lipid-extracted *Chlorella* sp.	NA	1	30	8	60 min	25.5	Zheng 2016
	Parachlorella sp.	100	1	35	7	NA	96.2	Dirbaz 2018
Cr (III)	*Chlorella miniate*	100	NA	25	5	24 h	41.12	Han 2006
	Spirogyra sp.	50	NA	25	5	3 h	30.21	Bishnoi 2007
	Chlorella sorokiniana	NA	1	25	4	NA	58.8	Akhtar 2008
Cr (VI)	Immobilized *Chlorella minutissima*	100	60% (w/v)	30	2	2160 min	57.33	Singh 2012
	Rhizoclonium hookeri #	1000	1	NA	2	45 min	67.3	Kayalvizhi 2015
	Immobilized *Spirulina platensis*	250	1	25	2	NA	49	Kwak 2015
	Chlorella vulgaris	147	1	25	2	240 min	63.2	Sibi 2016
	Spirulina platensis	500	NA	60	1	90 min	59.6	Nithya 2019
	Lipid-extracted *Spirulina platensis*	500	NA	60	1	90 min	45.5	Nithya 2019
Cu (II)	*Spirogyra neglecta*	100	0.1	25	5	0.16 h	115.3	Singh 2007
	Sargassum sp. #	NA	1	22	6	180 min	72.5	Karthikeyan 2007
	Fucus vesiculosus #	NA	NA	23	5	120 min	105.48	Mata 2008
	Cladophora sp. #	100	NA	25	5	60 min	13.7	Lee 2011

TABLE 16.1 (*Continued*)
Biosorption efficiency of different heavy metal ions using a number of macro and microalgal strains under optimal condition (#: seaweed/macroalgae)

Metal	Algal strain	Initial metal conc (mg/L)	Biomass (g/L)	Temp (°C)	Optimal pH	Time	Max. sorption (mg g^{-1})	References
Pb (II)	*Chaetoceros* sp.	20	1.5	25	6	180 min	8	Molazedah 2015
	Chlorella sp.	20	1.5	25	6	180 min	10.4	Molazedah 2015
	Phormidium sp.	10	4	25	5	40 min	2.305	Das 2016
	Rhizoclonium hookeri #	NA	NA	40	5	NA	81.7	Suganya 2016
Se (IV)	*Cladophora hutchinsiae* #	NA	8	20	5	60 min	74.9	Tuzen 2010
U (VI)	*Chlorella vulgaris*	23.8	0.8	NA	4	96 h	27	Vogel 2010
Ni (II)	*Sphaeroplea sp*	NA	1	33	6	60 min	199.55	Srinivasa Rao 2005
Zn (II)	*Scenedesmus obliquus*	75	0	25	6 to 7	24 h	836.5	Monteiro 2011
Hg (II)	Transgenic *Chlorella sp.*	8	0.3	30	NA	120 min	7.33	Huang 2006
	Spirogyra sp.	1	3	4	4	30 min	0.253	Rezaee 2006
	Chlorella vulgaris	48	2	20	5	120 min	17.49	Solisio 2019

satisfactory level within 120 minutes. It is also evident that low pH favours metal ion uptake. Functional groups present on the cell walls of algal biomass influences the effect of pH on metal uptake.

Besides pH, phycoremediation is also affected by a number of factors such as concentration of algal biomass and metal ions, temperature, presence of competing ions, etc. Temperature variations show diverse biosorption behaviours in different algal species with different metal ions. Moreover, wastewater polluted with numerous heavy metals shows competition between them for binding to the algal cell wall, which, in turn, is affected by the properties and amount of the metal ions present. Light intensity, amount of dissolved nitrates and growth rate also play key roles in phycoremediation of heavy metals. Furthermore, several chemicals and physical pretreatments of algal biomass enhance its heavy metal uptake capacity. Bioremoval of heavy metals using algal biomass immobilized in either natural biopolymers such as alginate and agar or artificial substances like polyacrylamide and silica gel shows increased biosorption relative to free algal cells. These techniques aid the reuse of algal biomass during consecutive sorption/desorption cycles, boost photosynthetic capability and also lessen the toxicity of a number of metals. Additionally, as numerous genes and cellular substances are involved in uptake, detoxification and tolerance of heavy metals, production of cost-effective green biosorbents by exploiting genetic and protein engineering approaches is an emerging area of research nowadays.

16.2.2 Bioremoval of Emerging Contaminants Using Algae

Emerging environmental contaminants present in wastewater are drawing significant awareness as they exhibit a number of bad qualities such as high polarity, ability to be bioaccumulated by aquatic organisms and resistance to biodegradation. They have adverse effects on the aquatic ecosystem and human health as well. Algae-based technologies have shown much efficiency in removing emerging contaminants at laboratory-scale experiments, some of which are shown in Table 16.2.

TABLE 16.2
Percent removal efficiency of emerging contaminants by algae at laboratory scale

Emergent contaminants	% removal	Algal strain	Experimental conditions	References
Pharmaceutical; ciprofloxacin	100	*Chlamydomonas* sp.	Synthetic wastewater medium, 250μmol m^{-2} s^{-1} light intensity, (12/12 light/dark cycle), 2%, CO_2 25± 1°C temperature, 5–6 days' time	Xie 2020
Pharmaceutical: sulfadiazine	54.53			
Pharmaceutical: sulfamerazine	84	*Haematococcus pluvialis*	Pre-sterilized synthetic wastewater medium, 12 h: 12 h dark/light cycle, 25± 1°C temperature, 40 days' time	Kiki 2020
Pharmaceutical: sulfamethoxazole	74			
Pharmaceutical: sulfamonomethoxine	75			
Pharmaceutical: acetaminophen	67	*Chlorella sorokiniana*	Mann and Myers medium, 25± 1°C temperature, pH 7.5 ± 0.5, 370 μ Em^{-2} s^{-1} light intensity (12/12 light/dark cycle), 144 hours' time	Escapa 2019
Pharmaceutical: sulfamethazine	31.4–62.3	*Scenedesmus obliquus*	Sterilized Bold's Basal medium, 27°C temperature, 45–50 μmol m^{-2} s^{-1} light intensity (16/8 light/dark cycle), 14 days' time	Xiong 2019
Pharmaceutical: sulfamethoxazole	27.7–46.8			

TABLE 16.2 (*Continued*)
Percent removal efficiency of emerging contaminants by algae at laboratory scale

Emergent contaminants	% removal	Algal strain	Experimental conditions	References
Pharmaceutical: carbamazepine	<21	*Chlorella vulgaris*	Synthetic wastewater medium, 22°C temperature, 90–160 μmol m^{-2} s^{-1} light intensity (16/8 light/dark cycle), 25 days' time	Larsen 2019
Pharmaceutical: ibuprofen	60			
Pharmaceutical: gemfibrozil	<27			
Personal care product:methylisothiazolinone	100	*Scenedesmus* sp.	BG11 medium, 25± 1°C temperature, 55–60 μmol m^{-2} s^{-1} light intensity (14:10 h light/dark cycle), 4 days' time	Wang 2020
Industrial chemicals:para-xylene (aromatic hydrocarbons)	100	*Rhodomonas* sp.	F/2 medium, 60 μmol m^{-2} s^{-1} light intensity, (14 h:10 h light/dark cycle), 20°C temperature, 6 days' time	Li 2020

Bioremediation of emerging contaminants by algae takes place in three steps such as bioadsorption, bioaccumulation and biodegradation or biotransformation. Biodegradation is the breakdown of the complex materials into environmentally acceptable simpler forms which takes place intra and/or extracellularly. It occurs via two key mechanisms, for example metabolic degradation and co-metabolism. Algae take up emerging contaminants as their carbon source during metabolic degradation. In co-metabolism, enzymatic breakdown of the contaminants takes place, and a threshold concentration of contaminants is required for the enzymatic activity. Several factors such as the type of algal strains, feature of the pollutant, enzymatic pathway and environmental conditions influence biodegradation in various ways. Moreover, algae, especially microalgae, boost up the degradation process by forming a consortial relationship with bacteria.

Algae-based bioremediation of emerging contaminants is affected by several factors like lack of nutrients in wastewater, competition between different contaminants for the binding sites, etc. Surprisingly, some specific contaminants seem to boost the removal rate of other contaminants as well. Co-metabolism is another mechanism that enhances the elimination efficacy of a variety of emerging contaminants. In contrast, some organic substrates have negative impact on phycoremediation of few contaminants. Temperature and pH play important roles in removal of emerging contaminants, and higher temperature usually enhances the whole process. Wastewater pH affects contaminant structure, biosorption and removal kinetics. Increased pH reduces the removal of acidic xenobiotics. Indeed, excessive high or low pH levels hinder algal growth, thereby lowering overall removal efficiency. It is also important to check effluent quality, which has been ignored since most of the studies are restricted to laboratory conditions only. Many studies reported the presence of higher amount of contaminants in the effluent in comparison to the influent. Moreover, the formation of by-products may turn out to be more dangerous in comparison to the primary compounds.

16.2.3 Nutrients Bioremediation by Algae

Wastewater, especially municipal and agricultural wastewater, contains a large amount of nutrients. Presence of excessive nutrients results in eutrophication of natural water bodies that occurs through the production of dangerous algal blooms and depletion of dissolved oxygen (hypoxia) due to decomposition of algal biomass. It ultimately disturbs the whole aquatic ecosystem and imparts a severe hazard to the aquatic life forms, which, in turn, may harm humankind as well.

Nutrients released from wastewater are getting considerable attention and have been strictly controlled throughout the world. Some of the recent studies on algal nutrient removal along with the production of valuable products are shown in Table 16.3.

TABLE 16.3
Nutrient removal capacity of various algal strains from different wastewater types showing high-value product synthesis (tn: total nitrogen, tp: total phosphorus, sfa: saturated fatty acids, ufa: unsaturated fatty acids, pufa: polyunsaturated fatty acids, mufa: monounsaturated fatty acids)

Algal strain	Wastewater type	Nitrogen (%)	Phosphorus (%)	Product/ co-product	Reference
Dunaliella	Anaerobically digested poultry litter wastewater	63.8 TN	87.2 TP	7.26 mg L−1 β-carotene	Han 2019
Chlorella sorokiniana	Palm oil mill effluent	98.6 TN	96 TP	Lipid content: 14.43%(NPBR)	Cheah 2020
Scenedesmus obliquus	Municipal wastewater	96 TN	80 TP	Lipid content: 56%	Qu 2020
Desmodesmus sp.	Piggery wastewater	79.2 TN	65.3 TP	Total fatty acid/ dryweight (%): 29.4 ± 0.1728.3 ± 0.21 SFA 39.9 ±0.93 MUFA 31.3 ± 1.74PUFA	Chen 2020
Isochrysis sp.	Sewage discharge	5.57 TN	84–94	63.0, 16.92% MUFA, 20.00% PUFA	Singh 2021

TABLE 16.3 (*Continued*)
Nutrient removal capacity of various algal strains from different wastewater types showing high-value product synthesis (tn: total nitrogen, tp: total phosphorus, sfa: saturated fatty acids, ufa: unsaturated fatty acids, pufa: polyunsaturated fatty acids, mufa: monounsaturated fatty acids)

Algal strain	Wastewater type	Nitrogen (%)	Phosphorus (%)	Product/ co-product	Reference
Chlorella vulgaris, Chlorococcum vitiosum, Chroococcus turgidus, Desmococcus olivaceus, Scenedesmus acutus, Scenedesmus dimorphus and Oocystis solitaria	Coke plant wastewater	42.7 (NH4+ N)	NA	NA	Nagi 2021
Scenedesmus sp.	Domestic wastewater	80 (NH4+ N), 99 (NO2- N), 86 (NO3- N)	66 (PO4- P)	43.3% SFA, 44.4% MUFA, 12.3% PUFA	Baldev 2021

The main source of nitrogen in wastewater is primarily fertilizers and human wastes containing ammonia (NH_4^+), nitrite (NH_2^-) and nitrate (NO_3-), and majority of phosphorus comes from synthetic detergents used in households and different industrial activities. Phosphate enters the algal cell actively by a symporter, with H^+ or Na^+ ions providing the driving force. Algae also hydrolyze organic phosphorus compounds with membrane-bound as well as free phosphatases, releasing bioavailable phosphorus that are subsequently taken by the algal cells. Among inorganic nitrogen sources, algae preferentially take up NH_4^+ because of its more energetically favourable assimilation and incorporation process. Algae uptake NH_4^+ by a group of membrane transporter proteins belonging to the ammonium transporter family. On the other hand, NO_{3-} and NO_{2-} are reduced to NH_4^+ by nitrate reductase and nitrite reductase respectively for intracellular uptake. Moreover, the entry of NO_{3-} inside the algal cell involves ATP hydrolysis. In addition to inorganic nitrogen, algae can also assimilate nitrogen from a broad array of organic sources. The incorporation of organic nitrogen inside the algal cell may occur in both autotrophic and heterotrophic conditions.

16.3 USE OF BACTERIA IN WASTEWATER TREATMENT

Biological treatment of wastewater is gaining importance and is considered as a cost-effective and eco-friendly process for mitigation of pollutant. Bioremediation of wastewater by bacteria can be done by different processes such as biodegradation, biomineralization, bioaugmentation, biosorption, bioreduction of pollutants into less toxic and harmless products. Mostly, a consortium of microbes is used for wastewater treatment that has high biodegradation efficiency and uses a wide range of different substrates. Several bacteria have been reported to produce a plethora of enzymes such as arsenate oxidase which has the ability of oxidation of toxic arsenic [As (III)] to less toxic [As (V)] (Sanyal et al. 2016). Bacteria like *Pseudomonas* sp. and *Lysinibacillus* sp. have been extensively studied for their ability to remove heavy metals and act as a carrier matrix of natural non-polymeric electrospun cyclodextrin fibers (CDF) (Safdari et al. 2018). Isolates like *Aeromonas hydrophila* have been used to decolorize triarylmethane dyes; however, the efficiency was largely dependent on temperature, pH and oxygenation state (Imran et al. 2015). The detailed information on usage of bacteria or bacterial consortium for removal of different types of pollutants are enlisted in Table 16.4.

TABLE 16.4
Microorganism species used in bioremediation of different types of contaminants

Contaminants in the wastewater	Microorganisms	References
Hydrocarbon organic compounds		
Petrol and diesel polycyclic aromatic hydrocarbons, toluene, polyurethane, benzene and xylene	*Achromobacter, AcinetobacterFlavobacterium, Pseudomonas* sp.*Pseudomonas putida, P. alcaligenes,Pseudomonas mendocina,Pseudomonas veronii*	Patel and Gupte (2016)
Phenanthrene, pyrene and benzo(a) pyrene	A synthetically microbial consortium	Zafra et al. (2017)
Striatum Pyrene, anthracene, 9-metil anthracene, dibenzothiophene lignin peroxidase	*Gleophyllum striatum*	Birolli et al. (2018)
Oil and grease contaminants		
Oil	*Alcaligenes odorans, Bacillus subtilis, Corynebacterium propinquum, Pseudomonasaeruginosa*	Paikhomba et al. (2017)
Diesel oil	*Arthobacter* sp. *Bacillus* sp., *Bacillus coagulans, Bacillus cereus, Pseudomonas aeruginosa, Pseudomonas cepacia, Pseudomonas putida*	Dorcas et al. (2018)

TABLE 16.4 (*Continued*)
Microorganism species used in bioremediation of different types of contaminants

Contaminants in the wastewater	Microorganisms	References
Azo dye wastewaters		
Textile azo dyes	*Micrococcus luteus* strain SSN2, *Providencia rettgeri* strain HSL1, *Pseudomonas* sp. SUK1, *Pseudomonas fluorescens*, *Staphylococcus* sp.	Ghosh et al. (2016), Sadeghi et al. (2019)
Heavy metals		
Cobalt, copper, chromium, lead and cadmium	*Bacillus safensis* (JX126862) *Bacillus safensis* (PB-5), *Bacillus safensis* (RSA-4) *Lysinibacillus sphaericus* (CBAM5)	Fauziah et al. (2017)
Manganese, iron, copper, uranium, zinc, lead	*Geobacter metallireducens*, *Geobacter* spp., *Pseudomonas syringae*, *Pseudomonas fluorescens*, *Pseudomonas aeruginosa*, *Vibrio harveyi*	Igiri et al. (2018), Choudhary et al. (2017)
Lead (Pb) and cadmium (Cd)	*Escherichia coli*	Liu et al. (2021a)
Pesticides		
Endosulfan, Coragen	*Achromobacter* sp. *M6 Bacillus*, *Klebsiella pneumonia*, *Klebsiella sp.*, *Staphylococcus aureus*, *Pseudomonas* sp., *Rhodococcus* sp. *M2*	Alvarez et al. (2017), Nwokem et al. (2017)
Bensulfuron-methyl (BSM)	*Methylomonas* sp. strain LW13	Liu et al. (2021b)
Organochloride pesticides, organophosphorus pesticides, carbamates and pyrethroid	*Escherichia coli* strain BL21	Yang et al. (2012)

Heterotrophic bacteria are known to degrade a wide range of biodegradable organic components, using them as electron donors. Under aerobic, anaerobic or anoxic condition, different substrates, such as oxygen, and different nutrients, nitrite and sulphates present in the wastewater are used as electron acceptors. Several respiratory products and biomass are produced, depending on the substrate present in wastewater. Autotrophic bacteria procure energy by oxidizing ammonia to nitrate or nitrite and also use organic substances as carbon source. Both phototrophs and chemotrophs use solar energy or chemical energy using both organic and inorganic substances.

16.4 MICROBES ASSOCIATED WITH WASTEWATER

Wastewater is mostly dominated with heterotrophic and autotrophic bacteria which lead to different biochemical reactions and subsequent changes in wastewater. Most traditional activated sludge processes are characterized by long retention time, good aeration and low biological oxygen demand (BOD) to maintain concentration of nitrifying bacteria. The mode of wastewater treatment, that is, oxic and anoxic batch reactors, is largely determined by the activity of indigenous nitrifying microbiota.

16.4.1 Role of Heterotrophic Bacteria

The heterotrophic bacteria are known to degrade readily biodegradable chemical oxygen demand (rbCOD); however, they are characterized by low efficacy for the conversions of less biodegradable and slowly hydrolyzable COD to rbCOD. Accumulation of nonbiologically degraded COD concomitantly results in high biosolids per pound and requires additional treatment, aeration and disposal cost. These heterotrophic bacteria use inorganic and organic macronutrients including inorganic phosphate and dissolved organics for growth and metabolism and bioremediation. Processes like nitrification and denitrification helps in the removal of ammonium nitrogen (NH4-N) and total nitrogen (TN) from the wastewater. Many competitive heterotroph species, such as *Pseudomonas* sp., are known to inhibit NH_4-N conversion into nitrate, thus preventing several metabolic processes and decreasing efficiencies. However, when a consortium of nitrifying and denitrifying bacteria was used, an increase in performance was noted.

16.4.2 Role of Autotrophic Bacteria

Autotrophic bacteria play a determining role in the nitrification process. Both ammonia-oxidizing (AOB) and nitrite-oxidizing bacteria (NOB) have been extensively studied by both physiological as well as molecular in situ techniques. However, their distribution pattern is highly dependent on the different biological and environmental conditions of the wastewater. Autotrophic bacteria play a crucial role in activated sludge systems; however, a detailed study on growth pattern, operational strategies, and the degradation products formed by the autotrophs is still required to clearly understand the entire process. However, in an activated sludge system, high nitrite concentration inhibits the growth of autotrophs and requires considerable amount of energy during maintenance.

16.5 THE UTILIZATION OF ALGAE-BACTERIA CONSORTIA IN VARIOUS WASTEWATER STREAMS

The use of algae for wastewater bioremediation incites the emergence of algae-bacteria consortia as an alternative strategy to the conventional biological methods to treat various kinds of wastewaters. Being photosynthetic, algae grow using CO_2 and light as their carbon and energy sources respectively and release oxygen. Oxygen evolved by algal cells is subsequently utilized by the heterotrophic bacteria to carry out the oxidation of the organic matter present in wastewater. Simultaneously, the heterotrophic bacteria provide CO_2 required for the photosynthetic activity of the algal cells. In addition, the oxygen released by algal biomass can be utilized by nitrifying bacteria to oxidize ammonium to nitrate, consuming CO_2 as a carbon source.

Community structure plays an important role in the wastewater treatment process. The dominant species significantly affect the nutrient removal process. For example, phosphorus removal is associated with the dominance of algae in an aerobic autotrophic environment with a very little effect of bacterial community on total dissolved phosphate removal. Moreover, the community structure directly changes with a change in the pH of the medium.

Wastewater treatment with algae-bacteria consortia is a cost-effective, energy-efficient, environmentally sustainable and eco-friendly approach producing valuable biomass as biorefinery feedstock. Shetty (2019) proposed a condition-dependent wastewater treatment process by mutualistic interaction between a green alga *Chlorella* and bacteria with combined biohydrogen generation. The complications in determining individual concentrations of each species in the association emphasize the requirement for further research involving consistent estimation techniques. However, recently, Sánchez-Zurano (2021) showed the use of genetic algorithms to calibrate the values of several parameters, including the relative proportions of algae and bacteria present in the reactor to find out the optimum conditions for efficient wastewater treatment.

16.6 APPLICATION OF THE ALGAL BIOMASS

Wastewater-grown algal biomass can be applied in various fields. Anaerobic digestion of both the intact and lipid-extracted biomass results in conversion of complex organic substances into biogas such as methane, which could be used for the generation of electricity, heat or as transportation fuel, potentially increasing energy revenues. Lipid fraction of the biomass could be exploited for the production of biodiesel. Additionally, glycerol is synthesized as a by-product in biodiesel fabrication. Polysaccharides are also extracted from the algal biomass for biocrude production. A number of valuable bioproducts can be extracted from the harvested biomass. For instance, amino acids are obtained from the algal biomass and subsequently utilized in industrial applications. Moreover, the extracted protein-based compounds may serve as a high-quality feedstock for the production of polyols that could be exploited as a raw material for the synthesis of polyurethane foams. Harvested algal biomass can be used as bio-fertilizers because of their property to enhance nutrient content in the soil in addition to various plant-stimulating compounds needed by the land plants. Treated algae are also used in the production of lactic acid, alcohol and bioplastics. Algae secrete exopolysaccharides, which could be used as thickener, gelling agent, stabilizer, biolubricants and anti-inflammatory agents in various industries. Owing to the high nutritional value, harvested algal biomass can also be utilized as food for animals and fish.

16.7 CHALLENGES

Even though algal bioremediation offers numerous attractive benefits, it still shows some disadvantages, especially in large-scale applications. The challenges and the proposed strategies to overcome some of them are discussed subsequently:

- Contaminations with protozoa and zooplanktons badly affect algal bioremediation in open reactors such as raceways or high-rate algal ponds. Grazing by these organisms may decrease algal population considerably (as high as 90%) in just a few days.
- Presence of soluble organic matter (protein, carbohydrates, lipid, vitamins, enzymes, pigment, hormonal substances, inhibitors and toxins) in the effluent released by algal biomass is also a major concern.
- Bioaccumulation of emerging contaminants results in oxidative damage of the algal cells, leading to death. As a result, unbound contaminants may be discharged back. Such occurrence cannot be regulated, especially in commercial-scale application.
- Fungus, bacteria and other microorganisms present in raw wastewater act as possible competitors for available nutrients, and perhaps they dominate because of the higher rate of growth. However, pretreatment practices (like autoclaving, chlorination, filtration, ozonation, UV irradiation, etc.) and use of algae-fungus, algae-bacteria and other consortiums could solve this problem.

16.8 FUTURE PERSPECTIVE

In view of the multiple advantages of phycoremediation, the future outlook of widespread use of this green technology looks promising. However, certain considerations are to be taken care of. First of all, screening of suitable algal strains with specific attributes (such as high tolerance, ability to produce valuable products, low nutrient requirements, high CO_2 capturing ability, robustness toward existence of other microorganisms, resistance to predation by grazers, having self-flocculation capability) is a key step. Hence, research work based on genetic engineering to raise suitable algal strains is a prospective area of research.

16.9 CONCLUSION

Algal strains show various tolerances, reactions and varied bioremediation capabilities towards diverse environmental pollutants. Algal bioremoval efficiency depends on a number of factors. This green technology, with a few drawbacks, is more advantageous than conventional physicochemical techniques in many ways. For large-scale implementation and concurrent formation of value-added products, a better understanding of influencing parameters, research to overcome the disadvantages associated with phycoremediation as well as techno-economic analysis and a life cycle assessment are desperately required.

REFERENCES

Akhtar, Nasreen. 2008. Biosorption characteristics of unicellular green alga *Chlorella sorokiniana* immobilized in loofa sponge for removal of Cr(III). *J. Environ. Sci.* 20, 2: 231–239.

Alvarez, Analia, Juliana Maria Saez, José Sebastian Davila Costa, Veronica Leticia Colin, María Soledad Fuentes, Sergio Antonio Cuozzo, Claudia Susana Benimeli, Marta Alejandra Polti, and María Julia Amoroso. 2017. Actinobacteria: current research and perspectives for bioremediation of pesticides and heavy metals. *Chemosphere* 166: 41–62.

Baldev, Edachery. 2021. Wastewater as an Economical and Ecofriendly green Medium for microalgal biofuel production. *Fuel* 294: 120484. doi: 10.1016/j.fuel.2021.120484

Birolli, Willian G., Darlisson de A. Santos, Natália Alvarenga, Anuska CFS Garcia, Luciane PC Romão, and André LM Porto. 2018. Biodegradation of anthracene and several PAHs by the marine-derived fungus Cladosporium sp. CBMAI 1237. *Mar. Pollut. Bull.* 129, 2: 525–533.

Bishnoi, Narsi R. 2007. Biosorption of Cr(III) from aqueous solution using algal biomass *spirogyra* spp. *J. Hazard. Mater.* 145: 142–147.

Cheah, Wai Yan. 2020. Enhancing microalga *Chlorella Sorokiniana* CY-1 biomass and lipid production in palm Oil Mill Effluent (POME) using novel-designed photobioreactor. *Bioengineered* 11: 61–69.

Chen, Zhihong. 2020. Nutrients removal from piggery wastewater coupled to lipid production by a newly isolated self-flocculating microalga *Desmodesmus* Sp. PW1. *Bioresour. Technol.* 302: 122806.

Choudhary, Madhu, Kumar Raman, Datta Asim, Nehra Vibha, and Garg Neelam. 2017. Bioremediation of heavy metals by microbes. In: Bioremediation of salt affected soils: an Indian perspective. Springer, Cham, pp. 233–255. doi: 10.1007/978-3-319-48257-6_12

Das, Debayan. 2016. Biosorption of lead ions (Pb2+) from simulated wastewater using residual biomass of microalgae. *Desalin. Water Treat.* 57, 10: 4576–4586.

Dirbaz, Mahboobeh. 2018. Adsorption, kinetic and thermodynamic studies for the biosorption of cadmium onto microalgae *Parachlorella* sp. *J. Environ. Chem. Eng.* 6, 2: 2302–2309.

Dorcas, Obafemi Yemisi, Isibor Patrick Omoregie, Adeyemi Alaba Oladipupo, and Taiwo Olugbenga Samson. 2018. Plasmid profiling of crude petroleum degrading bacterial strains isolated from polluted soils in Ota, Nigeria. *Annu. Res. Rev. Biol.*: 1–11.

Escapa, Carla. 2019. Acetaminophen removal from water by microalgae and effluent toxicity assessment by the zebra fish embryo bioassay. *Water* 11, 9: 1929.

Fauziah, S.H., et al. 2017. Assessing the bioaugmentation potentials of individual isolates from landfill on metal-polluted soil. *Environ. Earth Sci.* 76: 401. doi: 10.1007/s12665-017-6739-x

Ghosh, A., M.G. Dastidar, and T.R. Sreekrishnan. 2016. Recent advances inbioremediation of heavy metals and metal complex dyes: review. *J. Environ. Eng. USA* 142:C4015003. https://doi.org/10.1061/(ASCE)EE.1943–7870.00009 65

Han, Ting. 2019. Two-step strategy for obtaining *dunaliella* sp. biomass and β-carotene from anaerobically digested poultry litter wastewater. *Int. Biodeterior. Biodegradation* 143: 104714.

Han, Xu. 2006. Surface complexation mechanism and modeling in Cr(III) biosorption by a microalgal isolate, *Chlorella miniata. J. Colloid Interface Sci.* 303, 2: 365–371.

Huang, Chieh-Chen. 2006. Expression of mercuric reductase from Bacillus megaterium MB1 in eukaryotic microalga *Chlorella* sp. DT: an approach for mercury phytoremediation. *Appl. Microbiol. Biotechnol.* 72, 1: 197–205.

Igiri, Bernard, Stanley Okoduwa, Grace O. Idoko, Ebere P. Akabuogu, Abraham O. Adeyi, and Ibe K. Ejiogu. 2018. Toxicity and bioremediation of heavy metals contaminated ecosystem from tannery wastewater: a review. *J. Toxicol.* https://doi.org/10.1155/2018/2568038

Imran, Muhammad, Crowley David, Khalid Azeem, Hussain Sabir, Mumtaz Waseem Muhammad, and Arshad Muhammad. 2015. Microbial biotechnology for decolorization of textile wastewaters. *Rev. Environ. Sci. Biotechnol.* 14: 73–92. https://doi.org/10.1007/s11157-014-9344-4

Jena, Jayashree. 2015. Biological sequestration and retention of cadmium as CdS nanoparticles by the microalga *Scenedesmus*-24. *J. Appl. Phycol.* 27, 6: 2251–2260.

Karthikeyan, Sathrugnan. 2007. Evaluation of the marine algae *Ulva fasciata* and *Sargassum* sp. for the biosorption of Cu(II) from aqueous solutions. *Bioresour. Technol.* 98, 2: 452–455.

Kayalvizhi, K. 2015. Biosorption of Cr(VI) using a novel microalga *Rhizocloniumhookeri*: equilibrium, kinetics and thermodynamic studies. *Desalin. Water Treat.* 56, 1: 194–203.

Kiki, Claude. 2020. Dissipation of antibiotics by microalgae: kinetics, identification of transformation products and pathways. *J. Hazard Mater.* 387: 121985.

Kwak, Hyo Won. 2015. Preparation of bead-type biosorbent from water-soluble *Spirulina platensis* extracts for chromium (VI) removal. *Algal Res.* 7: 92–99.

Larsen, Christian. 2019. Mechanisms of pharmaceutical and personal care product removal in algae-based wastewater treatment systems. *Sci. Total Environ.* 695: 133772.

Lee, H.S. 2004. Effect of aluminum in two-metal biosorption by an algal biosorbent. *Min. Eng.* 17, 4: 487–493.

Lee, Yi-Chao. 2011. The biosorption of heavy metals from aqueous solution by *Spirogyra* and *Cladophora* filamentous macroalgae. *Bioresour. Technol.* 102, 9: 5297–5304.

Liu, Cong, Yu Liu, Chenglian Feng, Peng Wang, Lanping Yu, Daqing Liu, Shuhong Sun, and Fangkun Wang. 2021b. Distribution characteristics and potential risks of heavy metals and antimicrobial resistant Escherichia coli in dairy farm wastewater in Tai'an, China. *Chemosphere* 262: 127768.

Liu, Yongchuang, Haili Zhang, Xiangrong He, and Juan Liu. 2021a. Genetically engineered methanotroph as a platform for bioaugmentation of chemical pesticide contaminated soil. *ACS Synth. Biol.* 10, 3: 487–494.

Mata, Y.N. 2008. Characterization of the biosorption of cadmium, lead and copper with the brown alga *Fucus vesiculosus*. *J. Hazard. Mater.* 158: 316–323.

Mata, Y.N. 2009. Gold(III) biosorption and bioreduction with the brown alga *Fucus vesiculosus*. *J. Hazard. Mater.* 166: 612–618.

Molazedah, Parvin. 2015. Adsorption of lead by microalgae *Chaetoceros* sp. and *Chlorella* sp. from aqueous solution. *J. Commun. Health Res.* 4, 2: 114–127.

Monteiro, C.M. 2011. Biosorption of zinc ions from aqueous solution by the microalga *Scenedesmus obliquus*. *Environ. Chem. Lett.* 9, 2: 169–176.

Nagi, Gurpreet Kaur. 2021. Bioremediation of coke plant wastewater from steel industry with mixed activated sludge microalgal consortium in lab-scale semi-continuous mode. *J. Chem. Technol. Biotechnol.* 96: 2249–2256.

Nithya, K. 2019. Algal biomass waste residues of *Spirulina platensis* for chromium adsorption and modeling studies. *J. Environ. Chem. Eng.* 7, 5: 103273.

Nwokem, Nsidibeabasi Calvin, Calvin Onyedika Nwokem, Casmir Emmanuel Gimba, and Beatrice Nkiruka Iwuala. 2017. Biodegradation of endosulfan by mixed bacteria culture strains of Pseudomonas aeruginosa and Staphylococcus aureus. *Sci. World J.* 12, 1: 9–12.

Oswald, W.J. 1953. Algae symbiosis in oxidation ponds: III. Photosynthetic oxygenation. *Sewage Ind. Wastes* 25, 6: 692–705.

Paikhomba, Singha L., Kotoky Rhitu, and Pandey Piyush. 2017. Draft genome sequence of alcaligenes faecalis BDB4, a polyaromatic hydrocarbon-degrading bacterium isolated from crude oil-contaminated soil. *Microbiol. Resour. Announc.* 5, 48.

Patel, Yogesh, and Akshaya Gupte. 2016. Evaluation of bioremediation potential of isolated bacterial culture YPAG-9 (Pseudomonas aeruginosa) for decolorization of sulfonated di-azodye reactive red HE8B under optimized culture conditions. *Int. J. Curr. Microbiol. Appl. Sci.* 5: 258–272.

Qu, Fanqi. 2020. Nitrogen ion beam implantation for enhanced lipid accumulation of *Scenedesmus obliquus* in municipal wastewater. *Biomass Bioenerg.* 134: 105483.

Rezaee, Abbas. 2006. Biosorption of mercury by biomass of filamentous algae *Spirogyra* species. *J. Biol. Sci.* 6, 4: 695–700.

Saavedra, Ricardo. 2018. Comparative uptake study of arsenic, boron, copper, manganese and zinc from water by different green microalgae. *Bioresour. Technol.* 263: 49–57.

Sadeghi, M., et al. 2019. Biodecolorization of reactive Black5 and reactive Red120 azo dyes using bacterial strains isolated from dairy effluents. *Int. J. Environ. Sci. Technol.* 16: 3615–3624. https://doi.org/10.1007/s1376 2-018-1750-7

Safdari, M.S., et al. 2018. Development of bioreactors for comparative study of natural attenuation, biostimulation, and bioaugmentation of petroleum-hydrocarbon contaminated soil. *J. Hazard. Mater.* 342: 270–278. https://doi.org/10.1016/j.jhazm at.2017.08.044

Sánchez-Zurano, Ana. 2021. ABACO: a new model of microalgae-bacteria consortia for biological treatment of wastewaters. *Appl. Sci.* 11, 3: 998.

Sanyal, Santonu Kumar, Taslin Jahan Mou, Ram Prosad Chakrabarty, Sirajul Hoque, M. Anwar Hossain, and Munawar Sultana. 2016. Diversity of arsenite oxidase gene and arsenotrophic bacteria in arsenic affected Bangladesh soils. *AMB Express* 6, 1: 1–11.

Sarı, Ahmet. 2011. Equilibrium, thermodynamic and kinetic investigations on biosorption of arsenic from aqueous solution by algae (*Maugeotiagenuflexa*) biomass. *Chem. Eng. J.* 167, 1: 155–161.

Shetty, Prateek. 2019. Exploitation of algal-bacterial consortia in combined biohydrogen generation and wastewater treatment. *Front. Energy Res.* 7: 52.

Sibi, G. 2016. Biosorption of chromium from electroplating and galvanizing industrial effluents under extreme conditions using *Chlorella vulgaris*. *Green Energy Environ.* 1, 2: 172–177.

Singh, Alpana. 2007. Copper(II) and lead(II) sorption from aqueous solution by non-living *Spirogyra neglecta*. *Bioresour. Technol.* 98, 18: 3622–3629.

Singh, Alpana. 2012. Continuous metal removal from solution and industrial effluents using Spirogyra biomass-packed column reactor. *Water Res.* 46, 3: 779–788.

Singh, Pankaj Kumar. 2021. Utilization of wastewater as nutrient media and biomass valorization in marine Chrysophytes-*Chaetoceros* and *Isochrysis*. *Energy Convers. Manag.* 10: 100062.

Solisio, Carlo. 2019. Adsorption of inorganic mercury from aqueous solutions onto dry biomass of *Chlorella vulgaris*: kinetic and isotherm study. *Environ. Technol.* 40, 5: 664–672.

Srinivasa Rao, P. 2005. Comparison of biosorption of nickel (II) and copper (II) ions from aqueous solution by *Sphaeroplea* algae and acid treated *Sphaeroplea* algae. *Sep. Sci. Technol.* 40, 15: 3149–3165.

Suganya, S. 2016. Sequestration of Pb(II) and Ni(II) ions from aqueous solution using microalga *Rhizocloniumhookeri*: adsorption thermodynamics, kinetics, and equilibrium studies. *J. Water Reuse Desalin.* 7, 2: 214–227.

Sulaymon, Abbas H. 2013. Column biosorption of lead, cadmium, copper and arsenic ions onto algae. *J. Bioprocess. Biotechniq.* 3: 1.

Tuzen, Mustafa. 2010. Biosorption of selenium from aqueous solution by green algae (*Cladophora hutchinsiae*) biomass: equilibrium, thermodynamic and kinetic studies. *Chem. Eng. J.* 158, 2: 200–206.

Vogel, M. 2010. Biosorption of U(VI) by the green algae *Chlorella vulgaris* in dependence of pH value and cell activity. *Sci. Total Environ.* 409, 2: 384–395.

Wang, Xiao Xiong. 2020. Mechanism and kinetics of methylisothiazolinone removal by cultivation of *Scenedesmus* sp. LX1. *J. Hazard Mater.* 386: 121959.

Xie, Peng. 2020. Revealing the role of adsorption in ciprofloxacin and sulfadiazine elimination routes in microalgae. *Water Res.* 172: 115475.

Xiong, Jiu-Qiang. 2019. Combined effects of sulfamethazine and sulfamethoxazole on a freshwater microalga, *Scenedesmus obliquus*: toxicity, biodegradation, and metabolic fate. *J. Hazard. Mater.* 370: 138–146.

Yang, Jijian. 2012. Construction of a genetically engineered microorganism that simultaneously degrades organochlorine and organophosphate pesticides. *Appl. Biochem. Biotechnol.*, 166: 590–598.

Yang, Jin Shui. 2015. Lipid production combined with biosorption and bioaccumulation of cadmium, copper, manganese and zinc by oleaginous microalgae *Chlorella minutissima* UTEX2341. *Bioresour. Technol.* 175: 537–544.

Zafra, German, Ángel E. Absalón, Miguel Ángel Anducho-Reyes, Francisco J. Fernandez, and Diana V. Cortés-Espinosa. 2017. Construction of PAH-degrading mixed microbial consortia by induced selection in soil. *Chemosphere* 172: 120–126.

Zheng, Heshan. 2016. Biosorption of cadmium by a lipid extraction residue of lipid-rich microalgae. *RSC Adv.* 6, 24: 20051–20057.

17 Algae as a Source of Bioremediation and Resource Recycling

S. Senthil Kumar and Magapu Solomon Sudhakar

17.1 INTRODUCTION

Bioremediation is the technique of biologically degrading organic wastes under regulated settings to a benign state or quantities beneath regulatory concentration thresholds. Since microorganisms have inherent enzymes within that enable them to devour environmental toxins, they are well-suited to the function of contaminant eradicator. The goal of the said process is to get them to perform by providing them with the right nutrition and other chemicals they need to degrade/detoxify pollutants that are harmful to the surrounding environment and living beings. Enzymes such as oxidoreductases, hydrolases, lyases, transferases, isomerases, and ligases are involved in every metabolic process (**Vidali, 2001**). Because of their non-specificity and particular substrate affinity, numerous enzymes have a very extensive degradation capability. Microorganisms have the ability to digest pollutants enzymatically and convert them to beneficial molecules, allowing bioremediation to be effective. Bioremediation is frequently used to control environmental parameters to speed up microbial growth and destruction because it is only effective when environmental conditions allow for microbial growth and production (**Perelo, 2010**).

In such a pursuit, seaweeds, or otherwise known as marine macroalgae, can absorb macronutrients and metal pollutants from their surroundings and are thus extensively exploited in the bioremediation of polluted waters (**Neori et al., 2004**). In many regions of the world, the fast growth of seaweeds is a normal result of increased industrial activities near polluted water bodies or coastal waters (**Smetacek and Zingone, 2013, Rabalais et al., 2004**), where nutrient levels are significant and factors such as light and temperature circumstances are conducive for algae growth. Seaweeds are grown under regulated settings in the process of bioremediation; however, by establishing a structural framework for the culture surrounding the pollution source, where pollutants are localized and the amount of water to be treated is decreased, such as in wastewater treatment.

Because of their huge size, seaweed biomass can be easily harvested, directly removing the toxins contained in the biomass from saltwater. Another benefit is that the resulting seaweed biomass can be exploited for bio-products, as either biomasses or as a substrate for bio-based chemical extraction processes (**Jung et al., 2013; Neveux et al., 2014**). These characteristics make bioremediation processes commercially viable in many circumstances, yielding advantages from both wastewater

DOI: 10.1201/9781003297901-17

treatment and biomass valorization. As a result, seaweed-based bioremediation is envisioned as a way to remediate wastewater while also providing the biomass required for conventional uses like food and phycocolloids, as well as a rising bio-based chemical sector.

The present chapter provides an overview of the utilization of algae as a source of bioremediation and resource recycling.

17.2 MACROALGAE: A POWERHOUSE

Algae, both microscopic and macroscopic species, are present in practically every moist environment on land, in fresh water, and in the oceans. Multicellular, macro-sized, eukaryotic, and autotrophic creatures are known as marine macroalgae or seaweeds. Algae retain starch as well as other big molecular chain polysaccharides in the inside of their cells as a power source. Ulvan is produced by green algae and contains carotene, xanthophylls, and chlorophylls a and b as colors. Chlorophylls a and d, as well as carotenoids, are found in red algae, and their pigmentation is caused by the inclusion of phycoerythrin (pigment) in their cells. The pigments named fucoxanthin, chlorophylls a and c, and carotenoids, as well as lipids and polysaccharides (such as laminarin), are prevalent in brown algae (**Barsanti and Gualtieri, 2014**; **Vidotti et al., 2014**). Seaweeds, like plants on land, provide analogous ecological functions in aquatic environments. Some macroalgae varieties can act as bio-monitors of water quality, while others can perform bioremediation through processes of bio-absorption and bioaccumulation (**Yu et al., 2016; Neveux et al., 2018; Henriques et al., 2017a**).

17.2.1 Chemical Composition

Macroalgae have significant carbohydrate content, intermediate protein content, and lower fat content, with a varied mineral ash concentration (**Vassilev et al., 2010; Pereira, 2011**; **Wei et al., 2013**). Laminarin, mannitol, starch, and mannan are among the readily soluble polysaccharides found in the high carbohydrate percentage (**Jard et al., 2013**).

The lack of lignocellulosic substances aids microbial breakdown and makes biogas production easier (**Vergara-Fernandez et al., 2008; Horn, 2009**). Alginate is one of the principal structural polymers found in macroalgae, and it gives aquatic creatures exposed to moving water both stability and flexibility. Seaweed biomass contains alginate and a variety of other industrially important carbohydrate molecules (**Jard et al., 2013**). Fucoidan is a sulfated polysaccharide derived from *Fucus vesiculosus* that has therapeutic applications (**Rupérez et al., 2002**). As value-added by-products, glycerol, mannitol, and organic acids are extracted and employed in the pharmaceutical and food sectors (**Moen, 1997**; **Reith et al., 2005**). Macroalgae are important in human nutrition because they contain significant levels of minerals, which makes them interesting as a flavor enhancer. Seaweeds, unlike table salt, have a low sodium content and might thus be used as an alternative, attempting to manage one of the world's most pressing problems, majorly in the food industry.

17.3 APPLICATIONS OF MACROALGAE

a. Production of biofuels

Owing to the characteristics of rapid rates of growth, nitrogen sequestration ability, and significant solar energy conversion ratios, algae have piqued scientists' curiosity as a possible source of renewable bioenergy in current history (**Demirbas, 2010**). Algae develop and assimilate nutrients, transforming them into algal biomass, which may be recovered and utilized as a feedstock for various forms of bioenergy relying on the qualities of the biomass. Some microalgal species, for example, may store lipids at a rate of up to 80% per unit of dry weight (DW), which can later be turned into biodiesel (**Georgianna and Mayfield, 2012; Chisti, 2007**). Seaweeds, on the other hand, are high in retention of structural carbohydrates (**Yazdani et al., 2011; Schiener et al., 2015**), making them ideally suitable for bioethanol fermentation or anaerobic digestion for transformation to methane or biogas (**Adams et al., 2011; Horn et al., 2000**; **Jang et al., 2012; Yanagisawa et al., 2011**). Algal biofuels are now economically unappealing in contrast to fossil fuels due to their high manufacturing and processing costs. However, integrating fresh concepts or attempting to address key issues such as the usage of low-cost waste streams, the development of effective isolation and processing technologies, the implementation of a biorefinery methodology, and governmental incentives, such as carbon credits, may greatly boost the commercial realization of algal biofuel output (**Mathews, 2008**).

Macrocystis pyrifera, Laminaria, Gracilaria, Sargassum, Ulva, and other seaweeds have all been examined as possible energy crops. The great yield of these seaweeds is necessary for producing energy. Since it grows swiftly to big sizes and may be collected multiple times a year, Macrocystis pyrifera may be the most ideal species among the stated ones. The manufacture of ethanol from macroalgae is seen as a possible biofuel substitute to food crops. Several macroalgae species, such as Gracilaria sp. and Ulva sp., as well as ocean-based culture of bigger kelp genus, have been presented as potential options for bio-ethanol crops (**Kraan, 2010**). The technique to produce ethanol from macroalgae already exists, and switching from existing food crops to macroalgae needs to be simple (**Adams et al., 2009**).

17.3.1 Advantages of Macroalgae as a Potential Source of Biofuel

The various advantages offered by the use of macroalgae as a potential source of biofuel are described subsequently:

1. Macroalgae yields vary from 10 to 15 tons per 1000 m^2 per year, roughly ten times higher than present corn yields.
2. They can be farmed in the open ocean, which implies the fact that they do not struggle for land with other fuel and resource crops. As a result, macroalgae is a distinctive choice for countries with little land but abundant seawater.
3. Contrary to other energy crops, macroalgae thrive in marine environments. As a result, they do not contend for fresh water as compared to other crops.

4. Macroalgae helps in mitigating the issues associated with first- and second-generation energy crops. They are devoid of food and fuel issue since they do not have to struggle for both land and fresh water to produce food. As a result, planting macroalgae will not raise food-associated costs.
5. Because macroalgae have a higher carbohydrate load, they might be used to generate cellulosic ethanol.
6. Macroalgae, like microalgae, can retain carbon dioxide as they grow. With the exception of providing us with a viable biomass supply, macroalgae farming will also aid in reducing CO2 emissions.
7. Harvesting some buoyant macroalgae, such as Sargassum, is a simple operation in contrast to harvesting microalgae, which requires a lot of energy source.
8. Because of the vast amount of ocean land available, open ocean aquaculture can be developed up to any required level.
9. Cultivation costs are cheap when compared to microalgae. It is also feasible to use natural seaweed, which virtually reduces the cost of cultivation. Drift seaweeds are available periodically in several parts of the world. Furthermore, using these seaweeds lowers the cost of collecting and shipping.
10. Storehouse of natural products.

In recent decades, researchers have focused on the development of high-value substances for applications like food supplements, cosmetics, medicines, and nutraceuticals (**Smit, 2004; Borowitzka, 2013**). *Haematococcus pluvialis* and *Dunaliella* sp., for instance, are grown commercially for the synthesis of astaxanthin and carotene, which are both powerful antioxidants. Antioxidants in the diet may assist in the fight against free radical–related or degenerative diseases, as well as several other malignancies, liver disease, and heart diseases (**Reyes et al., 2014; Guerin et al., 2003**). Meanwhile, due to its high amount of vitamins, proteins, and pigments, the cyanobacterium spirulina is commercially manufactured as a food supplement or health supplement (**Borowitzka, 2013; Hoseini et al., 2013**).

Furthermore, micro-algal species like Chaetoceros, Chlorella, Isochrysis, and Tetraselmis are rich in proteins and HUFAs (**Hemaiswarya et al., 2011**). As a result, they are manufactured for animal feeds, and they are also used in the ornamental fish industry, aquaculture hatcheries for molluscan and crustacean larval nutrition, and live feeds (**Priyadarshani and Rath, 2012; Conceição et al., 2010**). Phycocolloids are also grown on a huge scale from red and brown seaweeds. In the food, drink, textile, and medical industries, phytocolloids are used as stabilizers, thickeners, sizing, emulsifiers, and gelling agents (**Tseng, 2001**; **McHugh, 2003**).

Seaweeds are also employed as a fertilizer, a soil conditioner, and a source of animal fodder. Macro-algae and its extracts are also a reservoir of higher-valued products which are sold commercially as bio-stimulants, cosmetics, antioxidants, and nutritional supplements (**Hafting et al., 2012**). WWT and CO2 mitigation are the two latest environmental approaches that could be combined with other features of algal production. Algal biomass generated on waste streams, on the other hand, would not be acceptable for personal consumption or employment in the pharmaceutical or cosmetic industries regarding the associated health and safety concerns and general perspective. Alternatively, it may be exploited for bioenergy or other WWT uses (**Wells et al., 2017**).

11. Algal bioremediation

Algae can remove nutrients as well as CO2 from water (**Hughes et al., 2012a, 2012b**). They may also absorb heavy metals from solutions, with some varieties able to accumulate significant quantities in their cells (**Villares et al., 2001**). As a result, algae have the ability to reduce the negative health, ecological, and environmental consequences of high amounts of nutrients and metals, such as eutrophication, which is driven by worldwide population growth and growing anthropogenic activities. As a result, algae are seen as a good candidate for inclusion in WWT (**Lawton et al., 2013**). Micro- and macroalgae have both been found to be effective at removing nutrients from wastewater.

For example, a conglomerate of five freshwater microalgae were reported to eliminate 99% of nitrogen and phosphorus from anaerobic digestate wastewaters following five days of development, with initial concentrations of 4500 and 690 M, respectively (**Silkina et al., 2017**). Ayre et al. employed a hybrid micro-algal culture to remediate anaerobically digested piggery effluent and achieved a maximum NH4 + -N removal rate. Similarly, macroalgae have been used to remove nutrients from aquaculture effluents in a number of investigations (**Ayre et al., 2017**). When grown in finfish effluents, the red macroalga *Gracilaria vermiculophylla*, for instance, showed significant productivity levels, with the resulting biomass having considerable economic potential as a phycocolloid.

Algae has been examined as a prospective low material for hazardous metal remediation in addition to nutrient accumulation. This is because of their widespread distribution and intriguing cell wall/membrane characteristics, which naturally bind metals (**Mehta and Gaur, 2005**). **Klimmek et al.** (2001) evaluated the capacity of thirty different strains of dried microalgae and cyanobacteria to extract mono-metallic solutions of Cd2+, Ni2+, Pb2+, and Zn2+ (**Klimmek et al., 2001**). The biosorption capability of the algae varied considerably; nevertheless, Cd2+ and Pb 2+ were often removed more efficiently than Ni2+ or Zn2+. Henriques et al. employed live *Ulva lactuta* in an investigation and accomplished >95 % elimination of monometallic Cd, Hg, and Pb solutions in about six days (**Henriques et al., 2017b**).

17.4 REMOVAL OF TOXIC HEAVY METALS USING MACROALGAE

The deterioration of ecosystem condition attributed to the prevalence of dangerous and damaging contaminants, such as heavy metals, is one of the most pressing environmental challenges (**Gautam et al., 2015; Naushad, 2014**). Heavy metals that end up in the environment are generally due to the outflow of grossly mistreated industrial effluents, which is why industrial operations are still regarded as one of the most significant sources of pollution with the said type of pollutant (**Wu et al., 2015a, 2015b; Zhang et al., 2018**).

Multiple toxic heavy metals such as (Pb (II), Cd II), Hg (II), Cr (VI), and others are extensively utilized in innumerable industrial processes (**Ghasemi et al., 2014; Qaiser et al., 2007**), and because they cannot be degraded and have an accrual propensity in natural water sources, the inclusion of heavy metals in natural water above a certain has an impact on human health and reduces the efficiency of other biological systems quantity (**Aklil et al., 2004; Aydin et al., 2008**).

Considering these factors, eliminating heavy metal ions from industrial wastewater remains a critical challenge, both environmentally and economically. Environmental pollution can be avoided, and the restoration of heavy metals that are recovered from wastewater into industrial operations guarantees the conditions long progress (**Judd, 2016**; **Syukor et al., 2016**).

In such a scenario, marine algae are best suited as they meet all of these criteria and are hence regarded as a remarkable biological resource since they are widespread, do not require additional growing conditions, and have a high level of productivity (**Hannachi et al., 2015; Kumar et al., 2007**). Algae are favored as biosorbents since they are readily available and have a lower cost of preparation involved. The availability of multiple negatively charged functional groups on the surface of cell walls gives marine algae biomass a good capability to bioaccumulate toxic metal ions from aqueous media (**Davis et al., 2003**). Because these functional groups are mostly derived from marine algal elements, their capacity to eliminate heavy metals is largely determined by their chemical makeup and, consequently, their type. Table 17.1 shows how algae can be used to remove heavy metals from a variety of sources.

TABLE 17.1
Utilization of algae in removal of heavy metals from various sources

Species name	Source	Heavy metal	Removal efficiency	References
Nannochloropsis oculata	Marine drainage	Cu	99·92 ± 0·04%	Martínez-macias et al., 2019
Gracilaria sp.	Leachate from landfills	As, Fe, Ni, and Cd	Fe = 100%Cd = 100% As = 100%Ni = 98%	Selvam et al., 2016
Chlorella vulgaris and *Scenedesmus spinosus*	Mining wastewater	Cu, Mo	Cu = 64·7%Mo = 99·9%	Urruntia et al., 2019
Sargassum sp.	Stimulated wastewater	Ni2+ and Cu2+	Cu2+ = 2.06 mmol/g Ni2+ = 1.69 mmol/g	Barquilha et al., 2017
Osmundea pinnatifida	Stimulated wastewater	Cu2+ and Cd2+	Cd2+ = 57.29% Cu2+ = 50.89%	El Hassouni et al., 2014
Scenedesmus sp.	Wastewater from market	Zn, Fe	Zn = 84·14%Fe = 65·76%	Jais et al., 2015

TABLE 17.1 (*Continued*)
Utilization of algae in removal of heavy metals from various sources

Species name	Source	Heavy metal	Removal efficiency	References
Nanochloropsis, Pavlova lutheri, Tetraselmis chuii, Chaetoceros muelleri	Municipal wastewater	Fe, Al	Fe = 95%Al = 95%	Richards and Mullins, 2013
Chondracanthus chamissoi	Aqueous solution	Pb (II) and Cd (II)	Pb(II) = 1.37 mmol/ gCd(II) = 0.76 mmol/g	Yipmantin et al., 2011
Scenedesmus incrassatulus	Artificial wastewater	Cr (VI), Cd (II), Cu (II)	25–78%	Pena-Castro et al., 2004

It is widely reported in the literature that there exist two main diverse strategies for metal bioremediation with seaweeds. The first one, being analogous to activated carbon, seaweed biomass can be suitably dried, adsorbed, and employed as a passive adsorbate for metals in a process known as biosorption (**Davis et al., 2003**). The other refers to a process known as bioaccumulation, wherein live seaweed can be grown in contaminated water to aggressively trap dissolved metals within living algae cells. The latter method of wastewater bioremediation has lately been acknowledged as a considerably long source of biomass for bio-products such as bioenergy and high-value compounds (**Saunders et al., 2012**). Biosorption is a technique of metal adsorption upon the dead biomass that absorbs seaweed, whereas bioaccumulation with live seaweeds is a mechanism of metal intake by a living tissue that builds seaweed biomass in the meantime.

17.5 BIOSORPTION

The existence of significant amounts of sulfated polysaccharides in the cytoskeleton is a distinguishing feature of seaweed. The existing polysaccharides are known to play a crucial part in seaweeds' capability to remove metals from a solution in a passive manner using a variety of methods known as biosorption. Polysaccharides in seaweed tissues can participate in a range of biosorption activities, including ion exchange adsorption and metal complexation (**Gadd, 2009**). Yet dissolved metals are taken from the water column and functionalized to the surface of the seaweed biomass in all of these circumstances, allowing the metals to be removed from the system.

The effectiveness of various seaweeds as biosorbents is complicated by the identification of the metals to be eliminated from the solution, their concentration in the solution, the availability of competing components in the water, and the physicochemical attributes of the water in which the metals are located (**Hashim and Chu, 2004; Murphy et al., 2008**). Furthermore, practically all biological materials have some propensity for dissolved metals and might be utilized as biosorbents, according to the substantial biosorption-based reported literatures (**Gadd, 2009**). As a result, it has been argued that in assessing the success of treatment, the features of the water to be processed are significantly more essential than the biological substrate utilized for biosorption process (**Gadd, 2009**).

Biosorption with natural biological matter has a number of drawbacks, including the fact that it is usually always nonspecific and is influenced by nontarget molecules prevalent in wastewater. Johansson et al., for instance, constructed a selenate biosorbent using waste Gracilaria biomass obtained from agar extraction factories. While the research indicated excellent removal of selenate from mock effluents, the inclusion of comparatively small quantities of sulfate, which typically co-occurs with selenate in industrial effluents, had a major effect on the biosorbent's efficiency (**Johansson et al., 2015**). Despite the acknowledged possibility for competitive interactions in multi-element systems, most metal biosorption studies focus on evaluating biosorbent effectiveness in idealistic single-element systems. Furthermore, investigations on complex effluents with several target metals for elimination show that as the pool of potential metals in a solution rises, the effectiveness of biosorption of each single metal decreases (**Kidgell et al., 2014a, 2014b**).

Another disadvantage of seaweed-based biosorption is that if seaweed biomass is initially processed with Fe to create binding sites for dissolved oxyanions, as in for selenate (SeO4 2), it is inefficient at sequestering these pollutants from the solution (Johansson et al., 2015). This necessitates the employment of numerous pretreatment biosorbents strategically deployed to treat complicated effluents comprising both dissolved cations (e.g., Cu2+) and oxyanions (SeO4 2) comprehensively (**Kidgell et al., 2014a, 2014b; Roberts et al., 2015**). Furthermore, one of the most prominent justifications for seaweed-based biosorption is that seaweed biomass is inexpensive and plentiful. However, there are established markets for seaweed biomass and their associated bioactive compounds, particularly new high-value industries, that allow a far more cost-effective use of seaweeds than consuming them in biosorption operations.

17.6 BIOACCUMULATION

In contrary to the usage of dried seaweeds as passive biosorbents of metals, the utilization of live seaweeds to retain metals as a bioremediation method has received less attention. The widespread perception that nonliving biomass has higher metal adsorption capacities than living biomass is one of the most compelling arguments for employing it as a biosorbent (**Zeraatkar et al., 2016**). Various studies have indicated considerable uptake of metals such as Al, Cu, Cd, Cr, and Zn from contaminated streams into seaweed environments (**Roberts et al., 2008**). Seaweeds gathered from heavily polluted estuaries have been proven to be much less appealing to marine herbivores and to impair young crustacean survivability when fed metal-contaminated seaweed. As a result, there is definitely a lot of promise to use live seaweeds' potential to absorb and integrate dissolved metals as a strategy to biofilter contaminated coastal ecosystems or to link seaweed production with coastal enterprises to absorb metals.

Yu et al. evaluated the growth and metal bioaccumulation trends of two species of the brown macroalgae Sargassum in effluent from a fish farm to neighboring benchmark locations. Sargassum cultivated in aquaculture effluents had a considerably faster growth than the standard locations, and the farmed biomass was noted to provide a two- to threefold higher content of metals such as Cu, Zn, Pb, and Cr than biomass from comparable areas (**Yu et al., 2014**). Metals were removed from the facility effluent in large amounts due to a confluence of enhanced seaweed growth and biomass concentrations of metals. Furthermore, when the kelp L. digitata was grown in the open ocean near salmon farms, the ensuing biomass was higher in Cu, Mn, and V, which all came from the nearby fish farm's supplies (**Ratcliff et al., 2016**).

Microalgae consortiums have gotten a lot of consideration as a way to extract nutrients from wastewater. Heavy metals such as Ni, Cd, and Pb, as well as the organophosphate insecticide malathion, were removed from a mixture of industrial and agricultural wastewater using a combination of microalgae and cyanobacteria, namely Spirulina platensis, C. vulgaris, and Scenedesmus quadricauda (**Abdel-Razek et al., 2019**). The consortia was reported to eliminate Ni (95%), Pb (89%), Cd (88%), and malathion (99%) from a water sample, according to the findings. It was also reported elsewhere that Cr (VI) was removed from synthetic wastewater using an algal-bacterial anaerobic granular sludge (**Yang et al., 2020**). Yang et al. investigated the removal of Cr (VI) using a similar strategy and found that loosely bound exopolysaccharides generated by algal-bacteria favored Cr (VI) removal (**Yang et al., 2021**).

17.7 BIOREMEDIATION OF SOIL

Bioremediation of soil is the act of living organisms accumulating or degrading chemical pollutants into lesser hazardous forms (**Juwarkar et al., 2010**). The potential of seaweed to bioremediate polluted soil is hampered by the presence of pollutants in the soil and the capability of algae to thrive in these settings.

Seaweeds were employed in a study to extract DDT from the ecosystem, which is particularly harmful (**Kantachote et al., 2004**). To counteract Hg's fatal impact on fenugreek development, a dried green seaweed known as *Codium iyengrii* was put to the soil, and it was observed that seaweeds demonstrated to significantly reduce mercury mobility only at reduced concentrations of metal doses (**Azmat and Askari, 2015**). Seaweeds were less efficient at higher concentration because the plants were unable to deal with the damaging impacts of the said metal.

Researchers also used dried seaweeds (*Ulva* sp. and *Gelidium* sp.) combined with soil not only to remediate soil contaminated with Cu, Mn, Pb, and Zn but also to analyze their implication on the growth of roots and shoots and composition of the explored elements in red radish (*Raphanus sativus* L.). Specific experiments demonstrated that adding seaweeds to soil significantly improved porosity, accessible water content, and organic matter composition, as well as increased the content of the elements studied in red radish (**Abd-Elhady and El-Zabalawy, 2014**).

Green seaweeds such as *Ulva* sp. have been utilized for soil bioremediation because they're widespread and have rapid growth, a fairly basic structure, and great resistance to metal pollution (**Abd-Elhady and El-Zabalawy, 2014; Farias et al., 2017**).

In the in situ assessment, the macroalga Ulva australis was employed to remove Zn (II) ions from the polluted soil. After 45 days, the zinc level in the biomass was found to be 700 mg/kg. Zinc accumulation had no effect on U. australis physiology such as photosynthesis effectiveness, photosynthetic pigment concentration, or growth rate over the first 12 days. Following 45 days, Zn buildup increased, leading to the conclusion that there was a time limit of around 20 days for the process of bioremediation.

17.8 APPLICATION OF SEAWEEDS IN BIOREMEDIATION OF NUTRIENTS

Eutrophication of the sea develops in numerous places across the world. Seaweed growth in coastal waterways may be beneficial in reducing eutrophication and controlling hazardous algae blooms that cause animal killing and financial damage (**Wu et al., 2015b**). Variables such as light, temperature, nutritional content, salinity, cultivation level, water movement, and the existence of epiphytes and herbivorous fish are all important aspects in the successful culture of algae (**Yang et al., 2015**). The nitrogen accumulation of seaweed structures grows when dissolved nitrogen levels in seawater rise during bioremediation of eutrophicated water. Biorefinery possibility remains in the manufacture of seaweeds with significant tissue protein and carbohydrate content (**Ross et al., 2018**). The macroalgal biomass can also be restored to agricultural land in a sustainable manner or utilized to make value-added products, achieving both economic and environmental goals. Huo and colleagues reported that *Gracilariopsis longissima*, a red seaweed, was found to remove nutrients (**Huo et al., 2011**).

Integrated multi-trophic aquaculture (IMTA) is a technique in which aquaculture effluents are employed as a source of nutrients for animal as well as seaweed biomass development. Seaweeds can be utilized for wastewater phytoremediation as biological filters to eliminate inorganic nutrients or as aquaculture feeding additions (**Arumugam et al., 2018**).

Farm animal manure decomposes, releasing a plethora of nutrients that damage the soil and the neighboring rivers (**Yang et al., 2015**). The principal excretory by-products of fish nitrogen metabolism are total ammonia nitrogen regarded as unionized ammonia, NH3, and ionized ammonia. Many fish and shrimp varieties are hazardous at concentrations of more than 1.5 mg NH3eN/L. The choice of the most appropriate macroalgal choice for the IMTA system should also be considered as a priority. Seaweeds that are appropriate should have a robust growth, yield, nutrient availability, and resistance to high ammonium concentrations (**Grote, 2016**).

The green algae Ulva species, as well as the red algae Gracilaria spp. and Porphyra spp., are well-known aquaculture varieties. When contrasted to other seaweed varieties, they have superior nutrient intake and growth rates (Grote, 2016). Palmaria palmata is a red alga that is NH4-sensitive (**Grote, 2016**). According to researchers, the green macroalga Ulva lactuca has been reported to eliminate nitrogen and phosphorus from shrimp wastewater from an integrated recirculation system, improving growth and carotenoids concentration.

17.9 BIOREMEDIATION OF WASTEWATER

Seaweed has been extensively researched, being used as an adsorbent in wastewater treatment to substitute functional activated carbon. Any procedure or operation produces wastewater as a by-product. Several research are being undertaken to treat various types of wastewater using various methodologies and technologies, and one among them is the employment of macroalgae. The presence of phenolic contaminants in wastewater generated from chemical plants, cookery facilities, or petroleum refineries is noted to be widespread (**Zhang et al., 2011**). Because of their associated toxicity, carcinogenicity, teratogenicity, and mutagenicity, there arises a need for their immediate removal.

Despite the fact that phenolic chemicals are challenging to remove using traditional treatment methods, the wastewater treatment industry has turned to bioremediation and adsorption, particularly employing macroalgae (**Rubin et al., 2006; Mahdizadeh et al., 2018**). Seaweed biosorption of phenolic chemicals has been characterized as hydrophobic and with donor-acceptor interactions. The sorption of chlorophenols on to the granular activated carbon exhibits an identical pattern, and it was proposed that these substances were immobilized at the carbonyl oxygens on the surface of activated carbon via a donor-acceptor complexation pathway (**Aksu and Yener, 2001**).

The majority of colors or dyes used in textile manufacturing are not biodegradable and are known to be resistant to aerobic digestion. The two prevalent dyes discovered in wastewater are known to be methylene blue and malachite green. While looking for more cost-effective and easily accessible materials, many types of lower-cost adsorbents were investigated, which were further utilized for removal of dye (**Bao and Zhang, 2012; Kushwaha et al., 2014**).

Eutrophication has been considered the most important water quality concern for most freshwater and marine ecosystems, as it affects coral reef health and community decline, raises the frequency of fish fatalities, and reduces water clarity (**Smith and Schindler, 2009**). The importance of seaweed aquaculture in ecosystem balance is not novel (**McVey et al., 2002**). Table 17.2 presents an overview of bioremediation potential of seaweed in wastewater treatment.

TABLE 17.2
Bioremediation potential of seaweed in wastewater treatment

Seaweed name	Name of pollutants	Treatment efficacy	Reference
Nizamuddinia zanardinii	Methylene blue	565.96 mg/g	**Daneshvar et al., 2017**
Ulva lactuca	Nitrogen and phosphorus	N = 22.7 mg N g DW−1 d−1P = 2.7 mg P g DW−1 d−1	**Sode et al., 2013**
Caulerpa racemosa	Methylene blue	98% (dye = 50 mg/L)	**Cengiz and Cavas, 2008**

(Continued)

TABLE 17.2 (*Continued*)
Bioremediation potential of seaweed in wastewater treatment

Seaweed name	Name of pollutants	Treatment efficacy	Reference
Gracilaria vermiculophylla	Nitrogen	N = 40.54 ± 2.02 gm–2 month–1	**Abreu et al., 2011**
Gracilaria caudata J. Agardh	Nitrogen and phosphorus	N = 49.6%, P = 12.3%	**Marinho-Soriano et al., 2009**
Sargassum muticum	Methylene blue	98.2% when treated with HCl	**Rubin et al., 2005**
Ulva lactuca	Malachite green	4.5% when maintained temperature = 25◦C and 93.8% when pH = 8.0	**Omar et al., 2018**
Gracilaria corticata	Malachite green	93.3% when temperature = 25◦C,92.5% when pH is maintained at 8.0 and96% when t = 150 min	**Omar et al., 2018**

17.10 CONCLUSION AND FUTURE DIRECTIONS

To accelerate the utilization of algae in the domain of biotechnology, eons of foundational research on algal chemistry and physiology must be exploited. The emergence of genome sequencing initiatives and functional genetic methods has lately aided initiatives to increase the comprehension of varied algal features, uncovering novel elements of algal metabolism important to industrial uses. Technological innovations will expedite the process of algal industrialization by offering techniques and skills for delivering highly productive algae-based solutions to a wide range of social problems. This involves a better understanding of algal biology, genetics, and metabolic characteristics, all of which will help to improve both the organisms and the environment in which they are grown. This will pave the way for impromptu highly efficient strains in the coming years, whether as innovative natural isolates or genetically designed strains, as well as practical culture systems with little ecological impact. Algae-based innovative technologies will gain use in practically every industrial area, including those critical to satisfying human society's growing requirements, such as food, pharmaceuticals, and chemical manufacturing, while minimizing environmental effect and lowering manufacturing costs. The establishment of incredibly effective algal biorefineries will allow for the co-sourcing of various products while limiting waste and maximizing productivity, hence boosting the profitability of operations that are now inefficient. As a result, it is expected that algae biotechnology will destabilize the present industrial terrain, leading to the establishment of a configurable, feasible, and productive algae-based bioeconomy that will be critical in meeting the obstacles and constraints that conventional agriculture will confront in the upcoming years.

REFERENCES

Abd-Elhady, E., El-Zabalawy, K. 2014. Remediation of A soil contaminated with heavy metals using some seaweeds. J. Soil Sci. Agric. Eng. 5(12), 1623–1633. https://doi.org/10.21608/jssae.2014.49817

Abdel-Razek, M.A., Abozeid, A.M., Eltholth, M.M., Fatma, A., El-Midany, S.A., Moustafa, N.Y., Mohamed, R.A. 2019. Bioremediation of a pesticide and selected heavy metals in wastewater from various sources using a consortium of microalgae and cyanobacteria. Slov. Vet. 56, 61–73.

Abreu, M.H., Pereira, R., Yarish, C., Buschmann, A.H., Sousa-Pinto, I. 2011. IMTA with Gracilaria vermiculophylla: productivity and nutrient removal performance of the seaweed in a land-based pilot scale system. Aquaculture 312, 77–87.

Adams, J., Toop, T., Donnison, I.S., Gallagher, J.A. 2011. Seasonal variation in Laminaria digitata and its impact on biochemical conversion routes to biofuels. Bioresource Technol. 102, 9976–9984.

Adams, J.M., Gallagher, J.A., Donnison, I.S. 2009. Fermentation study on *Saccharina latissima* for bioethanol production considering variable pre-treatments. J. Appl. Phycol. 21(5), 569–574.

Aklil, A., Mouflih, M., Sebti, S. 2004. Removal of heavy metal ions from water by using calcined phosphate as a new adsorbent. J. Hazard Mater. 112, 183–190.

Aksu, Z., Yener, J.A. 2001. Comparative adsorption/biosorption study of mono-chlorinated phenols onto various sorbents. Waste Manag. 21, 695–702.

Arumugam, N., Chelliapan, S., Kamyab, H., Thirugnana, S., Othman, N., Nasri, N.S. 2018. Treatment of wastewater using seaweed: a review. Int. J. Environ. Res. Public Health 15, 2851.

Aydin, H., Bulut, Y., Yerlikaya, C. 2008. Removal of copper (II) from aqueous solution by adsorption onto low-cost adsorbents. J. Environ Manag. 87, 37–45.

Ayre, J.M., Moheimani, N.R., Borowitzka, M.A. 2017. Growth of microalgae on undiluted anaerobic digestate of piggery effluent with high ammonium concentrations. Algal. Res. 24, 218–226.

Azmat, R., Askari, S. 2015. Improvement in the bioenergetics system of plants under Hg stress environment via seaweeds. Pak. J. Bot. 47(3), 851–858.

Bao, Y., Zhang, G. 2012. Study of adsorption characteristics of methylene blue onto activated carbon made by Salix psammophila. Energy Procedia, 1141–1146.

Barquilha, C.E.R., Cossich, E.S., Tavares, C.R.G., Silva, E.A. 2017. Biosorption of nickel (II) and copper (II) ions in batch and fixed-bed columns by free and immobilized marine algae Sargassum sp. J. Clean. Prod. 150, 58–64.

Barsanti, L., Gualtieri, P. 2014. Algae Anatomy, Biochemistry and Biotechnology, CRC Press, Boca Raton, FL.

Borowitzka, M.A. 2013. High-value products from microalgae—their development and commercialisation. J. App. Phycol. 25, 743–756.

Cengiz, S., Cavas, L. 2008. Removal of methylene blue by invasive marine seaweed: Caulerpa racemosa var. cylindracea. Bioresour. Technol. 99, 2357–2363.

Chisti, Y. 2007. Biodiesel from microalgae. Biotechnol. Adv. 25, 294–306.

Conceição, L.E.C., Yúfera, M., Makridis, P., Morais, S., Dinis, M.T. 2010. Live feeds for early stages of fish rearing. Aquac. Res. 41, 613–640.

Daneshvar, E., Vazirzadeh, A., Niazi, A., Sillanpaa, M., Bhatnagar, A. 2017. A comparative study of methylene blue biosorption using different modified brown, red and green macroalgae—effect of pretreatment. Chem. Eng. J. 307, 435–446.

Davis, T.A., Volesky, B., Mucci, A. 2003. A review of the biochemistry of heavy metal biosorption by brown algae. Water Res. 37, 4311–4330. https://doi.org/10.1016/S0043-1354(03)00293-8

Demirbas. 2010. Use of algae as biofuel sources. Energ. Convers. Manage. 51, 2738–2749.

El Hassouni, H., Abdellaoui, D., El Hani, S., Bengueddour, R. 2014. Biosorption of cadmium (II) and copper (II) from aqueous solution using red alga (Osmundea pinnatifida) biomass. J. Mater. Environ. Sci. 5, 967–974.

Farias, D.R., Hurd, C.L., Eriksen, R.S. et al. 2017. In situ assessment of Ulva australis as a monitoring and management tool for metal pollution. J. Appl. Phycol. 29, 2489–2502.

Gadd, G.M. 2009. Biosorption: critical review of scientific rationale, environmental importance and significance for pollution treatment. J. Chem. Technol. Biotechnol. 84, 13–28.

Gautam, R.K., Sharma, S.K., Mahiya, S., Chattopadhyaya, M.C. 2015. Contamination of heavy metals in aquatic media: transport, toxicity and technologies for remediation. In: Heavy metals in water: presence, removal and safety. The Royal Society of Chemistry, Cambridge, pp 1–24.

Georgianna, D.R., Mayfield, S.P. 2012. Exploiting diversity and synthetic biology for the production of algal biofuels. Nature. 488, 329–335.

Ghasemi, M., Naushad, M., Ghasemi, N., Khosravi-fard, Y. 2014. Adsorption of Pb(II) from aqueous solution using new adsorbents prepared from agricultural waste: adsorption isotherm and kinetic studies. J. Ind. Eng. Chem. 20, 2193–2199.

Grote, B. 2016. Bioremediation of aquaculture wastewater: evaluating the prospects of the red alga *Palmaria palmata* (Rhodophyta) for nitrogen uptake. J. Appl. Phycol. 28, 3075–3082.

Guerin, M., Huntley, M.E., Olaizola, M. 2003. Haematococcus astaxanthin: applications for human health and nutrition. Trends Biotechnol. 21, 210–216.

Hafting, J.T., Critchley, A.T., Cornish, M.L., Hubley, S.A., Archibald, A.F. 2012. On-land cultivation of functional seaweed products for human usage. J. Appl. Phycol. 24, 385–392.

Hannachi, Y., Rezgui, A., Dekhi, A.B., Boubaker, T. 2015. Removal of cadmium(II) from aqueous solutions by biosorption onto the brown macroalga (Dictyota dichotoma). Desalin Water Treat 54, 1663–1673. https://doi.org/10.1080/19443994.2014.891078

Hashim, M.A., Chu, K.H. 2004. Biosorption of cadmium by brown, green, and red seaweeds. Chem. Eng. J. 97, 249–255.

Hemaiswarya, S., Raja, R., Ravi Kumar, R., Ganesan, V., Anbazhagan, C. 2011. Microalgae: a sustainable feed source for aquaculture. World J. Microb. Biot. 27, 1737–1746.

Henriques, B., Lopes, C., Figueira, P., Rocha, L., Duarte, A., Vale, C., Pardal, M., Pereira, E. 2017a. Bioaccumulation of Hg, Cd and Pb by Fucus vesiculosus in single and multimetal contamination scenarios and its effect on growth rate. Chemosphere 171, 208–222.

Henriques, B., Rocha, L.S., Lopes, C.B., Figueira, P., Duarte, A.C., Vale, C., Pardal, M.A., Pereira, E. 2017b. A macroalgae-based biotechnology for water remediation: simultaneous removal of Cd, Pb and Hg by living Ulva lactuca. J. Environ. Manage. 191, 275–289.

Horn, S., Aasen, I., Østgaard, K. 2000. Ethanol production from seaweed extract. J. Ind. Microbiol. Biot. 25, 249–254.

Horn, S.J. 2009. Seaweed biofuels: production of biogas and bioethanol from brown macroalgae, VDM, Verlag Dr. Müller, Saarbrücken, p. 104.

Hoseini, S., Khosravi-Darani, K., Mozafari, M. 2013. Nutritional and medical applications of spirulina microalgae. Mini Rev. Med. Chem. 13, 1231–1237.

Hughes, A.D., Black, K.D., Campbell, I., Davidson, K., Kelly, M.S., Stanley, M.S. 2012a. Does seaweed offer a solution for bioenergy with biological carbon capture and storage? Greenh. Gases Sci. Technol. 2, 402–407.

Hughes, A.D., Maeve, S.K., Black, K.D., Stanley, M.S. 2012b. Biogas from macroalgae: is it time to revisit the idea? Biotechnol. Biofuels 86.

Huo, Y.Z., Xu, S.N., Wang, Y.Y., et al. 2011. Bioremediation efficiencies of *Gracilaria verrucosa* cultivated in an enclosed sea area of Hangzhou Bay, China. J. Appl. Phycol. 23, 173–182.

Jais, N.M., Mohamed, R.M.S.R., Peralta, H.B.M. 2015. Removal of nutrients and selected heavy metals in wet market wastewater by using microalgae Scenedesmus sp. App. Mech. Mat. 773, 1210–1214.

Jang, J.-S., Cho, Y., Jeong, G.-T., Kim, S.-K. 2012. Optimization of saccharification and ethanol production by simultaneous saccharification and fermentation (SSF) from seaweed, Saccharina japonica. Bioprocess Biosyst. Eng. 35, 11–18.

Jard, G., Marfaing, H., Carrère, H., Delgenes, J.P., Steyer, J.P., Dumas, C. 2013. French brittany macroalgae screening: composition and methane potential for potential alternative sources of energy and products. Bioresour. Technol. 44, 492–498.

Johansson, C.L., Paul, N.A., de Nys, R., Roberts, D.A. 2015. The complexity of biosorption treatments for oxyanions in a multi-element mine effluent. J. Environ. Manage. 151, 386–392.

Judd, S.J. 2016. The status of industrial and municipal effluent treatment with membrane bioreactor technology. Chem Eng. J. 305, 37–45.

Jung, K.A., Lim, S.R., Kim, Y., Park, J.M. 2013. Potentials of macroalgae as feedstocks for biorefinery. Bioresour. Technol. 135, 182–190.

Juwarkar, A.A., Singh, S.K., Mudhoo, A. 2010. A comprehensive overview of elements in bioremediation. Rev. Environ. Sci. Biotechnol. 9, 215–288.

Kantachote, D., Naidu, R., Williams, B., McClure, N., Megharaj, M., Singleton, I. 2004. Bioremediation of DDT-contaminated soil: enhancement by seaweed addition. J. Chem. Technol. Biotechnol. 79, 632–638.

Kidgell, J.T., de Nys, R., Hu, Y., Paul, N.A., Roberts, D.A. 2014a. Bioremediation of a complex industrial effluent by biosorbents derived from freshwater macroalgae. PLoS One 9(2), e94706.

Kidgell, J.T., de Nys, R., Paul, N.A., Roberts, D.A. 2014b. The sequential application of macroalgal biosorbents for the bioremediation of a complex industrial effluent. PLoS One 9(7), e101309.

Klimmek, S., Stan, H.-J., Wilke, A., Bunke, G., Buchholz, R. 2001. Comparative analysis of the biosorption of cadmium, lead, nickel, and zinc by algae. Environ. Sci. Technol. 35, 4283–4288.

Kraan, S. 2010. Mass cultivation of carbohydrate rich macroalgae, a possible solution for sustainable biofuel production. Mitig. Adapt. Strateg. Glob Change. https://doi.org/10.1007/s11027-010-9275-5.

Kumar, Y.P., King, P., Prasad, V.S.R.K. 2007. Adsorption of zinc from aqueous solution using marine green algae-Ulva fasciata sp. Chem. Eng. J. 129, 161–166. https://doi.org/10.1016/j. cej.2006.10.023

Kushwaha, A.K., Gupta, N., Chattopadhyaya, M.C. 2014. Removal of cationic methylene blue and malachite green dyes from aqueous solution by waste materials of Daucus carota. J. Saudi Chem. Soc. 18, 200–207.

Lawton, R.J., de Nys, R., Paul, N.A. 2013. Selecting reliable and robust freshwater macroalgae for biomass applications. PLoS One. 8, e64168.

Mahdizadeh, F., Aber, S., Dehghan, G. 2018. Removal of phenol from aqueous solution using the green macroalga Chara sp. Clean: Soil Air Water 46, 1800181.

Marinho-Soriano, E., Panucci, R.A., Carneiro, M.A.A., Pereira, D.C. 2009. Evaluation of gracilaria caudata J. Agardh for bioremediation of nutrients from shrimp farming wastewater. Bioresour. Technol. 100, 6192–6198.

Martínez-Macias, M.D.R., Correa-Murrieta, M.A., Villegas-Peralta, Y., Dévora-Isiordia, G.E., Álvarez-Sánchez, J., Saldivar-Cabrales, J., Sánchez-Duarte, R.G. 2019. Uptake of copper from acid mine drainage by the microalgae Nannochloropsis oculata. Environ. Sci. Pol. Res. 26, 6311–6318.

Mathews, J.A. 2008. How carbon credits could drive the emergence of renewable energies. Energ. Policy. 36, 3633–3639.

McHugh, D.J. 2003. A guide to the seaweed industry, Food and Agriculture Organization of the United Nations, Rome.

McVey, J.P., Stickney, R.R., Yarish, C., Chopin, T. 2002. Aquatic polyculture balanced ecosystem management: new paradigms for seafood production. In Responsible marine aquaculture, Stickney, R.R., McVey, J.P., Eds., CAB International, Oxon, pp. 91–104.

Mehta, S.K., Gaur, J.P. 2005. Use of algae for removing heavy metal ions from wastewater: progress and prospects. Crit. Rev. Biotechnol. 25, 113–152.

Moen, E. 1997. Biological Degradation of Brown Seaweeds. Ph.D. Thesis, Norwegian University of Science and Technology, Trondheim.

Murphy, V., Hughes, H., McLoughlin, P. 2008. Comparative study of chromium biosorption by red, green, and brown seaweed biomass. Chemosphere 70, 1128–1134.

Naushad, M. 2014. Surfactant assisted nano-composite cation exchanger: development, characterization and applications for the removal of toxic Pb2+ from aqueous medium. Chem. Eng. J. 235, 100–108.

Neori, A., Chopin, T., Troell, M., Buschmann, A.H., Kraemer, G.P., Halling, C., Shpigel, M., Yarish C. 2004. Integrated aquaculture: rationale, evolution and state of the art emphasizing seaweed biofiltration in modern mariculture. Aquaculture 231, 361–391.

Neveux, N., Bolton, J.J., Bruhn, A., Roberts, D.A., Ras, M. 2018. The bioremediation potential of seaweeds: recycling nitrogen, phosphorus, and other waste products. Blue Biotechnol. 1, 217–239.

Neveux, N., Magnusson, M., Maschmeyer, T., de Nys, R., Paul, N.A. 2014. Comparing the potential production and value of high-energy liquid fuels and protein from marine and freshwater macroalgae. Global Change Biol. Bioenergy 7, 673–689.

Omar, H., El-Gendy, A., Al-Ahmary, K. 2018. Bioremoval of toxic dye by using different marine macroalgae. Turk. J. Bot. 42, 15–27.

Pena-Castro, J.M., Martınez-Jerónimo, F., Esparza-Garcıa, F., Canizares-Villanueva, R.O. 2004. Heavy metals removal by the microalga Scenedesmus incrassatulus in continuous cultures. Bioresour. Technol. 94, 219–222.

Pereira, L. 2011. A review of the nutrient composition of selected edible seaweeds. In Seaweed: ecology, nutrient composition and medicinal uses, Pomin, V.H., Ed., Nova Science Publishers, New York, pp. 15–47.

Perelo, L.W. 2010. Review: in situ and bioremediation of organic pollutants in aquatic sediments. J. Hazard. Mater. 177, 81–89.

Priyadarshani, I., Rath, B. 2012. Commercial and industrial applications of micro algae–a review. J. Algal. Biomass Utln. 3, 89–100.

Qaiser, S., Saleemi, A.R., Ahmad, M.M. 2007. Heavy metal uptake by agro based waste materials. Electron. J. Biotechnol. 10, 409–416.

Rabalais, N.N., Atilla, N., Normandeau, C., Turner, R.E. 2004. Ecosystem history of the Mississippi River-influenced continental shelf revealed through preserved phytoplankton pigments. Mar. Pollut. Bull. 49, 537–547.

Ratcliff, J.J., Wan, A.H.L., Edwards, M.D., Soler-Vila, A., Johnson, M.P., Abreu, M.H., Morrison, L. 2016. Metal content of kelp (Laminaria digitata) co-cultivated with Atlantic salmon in an Integrated Multi-Trophic Aquaculture system. Aquaculture 450, 234–243.

Reith, J.H., Deurwaarder, E.P., Hemmes, K. 2005. Bio-offshore: Grootschalige Teelt Van Zeewieren in Combinatie Met Offshore Windparken in De Noordzee, Energy Research Centre of the Netherlands, Petten, p. 137.

Reyes, F.A., Mendiola, J.A., Ibañez, E., del Valle, J.M. 2014. Astaxanthin extraction from Haematococcus pluvialis using CO_2-expanded ethanol. J. Supercrit. J.M. 92, 75–83.

Richards, R.G., Mullins, B.J. 2013. Using microalgae for combined lipid production and heavy metal removal from leachate. Ecol. Modell. 249, 59–67.

Roberts, D.A., Johnston, E.L., Poore, A.G.B. 2008. Biomonitors and the assessment of ecological impacts: distribution of herbivorous epifauna in contaminated macroalgal beds. Environ. Pollut. 156, 489–503.

Roberts, D.A., Paul, N.A., Dworjanyn, S.A., Hu, Y., Bird, M.I., de Nys, R. 2015. Gracilaria waste biomass (sampah rumput laut) as a substrate for selenium bioremediation. J. Appl. Phycol. 27, 611–620.

Ross, M.E., Davis, K., McColl, R., Stanley, M.S., Day, J.G., Semião, A.J.C. 2018. Nitrogen uptake by the macro-algae Cladophora coelothrix and Cladophora parriaudii: influence on growth, nitrogen preference and biochemical composition. Algal Res. 30, 1–10.

Rubin, E., Rodriguez, P., Herrero, R., Cremades, J., Barbara, I., de Vicente, M.E.S. 2005. Removal of methylene blue from aqueous solutions using as biosorbent Sargassum muticum: an invasive macroalga in Europe. J. Chem. Technol. Biotechnol. 80, 291–298.

Rubin, E., Rodriguez, P., Herrero, R., de Vicente, M.E.S. 2006. Biosorption of phenolic compounds by the brown alga Sargassum muticum. J. Chem. Technol. Biotechnol. 81, 1093–1099.

Rupérez, P., Ahrazem, O., Leal, J.A. 2002. Potential antioxidant capacity of sulfated polysaccharides from the edible marine brown seaweed fucus vesiculosus. J. Agric. Food Chem. 50, 840–845.

Saunders, R.J., Paul, N.A., Hu, Y., de Nys, R. 2012. Sustainable sources of biomass for bioremediation of heavy metals in waste water derived from coal-fired power generation. PLoS One 7, e36470.

Schiener, P., Black, K.D., Stanley, M.S., Green, D.H. 2015. The seasonal variation in the chemical composition of the kelp species Laminaria digitata, Laminaria hyperborea, Saccharina latissima and Alaria esculenta. J. App. Phycol. 27, 363–373.

Selvam, S.B., Chelliapan, S., Din, M.F.M., Shahperi, R., Aris, M.A.M. 2016. Adsorption of heavy metals from matured leachate by Gracilaria sp. extract. Res. J. Pharm., Biol. Chem. Sci. 7, 15–20.

Silkina, A., Nelson, G.D., Bayliss, C.E., Pooley, C.L., Day, J.G. 2017. Bioremediation efficacy—comparison of nutrient removal from an anaerobic digest waste-based medium by an algal consortium before and after cryopreservation. J. Appl. Phycol. 29, 1331–1341.

Smetacek, V., Zingone, A. 2013. Green and golden seaweed tides on the rise. Nature 504, 84–88.

Smit, A.J. 2004. Medicinal and pharmaceutical uses of seaweed natural products: a review. J. Appl. Phycol. 16, 245–262.

Smith, V.H., Schindler, D.W. 2009. Eutrophication science: where do we go from here? Trends Ecol. Evol. 24, 201–207.

Sode, S., Bruhn, A., Balsby, T.J.S., Larsen, M.M., Gotfredsen, A., Rasmussen, M.B. 2013. Bioremediation of reject water from anaerobically digested waste water sludge with macroalgae (*Ulva lactuca*, Chlorophyta). Bioresour. Technol. 146, 426–435.

Syukor, A.R.A., Sulaiman, S., Siddique, M.N.I., Zularisam, A.W., Said, M.I.M. 2016. Integration of phytogreen for heavy metal removal from wastewater. J. Clean Prod. 112, 3124–3131. https://doi.org/10.1016/j.jclepro.2015.10.103

Tseng, C.K. 2001. Algal biotechnology industries and research activities in China. J. Appl. Phycol. 13, 375–380.

Urrutia, C., Yañez-Mansilla, E., Jeison, D. 2019. Bioremoval of heavy metals from metal mine tailings water using microalgae biomass. Algal. Res. 43, 101659.

Vassilev, S.V., Baxter, D., Andersen, L.K., Vassileva, C.G. 2010. An overview of the chemical composition of biomass. Fuel 89, 913–933.

Vergara-Fernandez, A., Vargas, G., Alarcan, N., Velasco, A. 2008. Evaluation of marine algae as a source of biogas in a two-stage anaerobic reactor system. Biomass Bioenergy 32, 338–344.

Vidali, M. 2001. Bioremediation: an overview. Pure Appl. Chem. 73, 1163–1172.
Vidotti, E.C., Rollemberg, M., Do, C.E. 2014. Algas: Da economia nos ambientes aquáticos à biorremediação e à química analítica. Quim. Nova 27, 139–145.
Villares, R., Puente, X., Carballeira, A. 2001. Ulva and enteromorpha as indicators of heavy metal pollution. Hydrobiologia. 462, 221–232.
Wei, N., Quarterman, J., Jin, Y. 2013. Marine macroalgae: an untapped resource for producing fuels and chemicals. Trends Biotechnol. 31, 70–77.
Wells, M.L., Potin, P., Craigie, J.S., Raven, J.A., Merchant, S.S., Helliwell, K.E., Smith, A.G., Camire, M.E., Brawley, S.H. 2017. Algae as nutritional and functional food sources: revisiting our understanding. J. Appl. Phycol. 29, 949–982.
Wu, H., Huo, Y., Han, F., Liu, Y., He, P. 2015a. Bioremediation using Gracilaria chouae co-cultured with Sparus macrocephalus to manage the nitrogen and phosphorous balance in an IMTA system in Xiangshan Bay, China. Mar. Pollut. Bull. 91, 272–279.
Wu, H., Huo, Y., Hu, M., Wei, Z., He, P. 2015b. Eutrophication assessment and bioremediation strategy using seaweeds co-cultured with aquatic animals in an enclosed bay in China. Mar. Pollut. Bull. 95(1), 342–349.
Yanagisawa, M., Nakamura, K., Ariga, O., Nakasaki, K. 2011. Production of high concentrations of bioethanol from seaweeds that contain easily hydrolyzable polysaccharides. Process Biochem. 46, 2111–2116.
Yang, X., Zhao, Z., Yu, Y., Shimizu, K., Zhang, Z., Lei, Z., Lee, D.-J. 2020. Enhanced biosorption of Cr (VI) from synthetic wastewater using algal-bacterial aerobic granular sludge: batch experiments, kinetics and mechanisms. Sep. Purif. Technol. 251, 117323.
Yang, X., Zhao, Z., Zhang, G., Hirayama, S., Van Nguyen, B., Lei, Z., Shimizu, K., Zhang, Z. 2021. Insight into Cr (VI) biosorption onto algal-bacterial granular sludge: Cr (VI) bioreduction and its intracellular accumulation in addition to the effects of environmental factors. J. Hazard Mater 414, 125479.
Yang, Y., Chai, Z., Wang, Q., Chen, W., He, Z., Jiang, S. 2015. Cultivation of seaweed Gracilaria in Chinese coastal waters and its contribution to environmental improvements. Algal Res. 9, 236–244.
Yazdani, P. Karimi, K. Taherzadeh, M.J. 2011. Improvement of enzymatic hydrolysis of a marine macro-alga by dilute acid hydrolysis pretreatment. In: World Renewable Energy Congress-Sweden, Linköping University Electronic Press, Linköping, pp. 186–191.
Yipmantin, A., Maldonado, H.J., Ly, M., Taulemesse, J.M., Guibal, E. 2011. Pb(II) and Cd(II) biosorption on Chondracanthus chamissoi (A red alga). J. Hazard. Mater. 185, 922–929.
Yu, Z., Robinson, S.M.C., Xia, J., Sun, H., Hu, C. 2016. Growth, bioaccumulation and fodder potentials of the seaweed Sargassum hemiphyllum grown in oyster and fish farms of South China. Aquaculture 464, 459–468.
Yu, Z., Zhu, X., Jiang, Y., Luo, P., Hu C. 2014. Bioremediation and fodder potentials of two Sargassum spp. in coastal waters of Shenzhen, South China. Mar. Pollut. Bull. 85, 797–802.
Zeraatkar, A.K., Ahmadzadeh, H., Talebi, A.F., Moheimani, N.R., McHenry, M.P. 2016. Potential use of algae for heavy metal bioremediation, a critical review. J. Environ. Manage. 181, 817–831.
Zhang, F., Li, M., Li, W., Feng, C., Jin, Y., Guo, X., Cui, J. 2011. Degradation of phenol by a combined independent photocatalytic and electrochemical process. Chem. Eng. J. 175, 349–355.
Zhang, L., Zhao, B., Xu, G., Guan, Y. 2018. Characterizing fluvial heavy metal pollutions under different rainfall conditions: implication for aquatic environment protection. Sci. Total Environ. 635, 1495–1506.

18 Blue-Green Algae in Wastewater Remediation

Susmita Mukherjee, Nivedita Chatterjee, Asmeeta Sircar, Shimantika Maikap, Abhilasha Singh, Sonali Paul

18.1 INTRODUCTION

When it comes to bioremediation, or rather remediation, of any kind of wastewater, bacteria and algae have always extensively been used throughout the history of bioremediation. In these processes, the parameters that influence the growth of these organisms have been found to be present in such systems, that is, the wastewater treatment plants; all these ways of treatment are economically more viable as well as much more eco-friendly (Mathew et al., 2022). Blue-green algae offer some of the best solutions for removing heavy metals and other organic toxic compounds from the environment (Gupta and Bux, 2019).

Algae are the most primitive life form on earth and are efficient in phytoremediation. They cover over 90% of earth's surfaces and supply 70% of earth's oxygen (Gupta and Bux, 2019). Blue-green algae is the most important among all the other species due to their high growth rate, because of which they are considered one of the most economic biomass-producing organisms to ever exist. Apart from being used in treatment of wastewater of different kinds, they have also been used in agriculture, food as well as other industries that have an emerging trend of using SCP apart from ways to treat wastewater without much side effects (Ebtesam, 2008).

18.2 BLUE-GREEN ALGAE

Blue-green algae are one of the most primitive photosynthetic prokaryotes on the planet and are important components of every ecosystem on Earth (Singh et al., 2019). They have a tendency to become overpopulated at certain temperatures, and that significant growth rate actually helps in remediation of natural habitats from heavy metals as well as unneeded and unwanted minerals and organic compounds (Das and Dash, 2021).

Blue-green algae have been used to plan biotechnological processes for remediated regions so sustainable development can be done. In addition to their fast growth, they also have various bioactive molecules with antagonistic properties that make them some of the best degradants of many pollutants. Cyanobacteria are often referred to as phytoplankton, but they actually belong to the bacterial domain, as their name implies, except they have both photo systems I and II and are capable of nitrogen fixation as well as playing an important role in nitrogen cycles in certain ecological systems. (Whitford and Duval, 2020).

DOI: 10.1201/9781003297901-18

This specific species is used for detoxification because it is cost-effective and eco-friendly (Singh et al., 2019).

They are excellent for bioremediation processes because they can get rid of heavy metals from polluted sites and can also degrade hydrocarbons and xenobiotics. In this review, we will be further discussing how bioremediation can take place through the use of blue-green algae (Das and Dash, 2021).

18.3 WASTEWATER DEFINITION: INDUSTRIAL AND HOUSEHOLD

Municipal and industrial wastewater have high levels of pollutants that can destroy ecological systems if not treated properly (Ranade and Bhandari, 2014).

Nearly 24% of wastewater cannot be purified completely through chemical processes, therefore affecting nearby ecological habitats. Industrial wastewater, especially from chemical as well as pharmaceutical industries, contains mostly hazardous substances that need to be treated before being discharged into a biological treatment plant for BOD processes (Leonhauser et al., 2014).

Generally, industrial wastewater is treated worldwide, but domestic wastewater is not treated as properly as industrial wastewater, which is why developing countries may face a water crunch (Cisneros, 2011) in water soon, as shown in Figure 18.1. Now this is extremely important because wastewater treatment is the only way we can regain fresh water for drinking as well as for other necessary activities in our lives (Muralikrishna and Manickam, 2017).

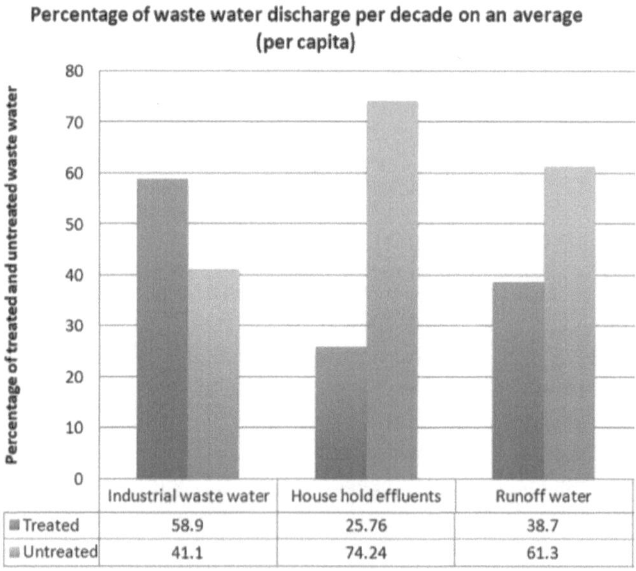

FIGURE 18.1 Percentage of wastewater that is treated and untreated per capita: industrial, household and runoff.

This is when bioremediation of wastewater comes into the big picture (Muralikrishna and Manickam, 2017). For bioremediation, there are many different techniques, but in this review we will talk about how blue-green algae plays an extremely important role in remediating wastewater, both industrial as well as domestic.

18.4 REMEDIATION

Remediation is a technique where reversal or stoppage of an environment contaminant takes place. Now when remediation uses biological components or agents during the removal of toxicity from an environmental area (lithosphere or hydrosphere), it is known as bioremediation (Kumar et al., 2010). It is one of the most effective management tools that helps in recovering both contaminated soil as well as water (Tang et al., 2007).

Bioremediation is a type of natural attenuation that takes place through biodegradation, where aerobic, anaerobic bacteria as well as small fungi come into play in a contaminated region (Strong and Burgess, 2008).

Bioremediation involves the use of blue-green algae to remove unneeded inorganic nitrogenous components, polynuclear aromatic hydrocarbons and even ketones from soil and water. In case of oil spills, usually there are very specific bacteria that are used to remediate those locations, but for industrial discharge in water as well as household discharge, blue-green algae, with the help of their enzymes, return the environment that had been altered by contamination to its original condition (Sutar and Das, 2012).

Remediation can be location-specific and is done by attacking specific containments in a specific geographical location depending on what kind of material is present in that region (Sutar and Das, 2012). Bioremediation is not a panacea, but a natural way to remediate pollution instead of partaking in processes like incineration, catalytic destruction or through absorbance and even through physical removal of pollutants (Kumar et al., 2010).

Using microorganisms with appropriate engineering can provide suitable growth to damaged environmental areas. However, accurate prediction of pollutants and degradation processes is necessary (Sutar and Das, 2012). Bioremediation is one of the best ways in pollution management, as it has less side effects to its environmental quotient.

18.5 HOW BLUE-GREEN ALGAE HELP IN REMEDIATION OF WASTEWATER

Blue-green algae has a multivariant range of usage in terms of biotechnological processes, to be more specific in agriculture biotech, natural products, the cosmetics industry as well as several bioremediation processes. When it comes to the potential of blue-green algae, it has been under attention globally for the numerous secondary metabolites that it produces, even after remediating a specific biological area (Gupta and Bux, 2019). They can be used for energy production like (bio)hydrogen and secondary products from nitrogen fixation, or by reversible activities of the hydrogenase enzyme (Singh et al., 2016).

Wastewater is heavily contaminated with pollutants of various kinds, including chemicals and physical contamination (Sharathchandra and Rajashekhar, 2011). Recent development has been noticed in research where it has been observed that in both genetic engineering as well as biotechnology, the modified blue-green algae has played a huge role when it comes to human welfare during water wastewater remediation, even in ancient times.

This is because there are a few factors that play into the process:

1. Firstly, it does not require external help for growth, as it starts growing in wastewater without any natural resistance and selectivity against the pollution.
2. Their sustainability and metabolic activities are not reduced at all even if the waste material in the water decreases; therefore, they play a huge role in utilizing the several organic as well as inorganic materials as a source of anoxygenic photosynthesis (Singh and Singh, 2013b).
3. Now what the biomass produces as a result is very easy to use in food after the water has been remediated and their by-products have been accumulated.

Blue-green algae can fix atmospheric nitrogen, which makes them completely self-sufficient for their growth and adaptability to survive in a polluted environment (Sharathchandra and Rajashekhar, 2011).

Blue-green algae have a high metal absorption capacity and a high doubling rate; therefore, they are very useful in the treatment of heavy metal–containing wastewater (Machado and Atsumi, 2012).

They are extremely adaptable to their environment, and, therefore, if there is a high concentration of heavy metals in effluent, they can increase the uptake of copper and iron by approximately 45%, and chromium and nickel by nearly 20%. Blue-green algae is not only used in remediation but also in energy production (Singh et al., 2016), and the usage in daily life of these specific species makes them commercially more valuable (pocket-friendly), and, therefore, they can be used frequently to help in environment-friendly treatment processes (Machado and Atsumi, 2012).

They have a huge role in remediation of various types of environmental contaminants, including pesticides, crude oil, naphthalene and even xenobiotics. Not only this, but they're even used in wastewater lagoons and are cultivated through these areas to check pollutants and pesticides and help in pollution reduction through processes like BOD and COD. They even help in promoting aquatic plant growth and even enhance seed germination since they have hormones such as cytokinins, oxin and abscisic acids, along with vitamins help in some potential development in plants in the surrounding ecosystem (Vijayakumar, 2012). They are capable of surviving in extreme environments like hypersaline waters or even in arid desert areas and even within hot springs, along with freezing environments (Nozzi et al., 2013; Prasanna et al., 2013); Algae can survive in pH levels lower than 5, which promotes the growth of other algae that are not harmful to freshwater ecosystems and check coliforms formation. All in all, blue-green algae in wastewater remediation has far too many advantages than disadvantages, which makes it one of the best ways of wastewater remediation (Rana et al., 2012).

18.5.1 Pathways/Mechanisms Involved

Blue-green algae not only remediates wastewater but creates useful secondary products that can then be used in different ways throughout industries dealing with animal husbandry, biofuel and even bioplastics (Al-Jabri et al., 2021), as stated in Figure 18.2. For biofuel and biohydrogen production, this specific species goes through a glycolytic as well as photosynthetic pathway (Mezzanotte et al., 2022; Kuppam et al., 2022), as shown in Figure 18.3.

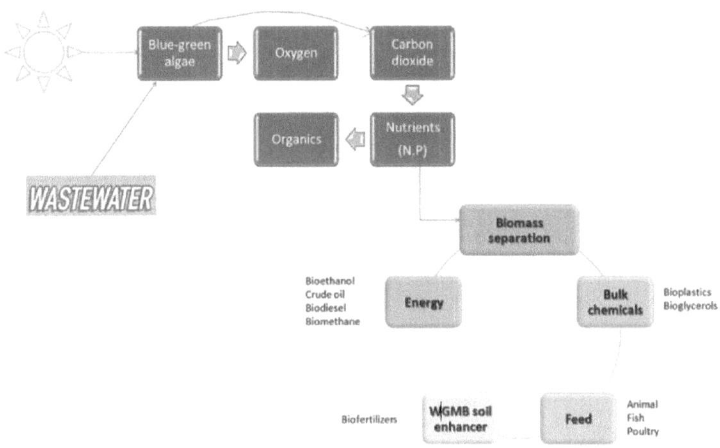

FIGURE 18.2 Pathway for wastewater treatment through blue-green algae schematic diagram.

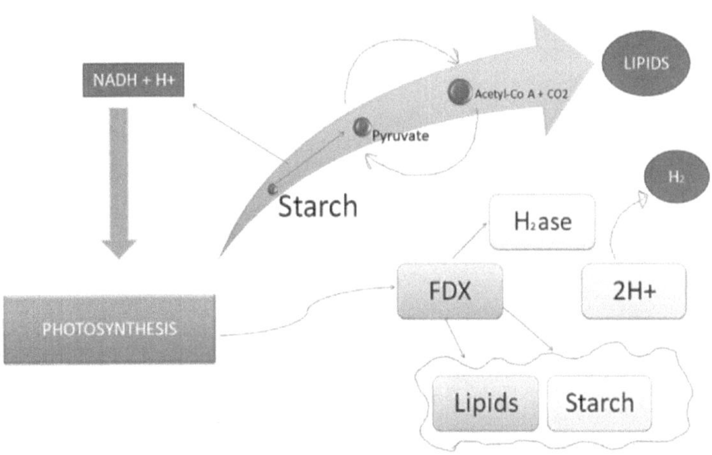

FIGURE 18.3 How glycolysis and photosynthesis in blue-green algae help in production of lipids and starch biomass.

18.5.2 Applications and Benefits

Blue-green algae has a lot of benefits, starting from how the biomass obtained from it can be manipulated with the help of several physiochemical treatments to attain the quality and quantity a customer requires while having good-quality biofuel products (El-Shafai et al., 2007; Vijayakumar, 2012).

1. The biofuel production through blue-green algae is also one of its benefits.
2. They have fast maturation as well as multiplication capability; therefore, they meet the huge demands for biofuel production as well as need for remediation.
3. Blue-green algae cultivation consumes less fresh water than any kind of wastewater or crop land (Pabbi, 2015).
4. Under properly evaluated carbon-enriched conditions, blue-green algae can produce biomass with high efficiency.
5. The production of greenhouse gases such as different kinds of nitrous and sulphur oxides can be lowered through the cultivation of blue-green algae for bio energy production (Iffat, 2022) as well as bioremediation (Figure 18.4).

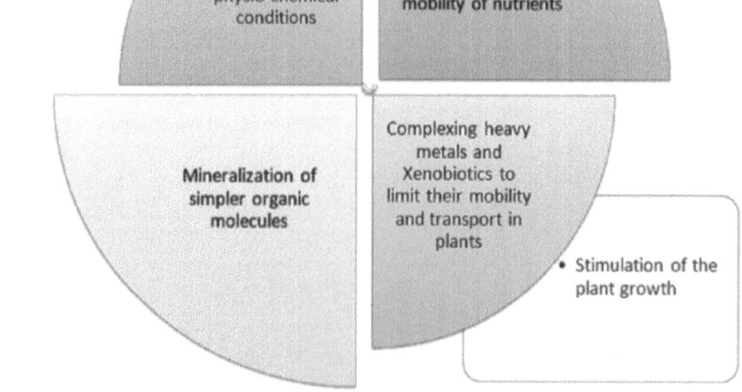

FIGURE 18.4 Various added benefits of blue-green algae.

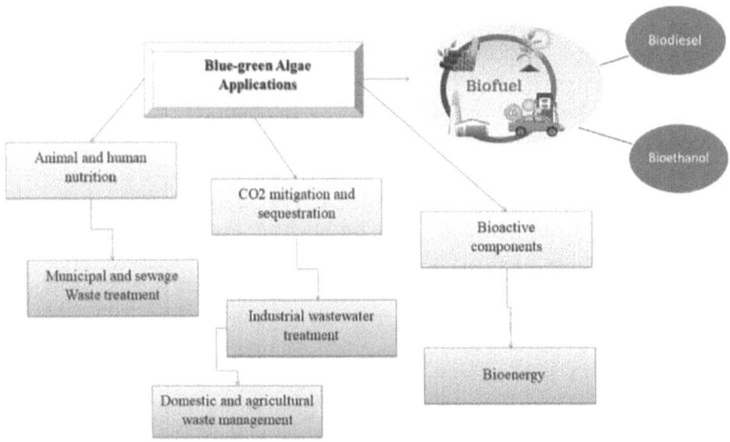

FIGURE 18.5 Applications of blue-green algae.

They are potentially more cost-effective, eco-friendly and much more sustainable and viable than any other conventional bioremediation source (Chu et al., 2018). They're economically sustainable and can overpower any kind of environmental challenge that needs to be remediated within a stipulated amount of time, along with providing high-quality biofuel (Ferreira et al., 2017; Zainith et al., 2021). Their applications range from nutrition for secondary consumers to sewage water treatment and even domestic wastewater treatment, as shown in Figure 18.5.

18.6 CONCLUSION

Blue-green algae have immense potential when it comes to treatment of wastewater and are not only cost-effective but also eco-friendly. They produce biomass as well as biofuel that can be used for several other important tasks in various industries; therefore, all in all, using blue-green algae for remediation of wastewater is something that enhances the removal of pollutants without affecting the habitat and the overall ecosystem while promoting growth to the plants as well as microorganisms that are beneficial in that location.

REFERENCES

Iffat Zareen Ahmad. 2022. The usage of Cyanobacteria in wastewater treatment: prospects and limitations. Lett. Appl. Microbiol. 75(4): 718–730. https://doi.org/10.1111/lam.13587

Kuppam Chandrasekhar, Tirath Raj, S. V. Ramanaiah, et al. 2022. Algae biorefinery: A promising approach to promote microalgae industry and waste utilization. J. Biotechnol. 345: 1–16.

B. T. T. Chu, M. L. Petrovich, A. Chaudhary, et al. 2018. Metagenomics reveals the impact of wastewater treatment plants on the dispersal of microorganisms and genes in aquatic sediments. Appl. Environ. Microbiol. 84: e02168–e02169.

B. J. Cisneros. 2011. Treatise on water science. Water-Quality Eng. 4: 147–200. https://doi.org/10.1016/B978-0-444-53199-5.00082-8

Ebtesam El-Bestawy. 2008. Treatment of mixed domestic–industrial wastewater using cyanobacteria. J. Ind. Microbiol. Biotechnol. 35(11): 1503–1516. https://doi.org/10.1007/s10295-008-0452-4

S. A. El-Shafai, F. A. El-Gohary, F. A. Nasr, et al. 2007. Nutrient recovery from domestic wastewater using a UASB-duckweed ponds system. Bioresour Technol. 98(4): 798–807.

A. Ferreira, B. Ribeiro, P. A. S. S. Marques, et al. 2017. Scenedesmus obliquus mediated brewery wastewater remediation and CO_2 biofixation for green energy purposes. J. Clean. Prod. 165: 1316–1327.

Al-Jabri Hareb, Das Probir, Khan Shoyeb, et al. 2021. Treatment of wastewaters by microalgae and the potential applications of the produced biomass. 13(1): 27. https://doi.org/10.3390/w13010027

Sutar Harekrushna and Das Chandan Kumar. 2012. A review on: bioremediation. Int. J. Res. Chem. Environ. 2: 13–21.

Muralikrishna V. Iyyanki and Manickam Valli. 2017. Industrial Wastewater Treatment Technologies, Recycling, and Reuse. In Environmental Management, pp. 295–336. https://doi.org/10.1016/B978-0-12-811989-1.00013-0

Leonhauser Johannes, Pawar Jyoti and Birkenbeul Udo. 2014. Industrial Wastewater Treatment, Recycling and Reuse. In Industrial Wastewater Treatment, Recycling and Reuse. Chapter 5 — Novel Technologies for the Elimination of Pollutants and Hazardous Substances in the Chemical and Pharmaceutical Industries. Elsevier Ltd, pp. 215–234.

A. Kumar, D. Chandel, I. Bala, et al. 2010.A Review on: Bioremediation. IUP J. Biotechnol. IV(2): 4556. *Int. J. Res. Chem. Environ.* Vol. 2, Issue 1, January 2012 (13–21).

Gupta Sanjay Kumar and Bux Faizal. 2019. Application of Microalgae in Wastewater Treatment (Volume 1: Domestic and Industrial Wastewater Treatment). In Potential of Blue-Green Algae in Wastewater Treatment, pp. 363–381. https://doi.org/10.1007/978-3-030-13913-1_17

I. M. P. Machado and S. Atsumi. 2012. Cyanobacterial biofuel production. J. Biotechnol. 162: 50–56. https://doi.org/10.1016/j.jbiotec.2012.03.005

Mathew Mammen Merwin, Khatana Kanchan, Vats Vaidehi, et al. 2022. Biological approaches integrating algae and bacteria for the degradation of wastewater contaminants—a review. Front. Microbiol. https://doi.org/10.3389/fmicb.2021.801051

N. E. Nozzi, J. W. K. Oliver and S. Atsumi. 2013. Cyanobacteria as a platform for biofuel production. Front. Bioeng. Biotechnol. 1: 7. https://doi.org/10.3389/fbioe.2013.00007

S. Pabbi. 2015. Blue Green Algae: A Potential Biofertilizer for Rice. In The Algae World. Springer, pp. 449–465.

Singh Pardeep, Kumar Ajay and Borthakur Anwesha. 2019. Abatement of Environmental Pollutants Trends and Strategies: Book. Science Direct.

R. Prasanna, V. Chaudhary, V. Gupta, et al. 2013. Cyanobacteria mediated plant growth promotion and bioprotection against Fusarium wilt in tomato. Eur J Plant. Pathol. 136(2): 337–353.

A. Rana, M. Joshi, R. Prasanna, et al. 2012. Biofortification of wheat through inoculation of plant growth promoting rhizobacteria and cyanobacteria. Eur. J. Soil. Biol. 50: 118–126.

Jay Shankar Singh, Arun Kumar, Amar N. Rai, et al. 2016. Cyanobacteria: a precious bioresource in agriculture, ecosystem, and environmental sustainability. Front. Microbiol. https://doi.org/10.3389/fmicb.2016.00529

K. Sharathchandra and M. Rajashekhar. 2011. Total lipid and fatty acid composition in some freshwater cyanobacteria. J. Alagal Biomass Utln. 2: 83–97.

J. S. Singh and D. P. Singh. 2013b. Plant Growth Promoting Rhizobacteria (PGPR): Microbes in Sustainable Agriculture. In Management of Microbial Resources in the Environment. Springer, pp. 307–319.

P. J. Strong and J. E. Burgess. 2008. Treatment methods for wine related ad distillery wastewaters: a review. Biorem. Jou. 12: 7087.

Zainith Surabhi, Saxena Gaurav, Kishor Roop, et al. 2021. Application of Microalgae in Industrial Effluent Treatment, Contaminants Removal, and Biodiesel Production: Opportunities, Challenges, and Future Prospects. Chapter 20. Bioremediation for Environmental Sustainability. Toxicity, Mechanisms of Contaminants Degradation, Detoxification, and Challenges. Elsevier, pp. 481–517.

Das Surajit and Dash Hirak Ranjan. 2021. Microbial Biodegradation and Bioremediation Techniques and Case Studies for Environmental Pollution (Second Edition). Science Direct.

C. Y. Tang, Q. S. Criddle, C. S. Fu, et al. 2007. Effect of flux (transmembrane pressure) and membranes properties on fouling and rejection of reverse osmosis and nanofiltration membranes treating perfluorooctane sulfonate containing waste water. J. Enviro. Sci. Tech. 41: 2008–2014.

Mezzanotte Valeria, Marazzi Francesca, Ficara Elena, et al. 2022. First results on the removal of emerging micropollutants from municipal centrate by microalgae. Environ. Clim. Technol. 26(1), 36–45. https://doi.org/10.2478/rtuect-2022-0004

S. Vijayakumar. 2012. Potential applications of cyanobacteria in industrial effluents—a review. J. Bioremed. Biodeg. 3: 2. https://doi.org/10.4172/2155-6199.1000154

Ranade V. Vivek and Bhandari M. Vinay. 2014. Industrial Wastewater Treatment, Recycling, and Reuse. Science Direct.

Whitford G. Walter and Duval D. Benjamin. 2020. Ecology of Desert Systems. Book (Second Edition). Science Direct.

19 Advances in Nano-Biotechnology for Bioremediation Applications

Titir Guha, Sugata Mitra,[#] Sumit Rez,[#] Rita Kundu
[#]Authors with equal contributions

19.1 INTRODUCTION

In recent years anthropogenic activities have alarmingly degraded the quality of our environment, specifically the lithosphere and aquatic ecosystem, which have been subjected to maximum exploitation. Water is not only an essential component of human civilization and progress but also houses an astonishing amount of wildlife and ecosystems. The heavy metals, organic compounds, industrial effluents, oil spills during exploration, transportation, storage and refining of crude oil (Kumari and Singh 2016), fertilizers, herbicides, pesticides and by-products that have resulted from the production and extraction of burning of fossil fuels are the principal environmental contaminants (Kapoor et al. 2021). The industrial contaminants also eventually find their way by leaching into the groundwater or invading streams and rivers. Other sources of contaminants include (but are not limited to) wastewater from dyeing textile mills, including different synthetic chemicals – salts of heavy metals, surfactants and mineral oils (Golob et al. 2005). Harmful substances from textile, cosmetic and paper processing industries could reach water bodies, leading to massive detriments to flora, fauna and human health (López-Miranda et al. 2019). Organic compounds produced in the form of by-products in petroleum industries have a detrimental effect on the environment, releasing huge amounts of BTEX (benzene, toluene, ethyl benzene, xylene), phenol and its derivatives (o, m, and p cresol). These pollutants are of a corrosive nature and cause degeneration of the nervous system (Singh et al. 2016). In addition to these, heavy metal contaminations are reportedly toxic for human health as well as plants (such as inhibition of pollen tube growth and aberrations in apical branching), affecting their growth and crop yield (Sawidis 2008).

In this regard, applications of innovative environmental remediation technologies are essential. Remediation involves the removal of contaminants from the soil, air, groundwater and water bodies to safeguard ecosystems and human health (Ingle et al. 2014). However, the need for developing a cost-effective and eco-friendly technology

 DOI: 10.1201/9781003297901-19

for any real remediation is critical, since many of these conventional physico-chemical methods have been found to be less economical in the long run, mainly due to generation of secondary by-products which, in turn, require additional remediation efforts (Kumari and Singh 2016). Although bioremediation techniques are relatively cheaper, they are extremely slow processes, and microbes fail to grow under very high concentrations of contaminants. In recent times, nanoparticles have also gained immense popularity as remediation tools. Several reports have suggested an alternative technique which involves integration of nanotechnology and bioremediation process which can enhance the degradation rate of environmental contaminants. The present chapter deals with the applicability and advantages of nano-bioremediation techniques as potential means of environmental remediation.

19.2 CONVENTIONAL TECHNOLOGIES FOR REMEDIATION

19.2.1 Physical Methods

Physical methods for the remediation of contaminated soils involve soil washing, soil vapour extraction, soil flushing etc. Contaminated sites can be cleaned up rapidly by these oldest remediation techniques, but due to expenses and risk of contaminant mobility, it is a remedial method not well suited for removal of heavy metals from a large area (Ingle et al. 2014).

Physical methods in water treatment involve precipitation, ion exchange, electro-coagulation, electro-dialysis, membrane filtration, etc. However, several of these methods are reportedly expensive.

19.2.2 Chemical Methods of Remediation

During a chemical remediation process, a highly toxic contaminant is generally converted to comparatively less toxic forms. However, its applicability is limited in cases of large industrial scales of remediation. The main critical methods involve remediation using chelators, surfactants (like rhamnolipids), chemical immobilization, critical fluid extraction, in situ catalyzed peroxide remediation and photo-degradation. These physico-chemical methods are also inefficient because of the high per capita cost of organizing equipment and workforce when the concentration of heavy metals is very low.

19.2.3 Biological Methods of Remediation

Bioremediation is an eco-friendly technology available for removal and recovery of various contaminants in wastewater and contaminated groundwater systems by using biological agents and their metabolism. Organisms that generally survive in highly contaminated areas are known to possess several detoxifying mechanisms such as biosorption (such as *Saccharomyces cerevisiae* on Zn (II) and Cd (II) via ion exchange), bioaccumulation (such as the bacteria *Cupriavidus metallidurans*, fungi *Fusarium* and plant *Noea mucronata* on Cd (II), Cu (II) and Pb (II) accumulation respectively), degradation of aromatic contaminants by bacteria like *Pseudomonas*,

Nocardia and *Mycobacterium*, bio-bleaching and de-colorization of dyes by basidiomycete *Phlebia* sp. and bio-conversion (such as by *Verticillum terrestre*on Hg (II) to non-toxic Hg), which can be exploited in ex situ or in situ forms (Dixit et al. 2015; Lim et al. 2003; Baker et al. 1996; Kelly et al. 2006). While in situ bioremediation involves on-site treatments, ex situ comprises transferring the contaminated material to be purified elsewhere. In phytoremediation, a number of processes such as phytoextraction, phytofiltration, phytostabilization, phytovolatilization and phytodegradation have been associated (Alkorta et al. 2004).

19.2.4 Limitations of Traditional Methods and the Need for Alternative Technologies

Contamination of oil and groundwater occurring due to different anthropogenic activities involve presence in these areas, heavy metals and organic compounds. The physical and chemical methods, which are usually in practice for remediation purposes, are time-consuming and significantly expensive. Bioremediation is also not the best way forward due to strict requirement of the continuous maintenance of optimized conditions for the maximal growth and metabolism of organisms. Consequently, bioremediation is not instantaneous. A pretreatment procedure for the removal of concerned inhibitory compounds or conversion of pollutants to forms that easily degrade might be required often, with the result being disturbance of the ecosystem.

In the field of remediation of environmental pollution, it is extremely important to find new and advanced techniques. The development of a cost-effective, fast and eco-friendly technology which can be effective in action and reach inaccessible areas such as crevices and aquifers is urgently required. Moreover, an additional ability to manipulate matter at its molecular levels, and develop tools that are specific for a certain pollutant, in turn increasing the affinity and selectivity of the technique, was equally desired. The advent of nanotechnology has led us to a common answer to most of today's practical problems on an industrial scale. It involves the designing, characterization, manufacture and application of particles by dictating their structure and size at nanoscale (Cameotra and Dhanjal 2010; Kumar and Gopinath 2017)

19.3 APPLICATIONS OF NANOTECHNOLOGY FOR REMEDIATION OF CONTAMINANTS

Derived from the Greek word "nanos" (meaning dwarfish), dimensions of nanoparticles generally range from 0.1–100nm (1nm = 10^{-9} m). Due to nanoparticles having extremely small size, they possess large surface areas relative to their volumes and have, consequently, unique chemical and biological reactivity. Nanostructures exist commonly as spheres, rods, wires and tubes. From naturally occurring nanoparticles as in volcanic ash, minerals and magnetosomes (inorganic crystals of magnetite or greigite of size 35–120 nm) in magnetotactic bacteria (term coined by Richard Blakemore) that exist ubiquitously, to anthropogenic nanoparticles arising due to diesel exhausts, coal combustion, mining and other human activities, there exist two main categories of nanoparticles. Recently, due to nanotechnological advances, the third category of chemically or biogenically engineered nanoparticles has come into play (Cameotra and Soniya 2010).

In recent years, a great consensus has been reached on the application of nano-structured materials as adsorbents or catalysts to purify soil and wastewater from toxic pollutants. Nanoparticles are more than ten times smaller than a typical bacterial cell (diameter around 1um = 1000nm). Effective transportation apart, they can remain in suspensions for long periods of time and, thus, can establish an in situ treatment zone (Agarwal and Joshi 2010). It is also possible to adhere nanoparticles to solid matrices like activated carbon, zeolites, etc. for improved treatment. The procedures and applications of which are discussed further subsequently.

19.3.1 Degradation of Polychlorinated Hydrocarbons

While polychlorinated hydrocarbons (PCH)/polychlorinated biphenyls (PCB) have wide uses as dielectric fluids in transformers, capacitors and coolants, the inherent toxicity and persistence of these organic pollutants have led to reports suggesting bimetallic nanoparticles for environmental clean-up of PCH due to small size of nanoparticles, leading to higher infusibility in contaminated areas where even micro particles cannot reach. Oxide-coated Fe^0 can form weak complexes with carbon tetrachloride (CTC) and consequently break CTC into methane, carbon monoxide or formate (Nurmi et al. 2005).

19.3.2 Degradation of Hydrophobic Compounds

Polymeric nanoparticles manufactured from poly(ethylene)glycol-modified urethane acrylate (PMUA) precursor. The interior regions of PMUA are hydrophobic, which lead to increased affinity between PAHs and outer hydrophilic surfaces. This increases amount of PAHs released into the aqueous phase. Eventually the released PAHs can be treated by natural attenuation.

19.3.3 Reductive Dechlorination by Oxidation of nZVI

Nano zero valent (nZVI), emulsified zero valent nanoparticles and bimetallic particles, are the foundations to remediation of contaminated waters in aquatic ecosystems as well as underground. nZVI (Fe^0) particles have the potential to get oxidized into +2/+3 states, consequently reducing surrounding organic and inorganic pollutants (act as electron donors).

$Fe^0 \rightarrow Fe^{2+} + 2e-$ (oxidation of nZVI)

$RCl + H^+ + 2e- \rightarrow RH + Cl^-$ (reductive dechlorination) (Agarwal and Joshi 2010)

An example of a stoichiometric equation involving reduction of C_2Cl_4 during oxidation of Fe^0 is shown: $C_2Cl_4 + 4Fe^0 + 4H^+ \rightarrow C_2H_4 + 4Fe^{2+} + 4Cl^-$ (Zhang 2003).

It is because of the lower standard reduction potential (E^0) of Fe^{2+}/Fe (~-0.44V) than organic pollutants and metals like Pb, Zn and Cr that these get reduced by ZVI nanoparticles.

It has been found out that bimetallic iron nanoparticles (Fe coupled with a less reactive metal like Pd, Ni, Pt or Ag) achieve higher rates of contaminant deterioration and reduce the amount of toxic side products released. Addition of surfactants or biodegradable vegetable oil with nZVI forms a surfactant-stabilized oil film around the iron particles in water (like micelles). This is an augmentation of nZVI technology called EZVI (emulsified zero valent iron) (Geiger et al. 2003).

The integrated bimetallic system can be divided into a catalyst (Pd, Pt, Ni) and an electron donor (Fe) which aids in the dechlorination of PCE, TCE, DCE and VC by surface-mediated redox reactions. It is notable that this dichlorination reaction proceeds at a pace higher than in cases where only iron nanoparticles have been used (Zhang et al. 1998).

19.3.4 Stabilization by Protective Coating on Nanoparticles

Due to exceedingly low stability of bare metal NPs which get oxidized easily on exposure to air, shells of carbon and silica coating the NPs show increased stability. Removal of dyes like methyl orange, etc. have been reported by the action of carbon-coated magnetic nanoparticles. Narrow size distribution of metallic nanoparticles helps dispersion stability, which is complimented by the large surface area of coated carbon shell. It has been suggested that for heavy metal remediation like Cr (VI), carbon shells effectively adsorb Cr (VI) from the solution due to presence of the carboxylic functional group. Moreover, the response of the Fe-C nanoparticles evidence the intact structure of the Fe core which is protected by the outer carbon shell from etching due to Cr (VI) (Zhang et al. 2010).

19.3.5 Nanoparticles in Water Purification

Resultant advanced methods of nanotechnology – nanosorbents (carbon nanotubes like SWCNTs and MWCNTs, graphene, fullerene), nanocatalysts (TiO_2), bioactive nanoparticles (MgO and Ag nanoparticles against spores of *Bacillus subtilis* and cell membranes of *Klebsiella pneumonia* respectively) and catalytic membranes for nanofiltration (alumina UF membranes, ceramic reinforced graphene oxide (GO)/attapulgite (ATP)-fused membranes) – would greatly improve the quality of potable drinking water (Singh et al. 2016; Liu et al. 2019; Sondi and Salopek-Sondi 2004). The development of novel nanoscale techniques for treatment of wastewater contaminated by toxic metal ions, radionuclides, pesticides, bacteria and viruses would be the major contribution of nanotechnology in the field of water nanoremediation.

19.3.6 Enzyme-Linked Magnetic Nanoparticles

Enzymes are proteins with biocatalyst function that modify the rate of biochemical reactions in living organisms. Enzymes can be extracted from cells by different advanced technologies and then implemented to catalyze a wide range of commercially significant processes, one of which being coupling with nanoparticles in their immobilized form. Enzymes are generally operational for long periods (merely acting as reaction catalysts and remaining as is at the end of the reaction) and can be recovered and hence reused. This lowers the overall cost of production. Immobilization of enzymes on a nanoparticle matrix can stabilize the enzyme such that it is functional in a broader range of pH and temperature as compared to free native enzymes. Without potential accretion of immobilized enzymes, an improved enzyme-substrate interface is observed (Dutta et al. 2013).

Wang et al. 2012 immobilized lipase with Fe_3O_4 nanoparticles via physical adsorption, providing increased surface area and higher enzymatic activity. The NPs were coated with a polydopamine film, and lipase was immobilized on the PD-Fe_3O_4. Darwesh et al. 2019 focussed on the environmental degradation of phenolic compounds like azo dyes, selecting peroxidase enzyme that acts as a catalyst in these reactions, along with Fe_3O_4 nanoparticles.

19.4 APPLICATION OF NANOPARTICLES ALONG WITH BIOREMEDIATION FOR REMOVAL OF HAZARDOUS MATERIALS

Application of nanoparticles, especially iron nanoparticles, have recently gained popularity as a remediation tool. However, the physiochemical interactions between soil particles and nanomaterials, which include the behavior of the nanoparticles in the soil pore, mineral adsorption tendencies and effects on the soil microorganisms, are not yet clear. Nanoparticles are known to have either negative or positive effects on the soil microorganisms. Considering several other challenges of nanoparticles such as aging-induced loss of reactivity and high clustering tendency upon in situ applications, the concept of nano-bioremediation can be regarded as a more efficient sustainable approach of remediation of hazardous materials from contaminated sites. In the following sections, several researches related to nano-bioremediation techniques which integrate the biotic and abiotic degradation of the toxic compounds have been discussed in detail.

19.4.1 Chlorinated Organic Contaminants

Chlorinated organic compounds are one of the most dangerous contaminants in the environment, because even in trace amounts, it can cause health risks. In industrial and urban areas, these contaminants are quite abundant. The chlorinated organic compounds can easily evaporate from the surface water but are retained for much longer duration in soils and groundwater (Hara 2012). Although several technologies have been explored by the researchers, during remediation of the chlorinated organic compound contaminated sites, cost-effectiveness, efficiency and sustainability must be evaluated, and in most cases, use of a single technology for this purpose has been proved to very expensive and less efficient (Němeček et al. 2016). Hence, a combination of two or more technologies must be encouraged (Koenig et al. 2016). Physico-chemical methods of remediation are costly but are indeed fast and efficient methods as an initial remediation step for very high concentrations of contaminants. This step can be followed by biological methods, which are relatively slower but cheap and a sustainable approach for in situ remediation purpose. Also, nano-bioremediation techniques can be an effective alternative in this regard.

The integration of bioremediation technique and nanoparticle-mediated degradation of chlorinated organic compounds was first proposed by Bokare et al. 2010. At first, 5 g/L triclosan solution was subjected to sequential degradation via Pd/Fe bimetallic nanoparticle mediated anaerobic dechlorination. Followed by this, the by-products were oxidized with the help of the laccase enzyme produced from *Trametes versicolor* (a common polypore mushroom). Complete dechlorination of the triclosan was achieved within 20 minutes by this nano-bioremediation technique.

Another study was conducted by Bokare et al. 2012, where Pd/Fe bimetallic nanoparticles were again used for mediating the dechlorination of 2,3,7,8-tetrachlorodibenzo-p-dioxin or dioxin, which is a toxic carcinogen present in herbicides which possess high persistence capacity in the environment and at the same time poorly biodegradable in nature. In an anaerobic environment, 100% dechlorination of dioxin was achieved via Pd/Fe bimetallic nanoparticle treatment for 10 hours. The residual by-product (dibenzo-p-dioxin) was not degraded by the nanoparticles, and hence further degradation was mediated by the action of *Sphingomonas wittichii*, which is an alphaproteobacteria.

Another nano-bioremediation approach has been reported by Singh et al. 2013, who conducted a combined biotic and abiotic degradation of soil polluted with Lindane or *gamma*-hexachlorocyclohexane (γ-HCH), a neurotoxin used as insecticides. At first it was found that CMC-Pd/Fe nanoparticles at the concentration of 40 mg/L can promote the growth of *Sphingomonas* sp. strain NM05, a Gram-negative chemoheterotrophic aerobic bacteria. Upon incubation of γ-HCH spiked soils (concentration: 25μg/g), with 40 mg/L CMC-Pd/Fe nanoparticles and NM05 cells for six days, 99% degradation of the contaminants was recorded, and overall this integrated approach showed 1.7–2.1 times higher efficiency compared to that of individual NMO5 or CMC-Pd/Fe nanoparticle treatments.

Le et al. (2015) reported a nano-bioremediation technique for degradation of Aroclor 1248, which is polychlorinated biphenyl (PCB), a neurotoxic compound often present in high concentrations in soils and sediments. Upon treatment of the Aroclor 1248 (PCB) solution with nZVI (1000 mg/L), 89% degradation of the contaminant was achieved, and biphenyls were obtained as a by-product after the nanoparticle-mediated dechlorination under anoxic condition. Followed by this, 90% decline of the biphenyls was found upon biodegradation with the bacteria *Burkholderia xenovorans*.

19.4.2 Petroleum Hydrocarbons

Bioremediation of petroleum hydrocarbons is an extremely slow process. Due to the insoluble nature of petroleum hydrocarbons, the bioavailability of crude oils gets reduced. In addition to this, petroleum hydrocarbons are toxic to the microbes in very high concentrations. However, application of nanotechnology can provide an alternative approach to combat these limitations. Since nanoparticles possess higher surface area, higher reactivity and other catalytic properties, they are often render to be more useful than their bulk counterparts and can help in the remediation of the contaminants at a much faster rate (Gupta et al. 2011; Sayan et al. 2013). Previous studies by Shao et al., 2010 and Gotovac et al. 2006 have already demonstrated that multi-walled carbon nanotubes can be employed for the remediation of organic pollutants like polycyclic aromatic hydrocarbons (PAHs) and polychlorinated biphenyls (PCB). Enzyme-conjugated nanoparticles are also being used for remediation purposes owing to the biocompatibility of the nanoparticles, and the nanoparticles can also provide an inert micro-environment which can help to preserve the native properties of the enzyme and thus aid to maintain their biological activity (Ansari and Husain 2012). Due to the presence of the magnetic properties of nanoparticles, the immobilized enzymes or proteins can be easily separated from the reaction mixtures, and this eliminates the need for other techniques such as centrifugation, filtration, etc. (Khoshnevisan et al. 2011; Ranjbakhsh et al. 2012). However, selection of suitable nanomaterials is a crucial process since it must be ensured that the nanomaterials themselves are not toxic to the microorganisms meant for the bioremediation process (Rizwan et al. 2014). Several studies have showed that nanoparticles can be utilized in order to ensure rapid destruction of the hydrophobic compounds via different methods which have been summarized subsequently:

19.4.2.1 Nanoparticles-Mediated Increased Solubility of Hydrophobic Compounds

For bioremediation, enhancement of the solubility of the hydrocarbons which are generally hydrophobic in nature is essential. Reportedly, this can be achieved by the use of biosurfactants which can decrease the surface tension and hydrophobicity of the hydrocarbons (Cameotra and Makkar 2010). The productions of these biosurfactants are often controlled by the presence of certain metal ions such as iron, which is a nutritional requirement for certain microorganisms (Haferburg and Kothe 2007). In this regard, iron nanoparticles have been proved to be an enhancer of the production of biosurfactants. Kiran et al. (2014) reported that iron nanoparticles increased the synthesis of the biosurfactant by 80% in *Nocardiopsis* sp. MSA13A, without any harmful effect on the growth or filamentous structure of *Nocardiopsis* sp. Liu et al. (2013), also found increase in growth of *Serratia* sp. by 57%, along with the enhancement in the synthesis of the biosurfactants by 63% upon treatment with 1mg/L of Fe nanoparticle.

19.4.2.2 Immobilization of Microbes and Enzymes with the Aid of Nanoparticles

According to Shan et al. 2005, the immobilization of microbial cells can remarkably enhance the microbial activity and longevity since it provides protection from adverse conditions, and in addition to this, it can also help in the reuse and recovery of the microbial cells. With the help of magnetic ferric oxide (Fe_3O_4), nanoparticles *Rhodococcus erythropolis* IGST8 were immobilized, and this resulted in enhanced rate of desulfurization of dibenzothiophene (DBT) in basic salt medium by 56% due to enhanced microbial activity (Ansari et al. 2009). Similarly, *Pseudomonas delafieldii* cells were also immobilized on Fe_3O_4 nanoparticles, which also contributed to increasing the desulfurization rate of DBT.

Also, the immobilization of the microbial enzymes which are responsible for the bioremediation process can be helpful in terms of reactivity, purity, selectivity, stability and reuse efficiency (Garcia-Galan et al. 2011). Ren et al. 2011 reported that immobilized lipase enzyme can be isolated easily from reaction mixture, and it lost only 30% of its initial activity after 21 cycles of reaction. According to Ansari and Husain 2012, enzymes that are immobilized on the surface of nanoparticles can be applied over a wide range of pH and temperature conditions due to increase in stability as compared to that of the native enzymes and nanoparticles acting like chaperones, which maintained the native structure of the protein (Singha et al. 2010). Also, immobilized enzymes had lower risk of aggregation and, hence, was more advantageous during the enzyme-substrate interactions (Dutta et al. 2013). Several reports highlight that utilization of nanoparticles for immobilization of the enzymes helped to increase enzyme activity levels. Several nanoparticles such as SnO_2, TiO_2, graphene, ZnO, CeO_2, Fe_3O_4, CNTs, etc. have been used for immobilization of enzymes, and this promoted the rate of electron transfer reaction between the enzyme active sites and substrate.

Acevedo et al. (2010) immobilized manganese peroxide enzymes, which was produced by *Anthracophyllum discolor*, on nanoclay, and that led to a reduction in the activation energy of the enzyme, which was 51.9KJ mol^{-1} for free enzyme and 34.4KJ mol^{-1} for the immobilized enzyme. Thus the immobilized enzyme showed higher efficiency in the degradation of PAH even in unfavorable conditions compared to that of free enzyme activity. Also, Chronopoulou et al. 2011 reported about an increase in the thermal stability of the lipase enzyme after its immobilization over nanoparticles.

According to Wang et al. 2012, in case of immobilization of enzymes on the nanoparticles, the magnetic nanoparticles are mostly used because of certain advantages like higher surface area, non-toxic nature and easy separation from reaction mixtures by means of magnetic field. Wang et al. 2012 also demonstrated that immobilization of lipase enzyme on the Fe_3O_4 nanoparticles helped to increase the enzyme activity by twofold due to higher surface area and increase in surface hydrophobicity.

19.4.2.3 Adsorption of PAH by Nanoparticles

Adsorption capacity of nanoparticles can indirectly help in bioremediation processes. When PAHs are adsorbed by the nanoparticles, the toxic effects of PAHs get reduced, which allows the microbes to grow even in high concentrations of the PAHs (Nowack and Bucheli 2007). According to reports by Gotovac et al. 2006; Yang et al. 2006, CNTs can adsorb PAHs.

Pavía-Sanders et al. 2013 reported that magnetic shell-cross-linked knedel-like nanoparticles, MSCKs, were designed by the co-assembling of amphiphilic block copolymers of hydrophobic poly (acrylic acid) 20-b-polysterene 280 (PAA_{20}-b-PS_{280}) and hydrophilic oleic acid stabilized Fe_2O_3 nanoparticles using tetrahydrofuran as hydrocarbon sequestering agents from crude oil, and these MSCKs showed higher hydrocarbon removal efficiency (Pavía-Sanders et al. 2013).

19.4.2.4 Chemical Oxidation of Petroleum Hydrocarbon by NP

Integration of bioremediation and application of nanoparticles can speed up the process of degradation of petroleum hydrocarbons, since nanoparticles can also help to oxidize the petroleum hydrocarbons and, thus, reduces its toxicity on the microbial growth. According to Northup and Cassidy 2008, CaO_2 nanoparticles can act as oxidants of hydrocarbons. Nador et al. 2010 also reported that cobalt (Co) and manganese (Mn) nanoparticles can assist in the reduction of poly-aromatic hydrocarbons.

19.4.3 Heavy Metals Detoxification through Biogenic Synthesis of Nanoparticles

Microorganisms which thrive in high metal ion concentrations are generally equipped with special defense mechanism. Metal toxicity is often reduced via reduction process, which changes the redox state of metals, and the microorganism can help in remediation of heavy metals via reduction of metal ions. Heavy metals can be also transformed into nanoparticles and thus help in the process of detoxification. Few microorganisms that thrive in arsenic-rich environments are reported to be capable of synthesizing arsenic nanostructures during reduction process. One such report made by Lee et al. (2007) showed formation of extracellular network of filamentous, AsS nanotubes which were 20 to 100 nm in diameter and around 30 μm in length with the help of *Shewanella* sp. HN-41. Oremland et al. (2004) and Baesman et al. (2007) first demonstrated the biogenesis of selenium nanomaterials via reduction of two chalcogenide oxyanion species, viz., selenite/nate and tellurite/rate, to elemental selenium and tellurium by two anaerobic bacteria, *Bacillus selenireducens* and *Sulfurospirillum barnesii*. Few microorganisms are also capable of synthesizing Cd nanoparticles. According to Bai et al. 2009, *Rhodopseudomonas palustris*, a photosynthetic bacteria, when incubated in $CdSO_4$ solution, can produce CdS nanoparticles with the help of cysteine desulfhydrase enzyme.

19.5 REMEDIATION USING NANOPARTICLES IN AQUATIC ENVIRONMENTS

Water is one the most essential resources for sustenance in this planet. Hence, removal of pollutants from water bodies is crucial. Industrial effluents are often released in water bodies without proper treatment, which reportedly causes great damage on the aquatic ecosystem. Additionally, marine pollution as a result of oil seepage or oil spills has huge detrimental impacts on the marine ecosystem and also on the economy, which is directly or indirectly associated with the marine ecosystem. Hence, application of a sustainable approach for the clean-up of these contaminants is essential.

19.5.1 Remediation of Industrial Wastewater with the Aid of Nano-Biotechnology

Removal of pollutants from the industrial effluents is indeed a challenging task. Although several remediation techniques like advanced oxidation processes, vaporization methods, etc. are capable of minimizing the toxicity of industrial effluent wastewaters, its application is not cost-effective for several industries (Gupta and Shukla 2020). Hence, alternative remediation strategies with the help of nanotechnology and along with that green synthesis of nanoparticles can provide a new avenue to develop sustainable and cheaper methods of remediation.

In general, in order to get rid of harmful industrial contaminants, nanoparticles have been already deployed as nano-adsorbents and nano-filtration membranes (Shukla 2020). Additionally, biofabrication of nanoparticles, along with the use of microbes, can make the application of nanotechnology more cost-effective, eco-friendly and sustainable. In the aqueous environment, the green synthesized nanoparticle can attain superior solidity due to co-precipitation or addition of proteinaceous elements on the nanoparticle outer surface.

According to Mahanty et al. 2020, iron oxide nanoparticles which were produced from *Aspergillus tubingensis* (STSP 25), which were obtained from the rhizosphere of *Avicennia officinalis* in the Sundarbans of India, demonstrated 90% ability of removal of toxic heavy metals like Pb, Ni, Cu, Zn from the wastewater following adsorption of the heavy metal ions on the nanoparticle surfaces via endothermic reactions.

Exopolysaccharide from the green algae *Chlorella vulgaris* was utilized to co-precipitate with iron oxide nanoparticles. The resultant nanoparticles with functional exopolysaccharide were capable of removing 85% of ammonia and 91% of phosphates from wastewater (Govarthanan et al. 2020).

Several biogenic nanoparticles have also reportedly helped to degrade azo dye, Naphthol Green B dye, etc. from textile effluents. Copper nanoparticles, which were produced from a copper-resistant bacteria, *Escherichia* sp. SINT7, helped in reduction of the concentrations of several harmful dyes like reactive black-5, congo red, direct blue-1 and malachite green (Noman et al. 2020). *Pseudoalteromonas* sp. CF10–13 has been utilized by Cheng et al. 2019 to synthesize iron-sulfur nanoparticles which were capable of degrading the Naphthol Green B dye.

There are several instances where microorganisms could further enhance the reactivity of nanoparticles by providing additional catalytic enzymes which can enhance the remediation capacity. According to the reports of Darwesh et al. (2019), peroxidase enzymes were immobilized on the surface of iron oxide nanoparticles, and those modified nanoparticles were capable of completely removing both green and red azo dyes within six hours of treatment. Fe_3O_4 and chitosan nanocomposite with immobilized laccase enzyme also helped to degrade 2, 4-Dichloro-Phenol and 4-Chloro-Phenol after 12 hours of treatment, according to the study by Zhang et al. 2020. Ranjan et al. 2018, 2019 reported that cyanate hydratase enzymes, which were immobilized on iron-oxide-filled magnetic multi-walled carbon nanotubes (MWCNTs), were capable of removing several heavy metals like Cu, Fe, Cr, and Pb, and 84% of cyanate was also degraded upon application of those modified nanomaterials.

19.5.2 Enhancement of Bioremediation of Oil Spills via Nanotechnology

In recent decades, researches regarding utilization of nanotechnology for remediation of oil spill, which is responsible for marine pollution, have been gaining momentum. Integration of nanotechnology and bioremediation methods have also been explored at great lengths, and results have shown significant efficiency. Reports have confirmed that nanoparticles can act as pickering emulsifiers which can contribute to increasing the growth of microorganisms responsible for remediation of oil spills. Clay nanoparticles are reported to act as pickering emulsifiers and can upregulate the process of bioremediation (Pete et al. 2021). According to Panchal et al. 2018, hydrophobic modified halloysite nanotube or HNT can increase the metabolic activity of *Alcanivorax borkumensis*, which is known to be marine petroleum oil–degrading bacteria.

Nanoparticles can be also used to deliver growth-limiting nutrients only in the vicinity of oil-water interface, which can be helpful to remove the non-oil-degrading bacteria. Mesoporous silica nanoparticles can serve as a carrier to deliver nitrogen, potassium and phosphorus, which are known to be growth-limiting nutrients. According to Pete et al. (2021), due to the modification of silica nanoparticles with long chain (C18) organosilane, a gated system can be created. This gate can open only when the nanoparticles come into the contact of lipophilic alkanes and, thus, release nutrients from the encapsulated nanoparticles. Application of this technology can help to deliver nutrients only at the region of oil-water interface and, thus, promote the growth of microorganisms which can help in oil degradation.

19.6 FUTURE PERSPECTIVES AND CONCLUSION

Nanotechnology holds immense prospects in the field of remediation of hazardous contaminants. Integration of nano-biotechnology has provided a new opportunity for sustainable and cheaper methods for remediation of industrial wastes, oil spills, chlorinated organic contaminants, hydrophobic compounds, heavy metals, etc. However, the big challenge lies in commercialization of these nanotechnological aspects, and according to Dwevedi, 2018, only 1% of these nanotechnological innovations are known to get commercialized. Additionally, biogenic nanoparticles have also provided a green approach for the management of hazardous contaminants. However, application of nano-biotechnology still requires further research regarding ecotoxicology and long-term exposure of the nanomaterials in the ecosystem.

REFERENCES

Acevedo, F., Pizzul, L., González, M.E., Cea, M., Gianfreda, L. and Diez, M.C., 2010. Degradation of polycyclic aromatic hydrocarbons by free and nanoclay-immobilized manganese peroxidase from Anthracophyllum discolor. *Chemosphere*, 80(3): 271–278.

Agarwal, A. and Joshi, H., 2010. Application of nanotechnology in the remediation of contaminated groundwater: a short review. *Recent Research in Science and Technology*, 2(6).

Alkorta, I., Hernández-Allica, J., Becerril, J.M., Amezaga, I., Albizu, I. and Garbisu, C., 2004. Recent findings on the phytoremediation of soils contaminated with environmentally toxic heavy metals and metalloids such as zinc, cadmium, lead, and arsenic. *Reviews in Environmental Science and Biotechnology*, 3(1): 71–90.

Ansari, F., Grigoriev, P., Libor, S., Tothill, I.E. and Ramsden, J.J., 2009. DBT degradation enhancement by decorating Rhodococcus erythropolis IGST8 with magnetic Fe_3O_4 nanoparticles. *Biotechnology and Bioengineering*, 102(5): 1505–1512.

Ansari, S.A. and Husain, Q., 2012. Potential applications of enzymes immobilized on/in nano materials: A review. *Biotechnology Advances*, 30(3): 512–523.

Baesman, S.M., Bullen, T.D., Dewald, J., Zhang, D., Curran, S., Islam, F.S., Beveridge, T.J. and Oremland, R.S., 2007. Formation of tellurium nanocrystals during anaerobic growth of bacteria that use Te oxyanions as respiratory electron acceptors. *Applied and Environmental Microbiology*, 73(7): 2135–2143.

Bai, H.J., Zhang, Z.M., Guo, Y. and Yang, G.E., 2009. Biosynthesis of cadmium sulfide nanoparticles by photosynthetic bacteria Rhodopseudomonas palustris. *Colloids and Surfaces B: Biointerfaces*, 70(1): 142–146.

Baker, A.J., McGrath, S.P., Sidoli, C. and Reeves, R.D., 1996. *The potential for heavy metal decontamination* (No. CONF-960592-). International Business Communications, Southborough, MA.

Bokare, V., Murugesan, K., Kim, J.H., Kim, E.J. and Chang, Y.S., 2012. Integrated hybrid treatment for the remediation of 2, 3, 7, 8-tetrachlorodibenzo-p-dioxin. *Science of the Total Environment*, 435: 563–566.

Bokare, V., Murugesan, K., Kim, Y.M., Jeon, J.R., Kim, E.J. and Chang, Y.S., 2010. Degradation of triclosan by an integrated nano-bio redox process. *Bioresource Technology*, 101(16): 6354–6360.

Cameotra, S.S. and Makkar, R.S., 2010. Biosurfactant-enhanced bioremediation of hydrophobic pollutants. *Pure and Applied Chemistry*, 82(1): 97–116.

Cameotra, S.S. and Soniya, D., 2010. Environmental nanotechnology: nanoparticles for bioremediation of toxic pollutants. In *Bioremediation technology*, ed. M.H. Fulekar, 348–374. Springer, Dordrecht.

Cheng, S., Li, N., Jiang, L., Li, Y., Xu, B. and Zhou, W., 2019. Biodegradation of metal complex Naphthol Green B and formation of iron–sulfur nanoparticles by marine bacterium Pseudoalteromonas sp CF10–13. *Bioresource Technology*, 273: 49–55.

Chronopoulou, L., Kamel, G., Sparago, C., Bordi, F., Lupi, S., Diociaiuti, M. and Palocci, C., 2011. Structure–activity relationships of Candida rugosa lipase immobilized on polylactic acid nanoparticles. *Soft Matter*, 7(6): 2653–2662.

Darwesh, O.M., Matter, I.A. and Eida, M.F., 2019. Development of peroxidase enzyme immobilized magnetic nanoparticles for bioremediation of textile wastewater dye. *Journal of Environmental Chemical Engineering*, 7(1): 102805.

Dixit, R., Malaviya, D., Pandiyan, K., Singh, U.B., Sahu, A., Shukla, R., Singh, B.P., Rai, J.P., Sharma, P.K., Lade, H. and Paul, D., 2015. Bioremediation of heavy metals from soil and aquatic environment: an overview of principles and criteria of fundamental processes. *Sustainability*, 7(2): 2189–2212.

Dutta, N., Mukhopadhyay, A., Dasgupta, A.K. and Chakrabarti, K., 2013. Nanotechnology enabled enhancement of enzyme activity and thermostability: study on impaired pectate lyase from attenuated Macrophomina phaseolina in presence of hydroxyapatite nanoparticle. *PLoS One*, 8(5): 63567.

Dwevedi, A., 2018. *Solutions to environmental problems involving nanotechnology and enzyme technology*. Academic Press, San Diego, USA.

Garcia-Galan, C., Berenguer-Murcia, Á., Fernandez-Lafuente, R. and Rodrigues, R.C., 2011. Potential of different enzyme immobilization strategies to improve enzyme performance. *Advanced Synthesis & Catalysis*, 353(16): 2885–2904.

Geiger, C.L., Clausen, C.A. and Brooks, K., 2003. Remediation of DNAPLs using emulsified zero-valent iron: laboratory and field results. *Scopus Export 2000s.1454*, 43(1): 939–944.

Golob, V., Vinder, A. and Simonič, M., 2005. Efficiency of the coagulation/flocculation method for the treatment of dyebath effluents. *Dyes and Pigments*, 67(2): 93–97.

Gotovac, S., Hattori, Y., Noguchi, D., Miyamoto, J.I., Kanamaru, M., Utsumi, S., Kanoh, H. and Kaneko, K., 2006. Phenanthrene adsorption from solution on single wall carbon nanotubes. *The Journal of Physical Chemistry B*, 110(33): 16219–16224.

Govarthanan, M., Jeon, C.H., Jeon, Y.H., Kwon, J.H., Bae, H. and Kim, W., 2020. Non-toxic nano approach for wastewater treatment using Chlorella vulgaris exopolysaccharides immobilized in iron-magnetic nanoparticles. *International Journal of Biological Macromolecules*, 162: 1241–1249.

Gupta, G.K. and Shukla, P., 2020. Insights into the resources generation from pulp and paper industry wastes: challenges, perspectives and innovations. *Bioresource Technology*, 297: 122496.

Gupta, K., Bhattacharya, S., Chattopadhyay, D., Mukhopadhyay, A., Biswas, H., Dutta, J., Ray, N.R. and Ghosh, U.C., 2011. Ceria associated manganese oxide nanoparticles: synthesis, characterization and arsenic (V) sorption behavior. *Chemical Engineering Journal*, 172(1): 219–229.

Haferburg, G. and Kothe, E., 2007. Microbes and metals: interactions in the environment. *Journal of Basic Microbiology*, 47(6): 453–467.

Hara, J., 2012. *Chemical degradation of chlorinated organic pollutants for in situ remediation and evaluation of natural attenuation*. IntechOpen, London.

Ingle, A.P., Seabra, A.B., Duran, N. and Rai, M., 2014. Nanoremediation: a new and emerging technology for the removal of toxic contaminant from environment. In *Microbial biodegradation and bioremediation*, ed. S. Das, 233–250. Elsevier, Amsterdam.

Kapoor, R.T., Salvadori, M.R., Rafatullah, M., Siddiqui, M.R., Khan, M.A. and Alshareef, S.A., 2021. Exploration of microbial factories for synthesis of nanoparticles–a sustainable approach for bioremediation of environmental contaminants. *Frontiers in Microbiology*, 12: 1404.

Kelly, D.J., Budd, K. and Lefebvre, D.D., 2006. The biotransformation of mercury in pH-stat cultures of microfungi. *Botany*, 84(2): 254–260.

Khoshnevisan, K., Bordbar, A.K., Zare, D., Davoodi, D., Noruzi, M., Barkhi, M. and Tabatabaei, M., 2011. Immobilization of cellulase enzyme on superparamagnetic nanoparticles and determination of its activity and stability. *Chemical Engineering Journal*, 171(2): 669–673.

Kiran, G.S., Nishanth, L.A., Priyadharshini, S., Anitha, K. and Selvin, J., 2014. Effect of Fe nanoparticle on growth and glycolipid biosurfactant production under solid state culture by marine Nocardiopsissp. MSA13A. *BMC Biotechnology*, 14(1): 1–10.

Koenig, J.C., Boparai, H.K., Lee, M.J., O'Carroll, D.M., Barnes, R.J. and Manefield, M.J., 2016. Particles and enzymes: combining nanoscale zero valent iron and organochlorine respiring bacteria for the detoxification of chloroethane mixtures. *Journal of Hazardous Materials*, 308: 106–112.

Kumar, S.R. and Gopinath, P., 2017. Nano-bioremediation applications of nanotechnology for bioremediation. In *Handbook of advanced industrial and hazardous wastes management*, 27–48. CRC Press (Taylor & Francis Group), London, UK.

Kumari, B. and Singh, D.P. 2016. A review on multifaceted application of nanoparticles in the field of bioremediation of petroleum hydrocarbons. *Ecological Engineering*, 97: 98–105.

Le, T.T., Nguyen, K.H., Jeon, J.R., Francis, A.J. and Chang, Y.S., 2015. Nano/bio treatment of polychlorinated biphenyls with evaluation of comparative toxicity. *Journal of Hazardous Materials*, 287: 335–341.

Lee, J.H., Kim, M.G., Yoo, B., Myung, N.V., Maeng, J., Lee, T., Dohnalkova, A.C., Fredrickson, J.K., Sadowsky, M.J. and Hur, H.G., 2007. Biogenic formation of photoactive arsenic-sulfide nanotubes by Shewanella sp. strain HN-41. *Proceedings of the National Academy of Sciences*, 104(51): 20410–20415.

Lim, P.E., Mak, K.Y., Mohamed, N. and Noor, A.M., 2003. Removal and speciation of heavy metals along the treatment path of wastewater in subsurface-flow constructed wetlands. *Water Science and Technology*, 48(5): 307–313.

Liu, J., Vipulanandan, C., Cooper, T.F. and Vipulanandan, G., 2013. Effects of Fe nanoparticles on bacterial growth and biosurfactant production. *Journal of Nanoparticle Research*, 15(1): 1–13.

Liu, W., Wang, D., Soomro, R.A., Fu, F., Qiao, N., Yu, Y., Wang, R. and Xu, B., 2019. Ceramic supported attapulgite-graphene oxide composite membrane for efficient removal of heavy metal contamination. *Journal of Membrane Science*, 591: 117323.

López-Miranda, J.L., Esparza, R., Rosas, G., Pérez, R. and Estévez-González, M., 2019. Catalytic and antibacterial properties of gold nanoparticles synthesized by a green approach for bioremediation applications. *3 Biotech*, 9(4): 1–9.

Mahanty, S., Chatterjee, S., Ghosh, S., Tudu, P., Gaine, T., Bakshi, M., Das, S., Das, P., Bhattacharyya, S., Bandyopadhyay, S. and Chaudhuri, P., 2020. Synergistic approach towards the sustainable management of heavy metals in wastewater using mycosynthesized iron oxide nanoparticles: biofabrication, adsorptive dynamics and chemometric modeling study. *Journal of Water Process Engineering*, 37: 101426.

Nador, F., Moglie, Y., Vitale, C., Yus, M., Alonso, F. and Radivoy, G., 2010. Reduction of polycyclic aromatic hydrocarbons promoted by cobalt or manganese nanoparticles. *Tetrahedron*, 66(24): 4318–4325.

Němeček, J., Pokorný, P., Lhotský, O., Knytl, V., Najmanová, P., Steinová, J., Černík, M., Filipová, A., Filip, J. and Cajthaml, T., 2016. Combined nano-biotechnology for in-situ remediation of mixed contamination of groundwater by hexavalent chromium and chlorinated solvents. *Science of the Total Environment*, 563: 822–834.

Noman, M., Shahid, M., Ahmed, T., Niazi, M.B.K., Hussain, S., Song, F. and Manzoor, I., 2020. Use of biogenic copper nanoparticles synthesized from a native Escherichia sp. as photocatalysts for azo dye degradation and treatment of textile effluents. *Environmental Pollution*, 257: 113514.

Northup, A. and Cassidy, D., 2008. Calcium peroxide (CaO_2) for use in modified Fenton chemistry. *Journal of Hazardous Materials*, 152(3): 1164–1170.

Nowack, B. and Bucheli, T.D., 2007. Occurrence, behavior and effects of nanoparticles in the environment. *Environmental Pollution*, 150(1): 5–22.

Nurmi, J.T., Tratnyek, P.G., Sarathy, V., Baer, D.R., Amonette, J.E., Pecher, K., Wang, C., Linehan, J.C., Matson, D.W., Penn, R.L. and Driessen, M.D., 2005. Characterization and properties of metallic iron nanoparticles: spectroscopy, electrochemistry, and kinetics. *Environmental Science & Technology*, 39(5): 1221–1230.

Oremland, R.S., Herbel, M.J., Blum, J.S., Langley, S., Beveridge, T.J., Ajayan, P.M., Sutto, T., Ellis, A.V. and Curran, S., 2004. Structural and spectral features of selenium nanospheres produced by Se-respiring bacteria. *Applied and Environmental Microbiology*, 70(1): 52–60.

Panchal, A., Swientoniewski, L.T., Omarova, M., Yu, T., Zhang, D., Blake, D.A., John, V. and Lvov, Y.M., 2018. Bacterial proliferation on clay nanotube Pickering emulsions for oil spill bioremediation. *Colloids and Surfaces B: Biointerfaces*, 164: 27–33.

Pavía-Sanders, A., Zhang, S., Flores, J.A., Sanders, J.E., Raymond, J.E. and Wooley, K.L., 2013. Robust magnetic/polymer hybrid nanoparticles designed for crude oil entrapment and recovery in aqueous environments. *ACS Nano*, 7(9): 7552–7561.

Pete, A.J., Bharti, B. and Benton, M.G., 2021. Nano-enhanced bioremediation for oil spills: a review. *ACS ES&T Engineering*, 1(6): 928–946.

Ranjan, B., Pillai, S., Permaul, K. and Singh, S., 2018. A novel strategy for the efficient removal of toxic cyanate by the combinatorial use of recombinant enzymes immobilized on aminosilane modified magnetic nanoparticles. *Bioresource Technology*, 253: 105–111.

Ranjan, B., Pillai, S., Permaul, K. and Singh, S., 2019. Simultaneous removal of heavy metals and cyanate in a wastewater sample using immobilized cyanate hydratase on magnetic-multiwall carbon nanotubes. *Journal of Hazardous Materials*, 363: 73–80.

Ranjbakhsh, E., Bordbar, A.K., Abbasi, M., Khosropour, A.R. and Shams, E., 2012. Enhancement of stability and catalytic activity of immobilized lipase on silica-coated modified magnetite nanoparticles. *Chemical Engineering Journal*, 179: 272–276.

Ren, Y., Rivera, J.G., He, L., Kulkarni, H., Lee, D.K. and Messersmith, P.B., 2011. Facile, high efficiency immobilization of lipase enzyme on magnetic iron oxide nanoparticles via a biomimetic coating. *BMC Biotechnology*, 11(1): 1–8.

Rizwan, M., Singh, M., Mitra, C.K. and Morve, R.K., 2014. Ecofriendly application of nanomaterials: nanobioremediation. *Journal of Nanoparticles*, 2014.

Sawidis, T., 2008. Effect of cadmium on pollen germination and tube growth in Lilium longiflorum and Nicotiana tabacum. *Protoplasma*, 233: 95–106.

Sayan, B., Indranil, S., Aniruddha, M., Dhrubajyoti, C., Uday, C.G. and Debashis, C., 2013. Role of nanotechnology in water treatment and purification: potential applications and implications. *International Journal of Chemical Science and Technology*, 3(3): 59.

Shan, G., Xing, J., Zhang, H. and Liu, H., 2005. Biodesulfurization of dibenzothiophene by microbial cells coated with magnetite nanoparticles. *Applied and Environmental Microbiology*, 71(8): 4497–4502.

Shao, D., Sheng, G., Chen, C., Wang, X. and Nagatsu, M., 2010. Removal of polychlorinated biphenyls from aqueous solutions using β-cyclodextrin grafted multiwalled carbon nanotubes. *Chemosphere*, 79(7): 679–685.

Shukla, P., 2020. Microbial nanotechnology for bioremediation of industrial wastewater. *Frontiers in Microbiology*, 2411.

Singh, R., Manickam, N., Mudiam, M.K.R., Murthy, R.C. and Misra, V., 2013. An integrated (nano-bio) technique for degradation of γ-HCH contaminated soil. *Journal of Hazardous Materials*, 258: 35–41.

Singh, T., Srivastava, N., Mishra, P.K., Bhatiya, A.K. and Singh, N.L., 2016. Application of TiO_2 nanoparticle in photocatalytic degradation of organic pollutants. *Materials Science Forum*, 855: 20–32.

Singha, S., Datta, H. and Dasgupta, A.K., 2010. Size dependent chaperon properties of gold nanoparticles. *Journal of Nanoscience and Nanotechnology*, 10(2): 826–832.

Sondi, I. and Salopek-Sondi, B., 2004. Silver nanoparticles as antimicrobial agent: a case study on E. coli as a model for Gram-negative bacteria. *Journal of Colloid and Interface Science*, 275(1): 177–182.

Wang, J., Meng, G., Tao, K., Feng, M., Zhao, X., Li, Z., Xu, H., Xia, D. and Lu, J.R., 2012. Immobilization of lipases on alkyl silane modified magnetic nanoparticles: effect of alkyl chain length on enzyme activity. *PLoS One*, 7(8): e43478.

Yang, K., Zhu, L. and Xing, B., 2006. Adsorption of polycyclic aromatic hydrocarbons by carbon nanomaterials. *Environmental Science & Technology*, 40(6): 1855–1861.

Zhang, D., Wei, S., Kaila, C., Su, X., Wu, J., Karki, A.B., Young, D.P. and Guo, Z., 2010. Carbon-stabilized iron nanoparticles for environmental remediation. *Nanoscale*, 2(6): 917–919.

Zhang, K., Yang, W., Liu, Y., Zhang, K., Chen, Y. and Yin, X., 2020. Laccase immobilized on chitosan-coated Fe_3O_4 nanoparticles as reusable biocatalyst for degradation of chlorophenol. *Journal of Molecular Structure*, 1220: 128769.

Zhang, W.X., 2003. Nanoscale iron particles for environmental remediation: an overview. *Journal of Nanoparticle Research*, 5(3): 323–332.

Zhang, W.X., Wang, C.B. and Lien, H.L., 1998. Treatment of chlorinated organic contaminants with nanoscale bimetallic particles. *Catalysis Today*, 40(4): 387–395.